HUMAN REPRODUCTION

THE CORE CONTENT OF OBSTETRICS, GYNECOLOGY AND PERINATAL MEDICINE

2nd Edition

ERNEST W. PAGE, M.D.

Professor and Chairman
Department of Obstetrics and Gynecology,
Rutgers Medical School, Piscataway, New Jersey

CLAUDE A. VILLEE, Ph.D.

Andelot Professor of Biological Chemistry,
Harvard Medical School, Boston, Massachusetts

DOROTHY B. VILLEE, M.D.

Assistant Professor of Pediatrics,
Harvard Medical School, Boston, Massachusetts

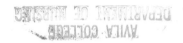
W. B. SAUNDERS COMPANY   Philadelphia   London   Toronto   1976

W. B. Saunders Company: West Washington Square
Philadelphia, PA 19105

1 St. Anne's Road
Eastbourne, East Sussex BN21 3UN, England

833 Oxford Street
Toronto, Ontario M8Z 5T9, Canada

Library of Congress Cataloging in Publication Data

Page, Ernest W

Human reproduction: the core content of obstetrics,
gynecology, and perinatal medicine.

Includes bibliographies and index.

1. Obstetrics.   2. Gynecology.   3. Human reproduction.
    I. Villee, Claude Alvin, 1917–   joint author.
    II. Villee, Dorothy B., joint author.   III. Title.
    [DNLM: 1. Obstetrics.   2. Reproduction. WQ100 P132h]

RG101.P27 1976   618   75–19851

ISBN 0–7216–7042–3

Human Reproduction: The Core Content of Obstetrics,
Gynecology and Perinatal Medicine   ISBN   0-7216-7042-3

Last digit is the print number: 9   8   7   6   5   4   3   2   1

PREFACE

The warm reception accorded the first edition of *Human Reproduction* prompted our publishers to encourage us to prepare a second edition. Our objective in this edition, as in the first, is to present the core content of obstetrics, gynecology and perinatal medicine within a single volume of modest size, and to present enough basic science in the fields of reproductive endocrinology, andrology, biochemistry, physiology and genetics to form a firm foundation for the clinical presentations. In preparing this edition, each chapter has been carefully reviewed and revised where necessary to bring the material up to date. There are new sections on prostaglandins, lactation, abortion, mechanisms of hormone action, and the male reproductive system. In addition, the clinical sections on obstetrics, gynecology and perinatal medicine have been expanded somewhat to render the text more complete for required and elective clinical clerkships. A new chapter (18) on gynecologic tumors has been added. We are in an era of continuing revision of therapy, which varies from one medical center to another and from year to year. Only the principles of therapy have been included; the details of medical and surgical treatment should be learned by precept during clerkship or residency training. Some disease entities and certain rarely encountered variants of specific disorders have been deliberately omitted. The body of knowledge included is, in our opinion, what every health professional involved with the health care of women should know.

We are indebted to many friends and associates who have read portions of the manuscript and offered valuable comments, and to the editorial staff of the W. B. Saunders Company for their suggestions and cooperation in preparing and producing this revision.

ERNEST W. PAGE, M.D.
CLAUDE A. VILLEE, PH.D.
DOROTHY B. VILLEE, M.D.

CONTENTS

Chapter 3

THE FEMALE REPRODUCTIVE CYCLE: THE FUNCTIONAL BASIS OF GYNECOLOGY 41

Chapter 8

PLACENTA AND GRAVID UTERUS: STRUCTURE AND
FUNCTION... 169

Chapter 15

Chapter 16

SOME MAJOR COMPLICATIONS OF PREGNANCY.................... 371

Chapter 17
COMMON GYNECOLOGIC DISORDERS OTHER THAN

Chapter 1

REPRODUCTIVE PROCESSES

If there is any one feature of a living system that qualifies as the "essence of life," it is the ability to reproduce and perpetuate the species. The survival of each species requires that its individual members multiply, that each generation produce new individuals to replace ones killed by predators, parasites, or old age. This can be contrasted to the processes of respiration, excretion, circulation, coordination and so on, which are needed for the day-to-day survival of the organism. Reproduction at the molecular level is a function of the unique capacity of the nucleic acids for self-replication, which depends on the specificity of the relatively weak hydrogen bonds between pairs of nucleotides. At the level of the whole organism reproduction ranges from the simple fission of bacteria and other unicellular organisms (a process which does not involve sex at all) to the incredibly complex structural, functional and behavioral processes of reproduction in the higher animals. Reproduction in man and higher animals involves not only the genetic transfer of biological information from one generation to the next but the endocrine regulation of the development of the genital tracts, oogenesis, ovulation and spermatogenesis, as well as the intricate behavior patterns that insure that eggs and sperm are released at the same time and the same place so that they can meet to form a fertilized egg or zygote (Fig. 1–1). This is followed by the complex processes of development and differentiation by which a zygote becomes an adult organism. The reproductive process is a cyclic one, and a study of reproduction could begin at any point in that cycle.

SEXUAL AND ASEXUAL REPRODUCTION

The details of the reproductive process vary tremendously from one kind of animal to another, but we can distinguish two basically different types—asexual and sexual. In asexual reproduction a single parent splits, buds or fragments to give rise to two or more offspring that have hereditary traits identical with those of the parent. Even higher animals may reproduce asexually; indeed, the production of identical twins in man by the splitting of a single fertilized egg is a kind of asexual reproduction.

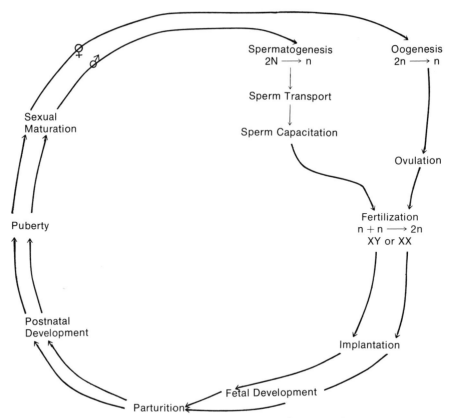

FIGURE 1–1.   The cyclic nature of the reproductive process in mammals.

In contrast, sexual reproduction involves two parents, each of which contributes a specialized gamete, an egg or sperm, which fuse to form the fertilized egg. The egg is typically large and nonmotile with stored nutrients to support the development of the embryo that results if the egg is fertilized. The sperm is typically small and motile, adapted to swim actively to the egg by the lashing movements of its long filamentous tail. Sexual reproduction has the biological advantage of making possible the recombination of the inherited traits of the two parents; thus, the offspring may be better able to survive than either parent. In contrast, in asexual reproduction the offspring have hereditary traits identical to those of the parent.

REPRODUCTIVE SYSTEMS

The basic components of the male reproductive system are the male gonad or testis, in which sperm are produced, and a sperm duct for the transport of sperm to the exterior of the body. Parts of the sperm duct or of areas adjacent to it may be modified for specific functions. A part of the duct may be given over for sperm storage — such a part is often called a seminal vesicle. There may be glandular areas for the production of seminal fluid, which serves as a vehicle or carrier for the sperm and may also activate, nourish or protect the sperm. The terminal part of the sperm duct may open onto or into a copulatory organ, a penis, which provides for the transfer of sperm to the female.

The basic parts of the female system are the female gonad, or ovary, and the oviduct, a tube for the transport of eggs to the exterior. The oviduct may also be modified in any of several ways in different species. It may have a glandular portion for the secretion of egg shells, or cases or cocoons. A section of the oviduct, a seminal receptacle, may be modified for the storage of sperm following their transfer from the male. Another portion of the oviduct may be modified as a uterus for egg storage or for the development of the fertilized egg within the body of the female. The terminal portion of the oviduct may be adapted as a vagina for receiving the male copulatory organ. The male and female systems of most animals do not possess all of these modifications, and they cannot be considered as any kind of progressive sequence of evolutionary change. Rather, the presence or absence of some specific adaptation of the reproductive system is correlated with the circumstances of reproduction, whether the animals live in the sea, in fresh water or on land, whether their fertilization is external or internal, whether the eggs are liberated singly into the water to develop or deposited onto the bottom within envelopes, and whether or not the developing individual passes through a larval stage.

Most of the 30,000 or so species of fish still utilize a primitive form of reproduction in which the female sheds enormous numbers of eggs which may or may not be fertilized by sperm from some passing male. Thereafter both male and female fish ignore the eggs. In such animals the uterus is missing and the male does not have a copulatory organ. In other fish, however, internal fertilization takes place before egg laying (oviparity). In sea catfish, for example, the male's anal fin is employed as an intromittent organ. The fertilized eggs are then deposited in a clump, which the male picks up and carries in his mouth during most of their incubation, presumably fasting and yielding not to temptation.

HORMONAL CONTROL OF REPRODUCTION

Reproduction in vertebrates is under three levels of hormonal control. The steroid hormones produced by the gonads compose the first level: testosterone produced by the Leydig cells of the testis and estradiol and progesterone synthesized by the ovary. Testosterone regulates the development of male genitalia and male features of the skeleton and muscular system, and estradiol regulates the development of the uterus, vagina, external genitalia, pubic hair, the changes in the structure of the pelvic bones, growth of the breasts, the proliferation of the glandular cells in the breasts and the characteristic deposition of adipose tissue in the hips and thighs of adult females. Progesterone has little effect on the development of the female sex characteristics, but it plays a major role in stimulating the development of the endometrium during the secretory phase of the menstrual cycle.

Ovarian structure and function and testicular structure and function are, in turn, under the second level of control: gonadotropins secreted by the anterior pituitary. In the male, follicle-stimulating hormone, FSH, stimulates the development of the testis and its seminiferous tubules, and luteinizing hormone, LH (also called interstitial cell-stimulating hormone, ICSH), stimulates the development of Leydig cells and their production of androgens. The complex events of the estrous and menstrual cycles are controlled by FSH and LH. A surge of LH causes the ripe follicle to rupture, releasing the egg, a process termed ovulation.

The production and release of the gonadotropins is regulated by releasing factors secreted by the hypothalamus. Whether there is a single releasing factor that brings about the release of both FSH and LH, or whether there are separate releasing factors for each gonadotropin is a problem that has not yet been resolved. Estradiol and progesterone also exercise both positive and negative feedback control of the secretion of these releasing factors. Whether the effect is positive or negative depends on the concentration of steroid in the blood and whether it is increasing or decreasing. Thus a small, rising concentration of estradiol may trigger the release of gonadotropin releasing factor, whereas a continuous high concentration will inhibit the secretion of LH releasing factor. This latter effect is the basis for the use of estrogens in oral contraceptives.

The endocrine control of reproduction, can, in turn, be altered or affected by certain environmental stimuli. A laboratory rat kept under normal conditions of day and night lighting ovulates early in the morning between 1:00 and 2:00 A.M. If rats are kept for two weeks or more under conditions in which the periods of light and dark are reversed, the time of ovulation is shifted 12 hours. If the rat is exposed to continuous light 24 hours a day she eventually stops ovulating and is in a state of continuous estrus, with persistent vaginal cornification.

The rhythm of the human menstrual cycle and probably of ovulation itself can be influenced by environmental factors. Nurses on night duty and airline hostesses who travel long distances east and west to different time zones frequently report changes in their menstrual cycles. Spontaneous ovulators have some sort of light-dependent hypothalamic clock that provides the neural stimulation for the release of the hypothalamic releasing factors. These are produced by neurons that end in the median eminence; they are then released from there and pass into the hypophysial portal vessels through which they reach the pituitary. This, in turn, stimulates the release of a surge of LH, which initiates ovulation.

The male pituitary also produces and secretes FSH and LH. The pituitary of a male rat transplanted into a hypophysectomized female rat will support a normal estrous cycle. An ovary transplanted into a castrate male will develop ripe follicles, but it will not undergo ovulation since there is no cyclic release of an LH surge from the male pituitary. The difference between the two sexes appears at a critical stage early in the development of the central nervous system. In the developing male, testosterone inhibits the development of the cyclic center. Young females, lacking testosterone, develop a hypothalamic cyclic center which subsequently regulates the rhythmic release of gonadotropins controlling the sexual cycles. If testosterone is injected into a female rat between the second and fifth day after birth, the activity of the cyclic center in the hypothalamus is permanently abolished. She never ovulates but remains in a state of constant estrus, like the rat kept under continual illumination.

EVOLUTION OF VIVIPARITY

The females of many aquatic invertebrate species, most insects, many lower vertebrates and all birds, lay eggs in which the young develop and from which they eventually hatch. Such animals are said to be oviparous — egg bearing. Female sharks, lizards, certain snakes and insects are ovoviviparous. They produce large yolk-filled eggs, which remain within the female reproductive tract for a considerable time after fertilization. The developing embryo usually forms no close connection with the wall of the oviduct or uterus and usually receives no nourishment from maternal blood.

In all mammals except the platypus and spiny anteater, the young are retained within the uterus for a considerable portion of development, and some specialized portion of the embryo comes into contact with the inner lining of the uterus, the endometrium. When these portions are utilized for the exchange of nutrients, gases, or fetal waste products, the structure is called the placenta. A placenta is an intimate apposition or fusion of fetal organs to maternal tissues for physiologic exchange. The tissue common to all placentas is the trophoblast, the outermost layer of cells of the developing embryo. When this is arranged as a membrane with a mesoderm layer it is termed the chorion. The chorion is nonvascular, but may become vascularized by the vitelline vessels to form the choriovitelline or yolk sac placenta found in many marsupials. In some species the vitelline vessels degenerate, leaving only a chorionic placenta. In all the higher mammals the allantoic sac fuses with and vascularizes the chorion, forming a chorioallantoic placenta. In these placentas there are mesodermal, vascular villi through which materials are exchanged.

If the fusion between fetal membranes and endometrium (termed the decidua) is very superficial and the placenta falls away at birth, leaving the maternal tissues intact, it is termed a nondeciduate placenta. If the chorion invades and intermingles with the maternal cells, and if maternal tissue is also shed at birth, it is termed a deciduate placenta. With the evolutionary development of the deciduate placenta, two complications arose. What limits the invasiveness of the trophoblast, and what immunologic mechanism permits the uterus to accept this homograft with a different genetic constitution?

Immunologic Considerations

In a deciduate placenta such as the human, the erosion of the endometrium continues until adequate uteroplacental circulation is established. Then the placenta becomes a docile, noninvasive organ. One might guess that the maternal tissue erects some sort of immunologic defense barrier against the fetal tissue, which contains paternal as well as maternal genes. This is not very likely, however, for if immunologic tolerance is induced in the female so that paternal skin transplants are accepted, then father-daughter matings do not result in more invasive placentas. The failure of maternal tissue to reject the placenta has puzzled immunologists and has led to a number of theories, none of which is completely satisfactory. In the deciduate placenta, the fetal layer in contact with maternal tissue is termed the syncytiotrophoblast. A widely held hypothesis is that the syncytiotrophoblast does not contain or exhibit transplantation antigens, and for this reason the placenta is not rejected.

Prolonging Pregnancy

In the evolution of viviparity another problem had to be solved. In most mammals the estrous cycle, the period of recurring heat, is much shorter than the period of gestation. With each cycle the endometrium either regresses or, as in the primate, is actually shed; hence, in evolution it was necessary to find some means of prolonging the pregnancy, preventing the shedding of the endometrium until the fetus was mature. The marsupials apparently did not solve this problem, for possums, kangaroos and wombats deliver an embryo about 25 mm. long, and the embryo crawls up from the cloaca into the marsupial pouch where it remains for seven or eight months. In the eutherian mammals the trophoblast has developed the property of producing a protein hormone with luteotropic properties, chorionic gonadotropin. This sustains the corpus luteum of the ovary, enabling it to continue to secrete progesterone, a key hormone for the maintenance of pregnancy. In the primates and in certain other animals the trophoblast has also developed a capacity for synthesizing progesterone and, in collaboration with the fetus, for synthesizing estrogens and other steroid hormones. With these developments the length of pregnancy became independent of ovarian function.

Nutrition of the Embryo

The trophoblast has also evolved remarkable abilities for the selective, facilitated transport of nutrients needed for the growth of the embryo and fetus. The trophoblast and embryonic tissues take up and concentrate glucose, amino acids, enzyme cofactors, trace minerals and a variety of essential nutrients.

Parturition

Finally, in the development of viviparity, some mechanism was needed to determine the end of the gestation period, that is, to initiate the birth process or parturition. The question of what signals parturition in the human is a fascinating puzzle that is not yet solved. The current speculation suggests that some signal from the fetus itself, from the fetal adrenal, acts perhaps via the placenta to initiate labor, the rhythmic uterine contractions that eventually expel the fetus. By comparison with many mammals the human infant is born in a rather premature state. However, in the course of human evolution, there has been a tendency toward increased brain size and head size. With the assumption of an erect posture the pelvic birth canal assumed such a size and shape that at many human births there is a problem of disproportion between the size of the baby's head and the mother's pelvis. Any significant further prolongation of pregnancy and increase in size of the baby's head would have led to the extinction of the human race.

SEXUAL CYCLES

In most animals there are periods of reproductive activity which recur in a rhythmic fashion and are affected by seasons of the year, by the length of daylight, and by other environmental factors. Increased light or increased length of day is ordinarily associated with increased pituitary activity and this, in turn, by way of gonadotropins affects sexual behavior and sexual receptivity. The females of some species ovulate reflexly — that is, ovulation is stimulated by copulation in a reflex fashion mediated through the hypothalamus. The stimulus may be a single copulation, as in the rabbit, but the female short-tailed shrew requires a minimum of 19 copulations per day to induce ovulation. The term estrus applies to the special period of sexual receptivity or desire on the part of the female, and the term "rut" applies to the period of male sexual activity. Human females are sexually receptive at any time in the ovarian cycle and at any season of the year. Human males after puberty are in constant rut.

The nature of the reproductive process determines to a considerable extent the organization of animal societies, their migrations and their behavior. It is interesting to contemplate what kind of human society might have evolved if women were sexually receptive only in April and October, or how the patterns of human behavior or even the family unit might have been altered if women emitted some visual or olfactory stimulus at the fertile period, or, heaven forbid, if they ovulated reflexly after each coitus!

BEHAVIORAL ASPECTS OF REPRODUCTION

Most animals breed only during relatively brief seasons of the year, and the production and release of eggs and sperm must be synchronized if fertilization is to occur. Typically, males and females are triggered to specific types of reproductive behavior by some environmental cue, such as a change in the photoperiod, the ambient temperature or the seasonal rainfall, or by specific relations of tidal and lunar cycles. The eggs and sperm of polychaetes, echinoderms and many other invertebrates are released into the sea water, where fertilization occurs. Many of these species have evolved behavioral mechanisms that insure the simultaneous release of eggs and sperm. By responding to certain rhythmic variations in the environment nearly all the males and females of a given species release their gametes at the same time. Seasonal cycles produce variations in the temperature, the length of day, and the amount of food available. Lunar cycles produce variations in the height of tides and the strength of currents, in the relation between time of tide and hour of day and in the amount of light at night. Diurnal cycles produce the great variations in light from day to night.

The Samoan palolo worms, which live on coral reefs in the South Pacific, shed their eggs and sperm within a single, two-hour period on one night of the year. The seasonal rhythm limits the reproductive period to November, the lunar rhythm limits it to a day during the last quarter of the moon when the tide is unusually low, and the diurnal rhythm limits it to a few hours just after complete darkness. The rear half of the palolo worm, loaded with gametes, breaks off from the rest of the body, swims backward to the surface, and eventually bursts, releasing the eggs or sperm so that fertilization may occur.

In some species the males and females must not only be brought to full sexual activity at the same time, but must be induced by specific environmental cues to move or migrate to specific mating and breeding grounds. The migration of salmon upstream to breed, the migration of eels to a specific breeding ground in the Central Atlantic, the migration of the gray whale to Baja California and the migrations of birds and sea turtles are examples of such movements. Many animals that live a solitary life during most of the year — seals, penguins, seabirds — come together for brief periods of mating at specific times and places.

Courtship

The reproductive synchronization of a particular male and a particular female frequently involves some sort of courtship behavior. The courtship, usually initiated by the male, may be a brief ceremony or, in certain species of birds, may last for many days. These precopulatory behavior patterns serve two important functions: they decrease the aggressive tendencies, and they establish species and sexual identification; that is, they identify a member of the same species but of the opposite sex. The members of many species normally fight whenever they meet, and some special cue is needed if two animals are to avoid fighting long enough to mate. Special structural and functional adaptations have evolved in some species in which aggression seems to be especially difficult to control. The male praying mantis is smaller than the female and is usually attacked by her. Mating in this species can continue even after the male's head has been bitten off by the female, since the nervous activities controlling copulation are centered in an abdominal ganglion. Indeed, removing the head enhances copulatory behavior by eliminating inhibitory impulses originating in the brain.

In some insects the male secretes a special substance on his back which is especially attractive to the female and diverts her attention during copulation. Males of certain flies present the females with little packages of food wrapped in silk threads. Other flies divert the female's attention by presenting her with an empty silk balloon. This balloon serves to decrease the female's aggressive tendencies and identifies the sex and species of the male presenting it.

The complex reproductive behavior of the stickleback fish has been studied intensively. After the male has migrated and staked out his territory he builds a nest and courts any egg-laden female who enters his territory. In the courting process he performs a special zig-zag dance. The female follows the male and he leads her to the nest, at which point he directs his head toward the entrance. The pointing stimulates the female to enter the nest and she, in turn, stimulates the male to tremble. His trembling stimulates the female to spawn, and then the male fertilizes the eggs. The presence of the eggs stimulates his release of sperm.

The bower birds that live in tropical jungles have a complex courtship behavior involving a fixed sequence of sign stimuli flashed from one partner to the other. In the special territory where mating takes place, the male sings and displays his brightly colored feathers in special short flight movements. Some male bower birds build and decorate an elaborate display area, attracting a female not only by his own color, but by the form and color of pebbles and berries that he has placed in his display area.

The song patterns of birds and the mating calls of certain fishes, frogs and insects provide effective cues for the discrimination of the species. Female frogs are attracted to calling males of their own species, but not to calling males of related species. The song patterns of each species of birds are distinct and function to keep other males of that species away from the male's territory and to attract a female of the species to the singing male.

Reproductive fighting directed towards reproductive rivals serves to space out the breeding pairs. Seldom are the combats mortal, and indeed the same purpose can be served by threat and threat display. Animals assume aggressive postures and display brightly colored or contrasting parts of their body. The result of this fighting is a more even distribution of breeding pairs, more even availability of food, more nesting and breeding space and the prevention of multiple copulations. This kind of spacing provides each pair with its own territory from which it subsequently drives trespassers. The size of a territory may range from the extensive hunting territory of a large carnivore to the very small domain of a nesting gull, which, like the territory of many other colonial nesting birds, is the diameter of a circle from within which one bird can reach out to peck another without having to stir from its nest.

In subsequent chapters we shall be dealing primarily with the phenomena of human reproduction, but these are frequently more readily understood against the background of their evolutionary development.

SUGGESTED SUPPLEMENTARY READING

Lamming, G. E., and Amoroso, E. C. (Eds.). Reproduction in the Female Mammal. Plenum Press, New York, 1967.
   A collection of papers with some discussion about the hypothalamic control of reproductive processes, gonadotropins, corpus luteum relationships and related topics in species other than man.
Mossman, H. W. Comparative biology of the placenta and fetal membranes. In Wynn, R. M. (Ed.). Fetal Homeostasis, Vol. II. N.Y. Acad. Sci., New York, 1967.
   This is the stenographic recording of a spirited discussion among experts about the evolution of placental types and their functions.
Villee, C. A. (Ed.). The Placenta and Fetal Membranes. Williams & Wilkins Company, Baltimore, 1960.
   In addition to introductory survey articles, this volume contains a classified and annotated bibliography of the literature from 1946 through 1958.
Wimsatt, W. A. Some aspects of the comparative anatomy of the mammalian placenta. Amer. J. Obstet. Gynecol., 84:1568, 1962.
   A classic, well illustrated description of the many morphological types of placentas that exist in mammals.

Chapter 2

DEVELOPMENT AND ANATOMY OF THE REPRODUCTIVE TRACTS

Inasmuch as structure traditionally precedes function, the anatomy of the reproductive organs should be thoroughly understood before we proceed with a description of their physiology or pathology. The anatomy, in turn, can be best appreciated by comprehending the embryological development of the several structures, which will also serve to explain the numerous congenital anomalies of the male and female reproductive organs encountered in clinical practice.

DEVELOPMENT OF THE REPRODUCTIVE SYSTEM

Ontogeny, the development of a complex individual from a single cell, the zygote, consists of a progressive revelation of a master plan which is stored in the genes, collectively known as the genome. The genome contains information which determines whether the zygote will become a mouse, a marmoset or a man. As the fertilized egg divides and becomes a solid ball, then a sphere, and then develops layers, it is the repression or the revelation of individual bits of genic information which results in the differentiation of tissues. The genetic basis for such a process will be discussed in Chapter 6 and aspects of early development of the embryo will be found in Chapters 10 and 15. For the present we will consider only the development of the urogenital system.

The Nephric Ducts

The urinary and reproductive tracts are remarkably intertwined as a result of the contributions of the various nephric ducts to the development of both systems. This explains why anomalies of the two tracts are frequently

11

associated. The paired pronephric tubules and ducts appear cephalad and are temporary structures. As the nephric ducts develop caudally, they drain the mesonephric tubules and form a collecting duct called the mesonephric or wolffian duct. More caudally the metanephros (permanent kidney) and its duct (the ureter) form bilaterally. A second pair of ducts,

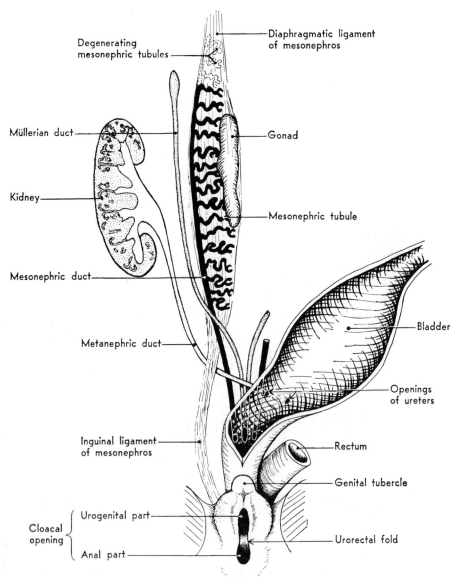

FIGURE 2–1.   Schematic diagram showing plan of the urogenital system at an early stage when it is still sexually undifferentiated. (From Human Embryology, by B. M. Patten. Copyright 1946, Blakiston Company. Used with permission of McGraw-Hill Book Company.)

the paramesonephric or müllerian ducts, develop independently beside the mesonephros. Thus human embryos of both sexes have at this "indifferent" stage (about two and one-half months) a double set of genital ducts, the mesonephric and paramesonephric ducts (Figure 2–1). The mesonephric ducts are retained in the male as the vas deferens but are reduced to vestiges in the female. The paramesonephric ducts disappear completely in subsequent male development but become the fallopian tubes, uterus and the uppermost portion of the vagina in the female.

Over the rather bulky mesonephros is a fold of peritoneum; its caudal end becomes fibrous and constitutes the inguinal ligament of the mesonephros (not to be confused with the inguinal or Poupart's ligament in the adult). In the male, this will become part of the gubernaculum testis and will be important in the descent of the testis into the scrotal sac. In the female, the inguinal ligament of the mesonephros will bend sharply, the upper portion becoming the ovarian ligament and the lower part becoming the round ligament of the uterus.

Urogenital Sinus

Just below the genital tubercle, the embryonic cloaca is divided into two sinuses by a urorectal fold (Figure 2–1), which will become part of the perineum. The cloacal membrane will be split so that the anal part will connect with the rectum, the ventral part of the urogenital sinus will be incorporated with the lower part of the bladder to form the urethra, and the dorsal part of the urogenital sinus will join with the caudal end of the fused paramesonephric ducts to form the vagina in the female.

Accessory Reproductive Organs in the Female

This, then, is the state of affairs in the sexually undifferentiated embryo. In the absence of functioning embryonic testes, and whether or not ovaries are present, the embryo will develop into a phenotypic female. This fact becomes important in certain cases of ambiguous sex differentiation. The events are shown diagrammatically in Figure 2–2. The mesonephros degenerates, leaving vestigial tubules called the epoophoron, within the broad ligament, and more caudally, remnants of ducts hug the lateral margins of the uterus, pass through the cervix and extend downward on either side of the vagina. These remnants, called Gartner's ducts, are clinically important because cysts may form there; they are known as paraovarian cysts when they arise within the mesosalpinx and as Gartner's cysts when they grow behind the vagina. In the cervix, a remnant may become malignant, giving rise to a mesonephric carcinoma. In Figure 2–3, the paramesonephric ducts are about to fuse, and the mesonephric ducts are shown within the embryonic myometrium. The folds of peritoneum which formerly covered the

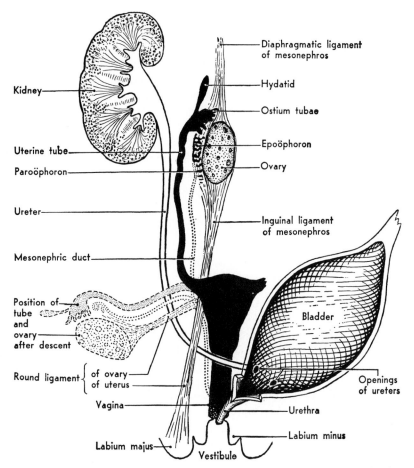

FIGURE 2–2. Schematic diagram showing plan of developing female reproductive system. Dotted lines indicate position of tube and ovary after their descent into the pelvis. (From *Human Embryology*, by B. M. Patten. Copyright 1946, Blakiston Company. Used with permission of McGraw-Hill Book Company.)

bulky mesonephros become the broad ligaments, leaving a small bay behind the uterus known as the cul-de-sac or pouch of Douglas.

Formation of Uterus and Vagina

The caudal portions of the paramesonephric (müllerian) ducts fuse to form the uterus and cervix. When the lower end of the fused ducts makes contact with the urogenital sinus, the cell cords are solid. They merge with the endodermal cells growing cephalad from the sinus to form a temporary barrier, which is called the müllerian tubercle. When canalized, the urogenital sinus forms almost the entire vagina and the tubercle becomes the lower portion of the uterus. Apparently the presence of the tubercle is necessary to induce the formation of the vagina, because when the uterus is missing the vagina does not form, except partially in the "testicular feminizing syndrome." When the ducts fail to fuse, the presence of two müllerian tubercles will induce the formation of a double vagina.

The funnel-shaped fimbriated ostia of the fallopian tubes do not form at the extreme cephalad tips of the paramesonephric ducts, but somewhat below this, opposite the gonads. The cephalad tips remain as little cysts which dangle from the ends of the tubes and are called hydatids (of Morgagni). As the broad ligaments and round ligaments form, the ovaries and tubes descend to lie laterally on either side of the uterine fundus (corpus uteri).

Anomalies

On occasion, the paramesonephric ducts may degenerate in the female fetus (as they do routinely in the male), resulting in an absence of the uterus

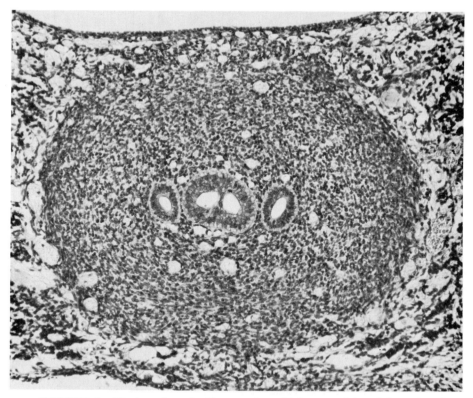

FIGURE 2–3. Uterine anlage of a 60 mm. embryo. Peritoneum of the cul-de-sac is at the top, the paramesonephric ducts are in the process of fusing, and the mesonephric ducts are still present on either side. (From Kraus, Gynecologic Pathology. C. V. Mosby Company, St. Louis, 1967.)

(and vagina). More commonly, the fusion, which always proceeds from below upward, is incomplete and results in a variety of anomalies, such as a bicornuate uterus (the normal situation in most lower mammals), a partial or complete septum dividing the uterine cavity into two lateral parts, a double uterus (didelphys) with double vagina, and many other permutations. If the vagina is present, some type of uterus is ordinarily present, but when the uterus is missing, so is the vagina.

In another type of anomaly, a functioning uterus is present but the lower portion of the vagina has not formed. With the onset of menstruation, the upper vagina and uterine cavity become distended with blood, and cor-

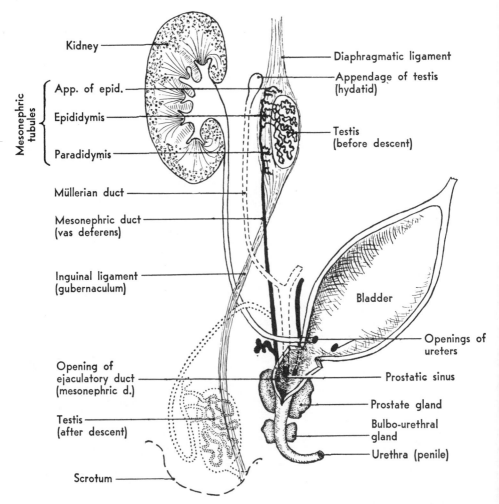

FIGURE 2–4. Schematic diagram showing plan of developing male urogenital system. (From Human Embryology, by B. M. Patten. Copyright 1946, Blakiston Company. Used with permission of McGraw-Hill Book Company).

rective surgery is required. The complete absence of the vagina creates serious psychological problems, but fortunately this can be remedied by the construction of an artificial vagina, using a skin graft from the buttocks sutured around an obturator and inserted into a space surgically dissected between the bladder and rectum. The optimal time to perform such surgery is between the ages of 16 and 19.

Accessory Reproductive Organs in the Male

When testes are present a different sequence of developmental events occurs (Figure 2–4). Sexual differentiation of the gonads begins about 40 days after fertilization of the ovum. About three weeks later the interstitial cells of the gonad undergo a marked hyperplasia and show histochemical reactions suggesting the production of steroid hormones. The seminiferous tubules become connected to the rete testis, a system of thin tubules which develop from the dorsal part of the gonad and form connections with the adjacent mesonephric tubules (which then differentiate into the epididymis) and through them to the wolffian ducts, which become the vas deferens, seminal vesicles and ejaculatory ducts. This development proceeds under the stimulus of testosterone secreted by the interstitial cells of the developing testis. The testes appear to secrete a second hormone, probably a peptide, which inhibits the development of the müllerian duct so that no fallopian tubes or uterus appear. The urethra of the male serves both excretory and reproductive functions, transporting both urine and semen. The external opening of the urogenital sinus is flanked on both sides by elongated thickenings, the genital folds, which meet anteriorly in front of the sinus and form a median outgrowth, the genital tubercle, the rudiment of the phallus. Under the stimulation of testosterone the rudiment of the phallus grows and becomes the penis, and outgrowths of the urethra form the prostate. The genital folds fuse on the posterior surface of the penis forming a tube, the penile urethra, continuous with the urethra leading from the urinary bladder. The outer genital folds become the scrotum.

Testicular Feminization Syndrome
(Androgen Insensitivity Syndrome)

The developmental effects of the two testicular hormones are strikingly evident in the curious type of male pseudohermaphroditism termed "testicular feminization syndrome." Individuals with this condition are genetic males, XY, and the testes, although intraabdominal or inguinal in location, secrete androgens in the same quantities as the testes of normal males. Like those of normal males, these testes also secrete estrogens and the individual develops as a phenotypic female lacking a uterus or tubes but with a shortened vagina formed by the urogenital sinus. In these individuals the basic defect appears to be a lack of response of the somatic cells to testosterone, in most cases caused by the absence of the cytoplasmic androgen receptors (p. 60). During embryonic development the testes elaborate testosterone as usual, but since the cells of the mesonephric duct, genital tubercle and urethra cannot respond, the accessory sex organs of a normal male cannot develop. The paramesonephric ducts do respond to the müllerian duct inhibiting hormone of the testis, and so degenerate, hence the absence of the uterus and tubes. In about half of the individuals the testes descend partially, leading to the proper diagnosis in early childhood when testes are found in bilateral inguinal hernias. In the remainder the diagnosis may not be made until after puberty when menstruation fails to occur. Breast development occurs at puberty under the influence of estrogens, but this development regresses if the testes are removed. The individuals are chromatin negative (p. 126) phenotypic females with little or no axillary or pubic hair, no sebum or acne but with typically feminine proportions, voice and habitus. The condition is usually treated by removing the testes (which are prone to develop neoplasia) and maintaining the patient on daily estrogen therapy.

Development of the Gonads

The primordial germ cells become segregated from the somatic cells early in embryonic development. Ten days after fertilization, they appear in the yolk sac as large cells with large vesicular nuclei and clear cytoplasm, staining deeply with an alkaline phosphatase reaction. Two days later, they have migrated to the hind gut, and three days after this the cells, by ameboid movements, have found their way into the mesonephric folds.

In the meantime, a genital ridge has appeared over the ventral part of the mesonephros on each side. At first, it consists of mesenchyme alone, but soon finger-like accumulations of cells move in to form sex cords. Then the germ cells move into the genital ridges (Figure 2-5), and later (day 33) into the sex cords to form the cortex of the gonad. By day 37, the male gonad has begun to organize itself into a testis. The sex cords condense to form seminiferous tubules, a capsule begins to form between the coelomic covering and tubules, the primordial germ cells become converted to spermatogonia and cells of the sex cords become Sertoli cells.

It would appear that unless the germ cells contain two sex chromosomes, they cannot persist in the gonads. When only one X chromosome is present (an XO, or Turner's syndrome), the gonads in later life are represented only by two streaks of mesenchyme with a complete absence of germ cells. Such individuals become phenotypic females and frequently have associated anomalies (see Chapter 6). In any individual with a Y chromosome (even XXXXY) a testis develops; thus the Y chromosome imposes the male pattern of development on the gonad. Evidence from chimeras (p. 140) produced experimentally and from Klinefelter's syndrome (XXY) suggest that XX germ cells die if present in a testis. Individuals with

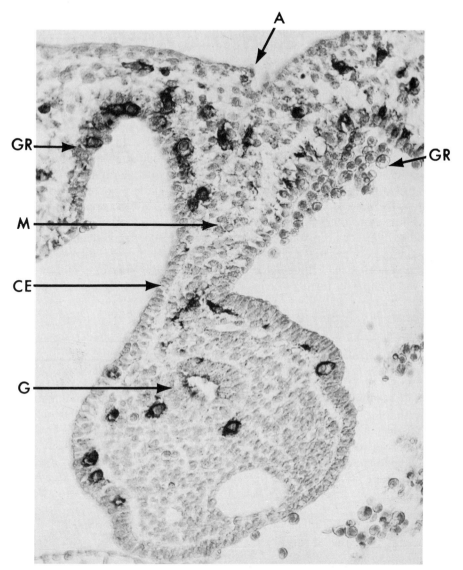

FIGURE 2–5. Transverse section from a 5 mm. human embryo showing migration of the germ cells (stained black with alkaline phosphatase) from the primitive gut through the mesentery and into the gonadal folds on either side. The space at the top is the aorta. *G*, gut; *M*, mesentery; *CE*, coelomic epithelium; *GR*, genital ridge; *A*, aorta. (From McKay *et al.*, Anat. Rec., *117*:201, 1953.)

Klinefelter's syndrome are phenotypic males but are sterile and have few or no germ cells in the testes.

The development of an ovary takes place much more slowly. At the embryonic age of 6 weeks, the sex cords break up into separate clumps of cells and the central part of the gonad becomes filled with loose mesenchyme permeated by blood vessels. By 16 weeks, these disorganized clumps become primary follicles, incorporating the germ cells, which now swell up and are called oogonia. The oogonia undergo furious mitotic activity, so that by age 20 weeks their number reaches a maximum of about 7 million. Thereafter, mitoses cease, and it is believed that no further ova are produced either later in fetal life or after birth. The term "germinal epithelium," which is used to denote the cortex of the adult ovary, is misleading and perhaps should be abandoned. Indeed, from mid-fetal life on for some 50 years, the ova undergo atresia (or are ovulated) until ovarian senility occurs at the menopause.

The oogonia, surrounded by follicle cells (which become granulosa cells), are protected from the advancing mesenchyme which forms the ovarian stroma, otherwise they would degenerate. The stroma thickens at the periphery to form a capsule-like tunica albuginea, and the original germinal epithelium thins out to a single layer of cuboidal epithelium which covers the adult ovary. By the time of birth, the oogonia have begun the first meiotic division and are called primary oocytes. They remain in the prophase until ovulation occurs. At birth there are about one million oocytes in the ovaries, and the number is further reduced to a half million or so by puberty. Of these, only about 450 are destined to ovulate, and a very few may eventually become fertilized. Such is the profligacy of Nature.

Excurrent Ducts in the Male

The sperm leaving the testis are not fertile but must undergo a maturation process in the epididymis, which enables them to become fertile. The epididymis can be divided into initial, middle and terminal segments, which are histologically quite distinct (Fig. 2–6). The initial segment is highly vascular and contains very tall columnar cells lacking cilia but having long

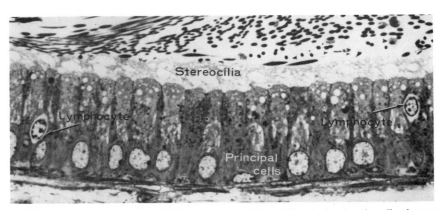

FIGURE 2–6. Photomicrograph of a portion of the epididymis showing the tall columnar principal cells with long stereocilia extending into the lumen of the epididymis, which contains many sperm. (Photograph courtesy of Dr. Martin Dym.)

microvilli that apparently function to concentrate the seminal fluid and remove water and electrolytes. The middle segment is lined with shorter columnar cells with shorter microvilli. Some of the sperm become trapped in the microvilli. The supranuclear cytoplasm of the cells in the middle segment is quite vesiculated and frothy in appearance. The terminal segment of the epididymis is lined with clear cells. The sperm are moved through the epididymis not by their own motility but by the contraction of the muscular coats in the wall. The cells of the epididymis have an extensive Golgi apparatus with an unknown function. The principal cell present in the terminal segment of the epididymis has an abundance of rough endoplasmic reticulum, which suggests that it may be synthesizing proteins for export. The fluid in the epididymis contains high concentrations of a number of enzymes, such as acetyl glucosaminidase and α-mannosidase. The epididymis also actively secretes carnitine into the epididymal lumen. Glutamate and inositol are found in high concentration in certain parts of the epididymis. An Androgen Binding Protein is synthesized in the Sertoli cells of the testis in response to FSH, and serves as an intraluminal carrier for transporting androgens, testosterone and dihydrotestosterone synthesized in the testis to the epididymis, an androgen target organ.

The Sertoli cells of the testis are unusual and interesting cells which provide support and nutrition to the sperm (Fig. 2–7). A considerable number of the germ cells, as many as one-third, undergo degeneration and are phagocytized by the Sertoli cells. They also phagocytize the residual parts of the spermatid that are cast off as the spermatid matures to a spermatozoon. The Sertoli cells also carry out certain aspects of steroid synthesis and transform androgens into estrogens, a process regulated by FSH. Adjacent Sertoli cells have tight junctions that provide the basis for a "blood testis barrier" which prevents the passage of albumin and other proteins from the blood into the seminal fluid. Germ cells are autoantigenic and can initiate the production of antibodies. Perhaps this blood testis barrier serves to protect the sperm against such antibodies.

From each epididymis a duct, the vas deferens, passes from the scrotum through the inguinal canal into the abdominal cavity, and then over the urinary bladder to the lower part of the abdominal cavity, where it joins the urethra. The terminal portion of the vas deferens is dilated to form the ampulla, which joins the seminal vesicle. The latter is derived embryologically from the vas deferens. The vas deferens empties into the ejaculatory duct, which empties, in turn, into the urethra. The seminal vesicle and the terminal part of the vas deferens are similar spongy tissues surrounded by muscle fibers. The secretions of the prostate and the seminal vesicle have no proven role in fertility, for sperm obtained from the terminal segment of the epididymis will fertilize eggs. However, these secretions do contain compounds that are important for sperm survival. The seminal vesicle, composed of columnar epithelium arranged in fronds, secretes fructose, ascorbic acid, some proteins, bicarbonate and potassium. The prostate is a small, chestnut shaped gland derived from the urethra and lying around it.

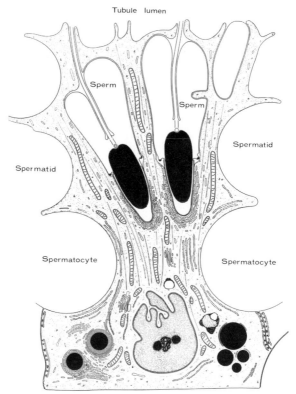

FIGURE 2–7. Diagram of a Sertoli cell showing nucleus and lipid droplets in the basal portion of the cell. Mitochondria and endoplasmic reticulum can be seen extending toward the distal portion of the cell. Imbedded in the cytoplasm of the Sertoli cell are several spermatocytes and spermatids, together with two older spermatids that have nearly completed the transformation into sperm. (Courtesy of Dr. Martin Dym.)

The prostatic secretions enter the urethra by many fine ducts in the ejaculatory duct. This secretion is acidic and contains citric acid as a major component. It has no reducing sugars but contains zinc, proteolytic enzymes and acid phosphatase. Nearly every man over the age of 50 has some degree of benign prostatic hypertrophy.

The prostate is under androgenic control. The bulbo-urethral glands empty into the urethra at the base of the penis; their secretion contains very viscous mucoproteins. The first portion of the human ejaculate to be released in coitus is provided by the bulbo-urethral glands, which is then followed by the prostatic secretion and finally by the seminal vesicular secretion. When first deposited, the human ejaculate coagulates, then about 20 minutes later the fibrinolysin contained in the ejaculate liquifies it.

The basic question in andrology is whether sperm have an intrinsic developmental sequence that takes them to maturity, or whether the epididymis or other structures in the male tract play some obligatory role in bringing them to maturity. Sperm taken from the upper part of the epididymis move in a rather curious circular motion, whereas those from the lower end of the epididymis move in a more direct, straight line. Sperm metabolize glucose or fructose to lactate and carbon dioxide, thereby generating ATP to drive the beating of the tail. The biophysical basis of the beating of the tail is not fully understood, but it appears to involve the sliding of filaments one on another, which is analogous to the sliding of filaments involved in muscular contraction.

PELVIC ANATOMY IN THE ADULT FEMALE

Our purpose in describing the gross and microscopic anatomy of the female generative system is to relate structure to function, both normal and dysfunctional. Unlike the format of textbooks of anatomy (which should be consulted for finer details), in this book the more important gross and microscopic descriptions will be interspersed with their clinical implications. A student of medicine, for example, must be thoroughly familiar with normal female anatomy before he can interpret variations which may be termed abnormal. Furthermore, any student of human biology must understand both form and function for a comprehension of human sexuality and reproduction.

Internal Genitalia

The appearance of the pelvic organs when exposed through a lower abdominal incision is shown in Figure 2–8. The uterine fundus is held forward to expose the cul-de-sac, which is bounded laterally by the uterosacral ligaments and posteriorly by the rectum. In either the supine or the erect posture, the cul-de-sac is the most dependent portion of the peritoneal cavity; hence any free fluids, such as blood or ascites, tend to gravitate into this pouch. Such fluids may be readily sampled by inserting a needle through the posterior fornix of the vagina, or the patient may be put in a knee-chest position, to permit the small bowel to fall away, and a culdoscope may be inserted through the same area to obtain a view of the pelvic organs.

The round ligaments, which are attached to the uterus at a point just anterior to the insertion of the tubes, form the upper margin of the broad ligaments, and they extend through the internal inguinal ring on either side, terminating in the labia majora, which are the counterpart of the scrotal sacs in the male. The broad ligament on each side is analogous to the mesentery or mesometrium of the uterine horn in those mammals which regularly possess a bicornuate uterus. The upper two-thirds consists of two thin layers of peritoneum, between which lie the tubular remnants of the mesonephros. The lower third consists of markedly thickened, dense connective tissue, which is referred to as the cardinal ligament because it is the main supporting structure for the uterus. Through this cardinal ligament course the uterine

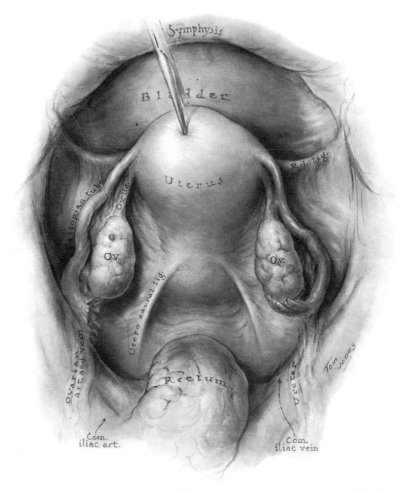

FIGURE 2–8. Female pelvis viewed from above. (Tom Jones, courtesy of S. H. Camp and Company.)

artery and vein, and just beneath these vessels lies the ureter, which wends its way through a tunnel just before inserting into the trigone of the bladder. One must be very careful near the base of the broad ligament when performing a hysterectomy, because the ureter may be injured while one attempts to control bleeding from the uterine vessels.

Just behind the insertion of the tube is the short ovarian ligament which suspends the ovary. During the reproductive period each ovary is a white, almond-shaped structure, 3 to 5 cm. in length and 5 to 8 gm. in weight. Small, clear follicular cysts and corpora lutea in various phases of formation and regression are commonly seen grossly. The true mesentery of the ovary, coursing from the ovarian hilum to the lateral pelvic wall, is called the infundibulo-pelvic ligament, and contains the ovarian vein and artery. These vessels anastomose with the uterine vessels, so that in essence the uterus has a dual blood supply on each side. Once again we find the ureter coursing just medial and posterior to the ovarian vessels at the lateral origin of the infundibulo-pelvic ligament, another area of potential danger when the tube and ovary must be removed. The ovarian arteries arise directly from the aorta at a point just below the renal vessels (where the gonads used to lie in embryonic life). The right ovarian vein empties into the vena cava; the left one is joined to the left renal vein.

The arrangement of the blood vessels in the pelvis is shown in Figure 2–9. Note again how the ureter runs its course beneath the bridges formed

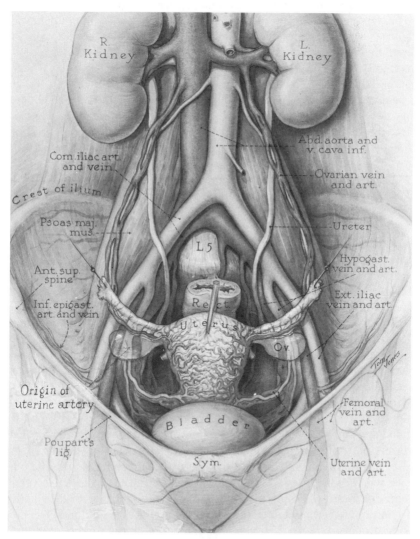

FIGURE 2–9.   Blood vessels supplying the pelvis. (Tom Jones, courtesy of S. H. Camp and Company.)

by the main blood vessels supplying the internal genitalia. ("Water flows under the bridge.")

The pelvic organs are richly supplied by lymphatic vessels interspersed with lymph nodes. In general, the lymphatic system follows the overall pattern of the major blood vessels. On either side of the cervix, for example, the lymphatic vessels extend laterally through the parametrial tissues to reach the retroperitoneal nodes around the bifurcation of the common iliac artery, and thence to the aortic nodes. This is the route commonly traveled by malignant cells arising from cervical cancer, the most common malignancy of the pelvic organs in women.

NERVE SUPPLY.   The internal genitalia, like other viscera, are richly supplied by afferent and efferent autonomic nerves, both motor and sensory. The parasympathetic fibers arise from the sacral nerves, and are believed to produce vasodilatation and an inhibition of muscular contraction. The efferent sympathetic motor nerves originate from the ganglia of T-5 to T-10, converge over the sacrum, and reach the uterus through ganglia which lie near the base of the uterosacral ligaments, as shown in Figure 2–10. They are believed to cause vasoconstriction and muscular contraction. The regulatory action of these autonomic nerves is superimposed upon an intrinsic uterine motility. Whereas a high spinal anesthesia up to T-5 tends to abolish the fundal contractions of labor, cord transections above this level do not necessarily prevent labor and delivery.

Of more immediate clinical significance, perhaps, are the afferent sensory fibers which carry pain sensation. From the uterus, these converge in the paracervical areas and travel upward through the superior hypogastric plexus, which is located just below the bifurcation of the aorta, and thence into the cord at T-11 and T-12. Thus it can be seen that pain of uterine origin can be interrupted by local anesthetics in the paracervical regions, by surgical extirpation of the superior hypogastric plexus, or by spinal, epidural or caudal anesthesia which reaches T-11, without interfering with the efferent autonomic motor functions.

Sensory fibers originating in the distal portions of the tubes or from the ovaries follow the course of the ovarian vessels to merge with the renal plexus and from there to the T-11 and T-12 segments. Thus pain of ovarian origin may mimic pain of ureteral origin, both of which may be referred to the flank and downward to the inguinal and vulvar areas.

In hollow viscera such as the uterus and tubes, the stimulus for pain is either stretch or ischemia. A biopsy of the cervix may be performed with minimal discomfort, but stretching the internal os of the cervix to any degree (as in the performance of a dilatation and curettage) requires some form of anesthesia. A torsion of the tube or ovary which interferes with the blood supply will produce severe pain. Peritoneal surfaces, which also receive somatic innervation, give rise to pain as the result of stretch or direct chemical irritation, as with inflammation or contact with clotted blood. The blood clotting process apparently liberates a compound, quite possibly serotonin, which is highly irritating.

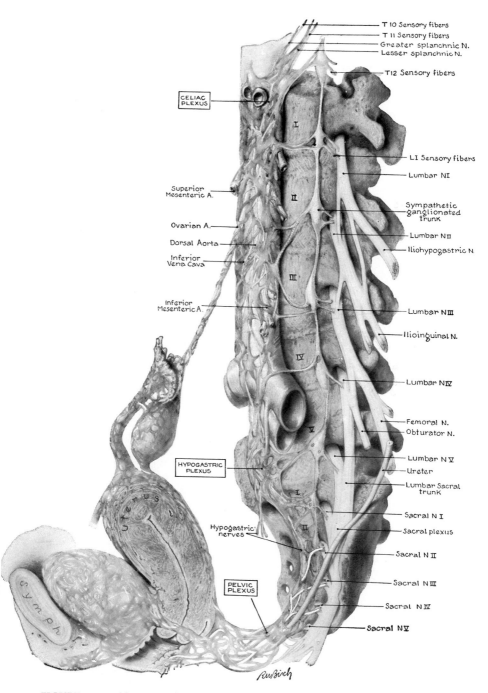

FIGURE 2–10. Nerve supply to the pelvic organs. (From Hellman and Pritchard, Williams' Obstetrics, 14th Ed. Appleton-Century-Crofts, New York, 1971.)

PELVIC SUPPORTS.   The pelvic organs as seen in a nearly sagittal section are shown in Figure 2–11. The anterior and posterior walls of the vagina are normally in contact with each other. In the illustration, they are being held apart by a vaginal speculum used to expose the cervix, which is shown in the inset. The uterine corpus is pictured in its usual anterior position, which in the standing posture is virtually horizontal; but the uterus is a rather mobile organ and it may either swing backward (retroversion) or bend backward (retroflexion) on its fulcrum, which consists of the cardinal ligaments. These retrodisplacements are so common as to be considered normal variations.

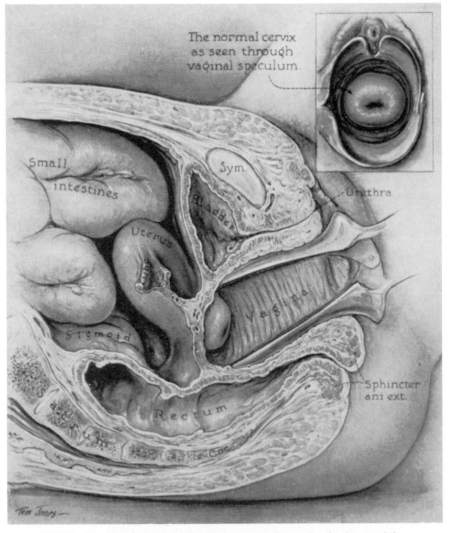

FIGURE 2–11.   Near-sagittal section showing vaginal speculum in place, and the appearance of a normal cervix. (Tom Jones, courtesy of S. H. Camp and Company.)

The bladder is loosely attached to the lower portion of the uterus. Once its peritoneal reflection is incised, it may be readily pushed off the uterus by blunt dissection in the midline, although laterally there are vesico-uterine ligaments which must be carefully dissected.

UROGENITAL DIAPHRAGM.   Attached to the under border of the symphysis is a condensation of fibromuscular tissue which surrounds the upper part of the urethra and lower part of the vagina to form the urogenital diaphragm (Figure 2–12). From the standpoint of pressure relationships, the urogenital diaphragm supports the upper part of the urethra as it slants

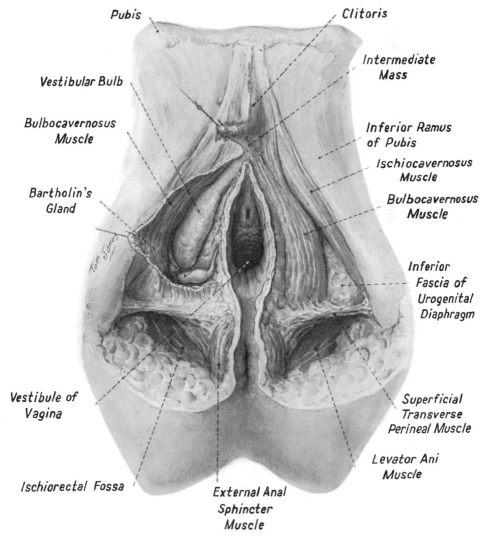

FIGURE 2–12.   Anterior and posterior components of the female perineum. (From Reid, A Textbook of Obstetrics. W. B. Saunders Co., Philadelphia, 1962.)

upward from the bladder. When the intra-abdominal pressure is suddenly increased, as with coughing or sneezing, the pressure is transmitted not only to the bladder but to the upper urethra as well, so that urine does not escape. If, however, the supporting tissues are torn or stretched by childbirth, the bladder neck and urethra may sag downward (*urethrocele*). Increases of intra-abdominal pressure push on only the bladder, forcing urine to escape. This is called *stress incontinence,* a common and most annoying symptom, but one which may be corrected by a surgical procedure designed to replace the bladder neck behind the symphysis where it belongs (see Chapter 17).

PELVIC RELAXATIONS.   There are other forms of pelvic relaxations

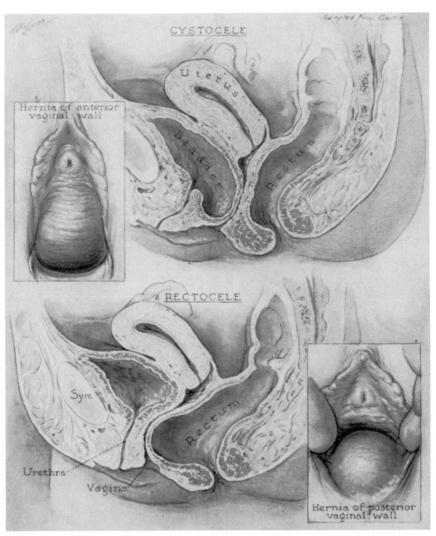

FIGURE 2–13.   Cystocele and rectocele. (Tom Jones, courtesy of S. H. Camp and Company.)

which are common in multiparous women. When the musculofascial portion of the anterior vaginal wall is stretched so that the bladder pushes the vagina through the introitus, the resulting bulge is a *cystocele*. (The suffix *-cele* means hernia.) A comparable situation with the posterior vaginal wall gives rise to a *rectocele*. These are illustrated in Figure 2–13, as observed both in sagittal section and clinically. When either herniation becomes troublesome, a vaginal repair may be indicated (see Chapter 17).

An additional herniation, somewhat more serious, may occur when the cul-de-sac, filled with small bowel, dissects downward between the vagina and the rectum, forming an *enterocele*. When the cardinal ligaments become stretched, the entire uterus descends like a plunger, producing various degrees of descensus. Modern obstetrical practices, including the rather routine use of episiotomy and the substitution of cesarean section for difficult mid-forceps delivery, are steadily decreasing the frequency with which these many forms of pelvic relaxations are encountered.

The External Genitalia

The external structures of a parous woman are shown in Figure 2–14. In the nulliparous female, the labia majora are opposed, usually concealing the labia minora. In most women the clitoris, the homologue of the penis, is completely covered by the prepuce. The labia majora consist of fatty tissue, covered by skin richly endowed with hair follicles and sebaceous glands. The labia minora are thin, pink, and covered with stratified epithelium resembling the vagina. They are devoid of hair, but contain many sebaceous follicles and a variety of nerve endings. The hymen is comprised of elastic and collagenous connective tissue, and presents many variations of form. In virgins, the hymenal opening may be only a few millimeters in diameter, but more commonly will just admit the tip of one finger, although occasionally two. When a young lady is seen for premarital examination and a virginal introitus is found, it is wise to advise her to perform self-dilatation over the next two or three weeks and perhaps to supply her with a suitably shaped dilator to facilitate the procedure. Otherwise, the initial coitus may result in a gross tearing of the hymen, which is not only painful but predisposes to the development of bladder infection ("honeymoon cystitis").

Above the hymenal orifice, the vaginal tube constitutes an excretory duct for the uterus, a repository for semen and a birth canal. It is relatively insensitive, and is not a sexual organ for the female in the erotic sense. The sexually responsive organs of women are the prepuce and clitoris, with its lateral crura (the corpora cavernosa), the erectile vestibular bulbs, and the labia minora. Taken collectively, these structures constitute what Masters and Johnson call the "sexual platform". They are shown in Figure 2–12 as they would be seen in a superficial dissection which also reveals some of the important muscles making up the pelvic floor. The Bartholin's glands are compound mucus-secreting glands; their ducts, each 1 to 2 cm. long, open laterally just outside the vaginal orifice. They are subject to gonorrhea and

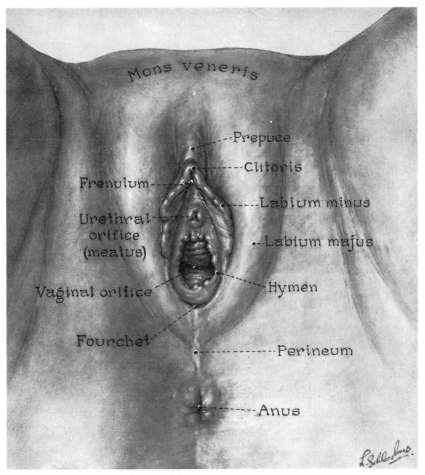

FIGURE 2–14.   External genitalia of the normal female. (From Hellman and Pritchard, Williams' Obstetrics, 14th Ed. Appleton-Century-Crofts, New York, 1971.)

other types of infection, resulting in abscesses or subsequent cysts. Two other ducts that are often subject to chronic infection are the paraurethral or Skene's ducts, which open lateral to the urethral meatus.

The blood supply to the upper vagina is largely from the descending cervico-vaginal branch of the uterine artery. The inferior vesical arteries supply the middle third; the lower portion and the perineal body are supplied by branches of the internal pudendal artery.

The lymphatic drainage of the labia and lower vagina is to the inguinal lymph nodes on both sides. Thus a malignant lesion involving one labium majus requires not only radical vulvectomy but also bilateral removal of all superficial and deep inguinal nodes.

The bony pelvis will be discussed in conjunction with the principles of antenatal care (Chapter 13).

MICROSCOPIC ANATOMY

Ovary

As mentioned earlier, the majority of ova in the cortex become atretic after early follicular development. Those which are destined for ovulation undergo a predictable course which is illustrated schematically in Figure 2–15. In the early phase, the follicle cells become cuboidal and the stromal cells around the follicle become quite prominent. The ovum itself enlarges and develops a prominent nucleolus, and the surrounding granulosa cells multiply rapidly (Figure 2–16).

The innermost three or four layers of granulosa cells become adherent to the ovum and are shed with it at ovulation to form a crown, the corona radiata. Now a fluid-filled antrum forms, and clear gelatinous material gathers around the ovum to form the zona pellucida, through which ultra-microscopic canals are found. The eccentrically placed ovum and its mound or cumulus of granulosa now give the picture of the classic graafian follicle (Figure 2–17). Surrounding the granulosa cells is a thin basement membrane,

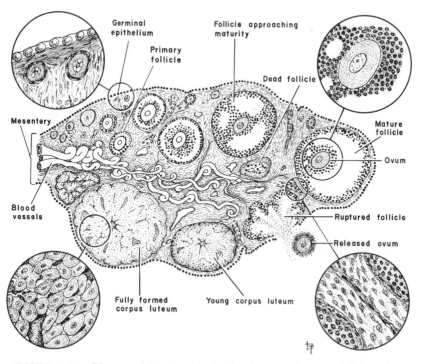

FIGURE 2–15. Diagram of the stages in the development of an egg, follicle and corpus luteum in a mammalian ovary. Successive stages are depicted clockwise, beginning at the mesentery. The insets show the cellular structure of the successive stages. (From Villee, C. A. Biology, 6th Ed. W. B. Saunders Co., Philadelphia, 1972.)

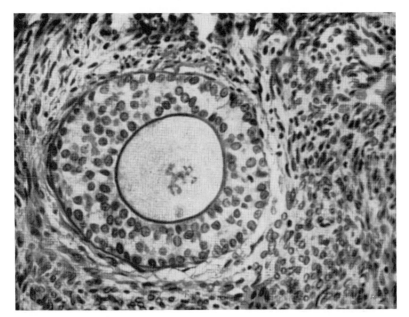

FIGURE 2–16. Follicle with two or three layers of granulosa cells separated by fluid-filled spaces which will later coalesce to form an antrum. Outside the basement membrane of the follicle are spindle-shaped cells of the theca interna. (From Shettles, Ovum Humanum. Hafner Publishing Company Inc., New York, 1960.)

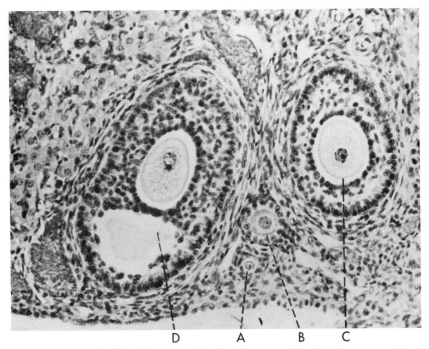

D   A   B   C

FIGURE 2–17. Portion of rat ovary showing (A) oocyte, (B) enlarging oocyte with early follicle, (C) nearly full-grown oocyte with zona pellucida and distinct theca and (D) fully developed graafian follicle with antrum formation. (From Witschi, Development of Vertebrates. W. B. Saunders Co., Philadelphia, 1956.)

and outside of this, the connective tissue cells have organized themselves into two coats, the theca interna and externa (Figure 2–18).

As the enlarging follicle begins to bulge from the surface of the ovary, a thin portion almost devoid of capillaries forms. This is the stigma through which the egg will escape. At the moment of ovulation (Figure 2–19), the ovum, together with its surrounding corona, is discharged into the peritoneal cavity, hopefully to be swept into the fingered ostium of the tube.

The cavity of the follicle often fills with blood, some of which may escape into the cul-de-sac, producing pain. During the next three days, there is an extensive invasion of the granulosa by capillaries, and both the granulosa and the theca interna cells become luteinized, that is, filled with a yellow carotenoid material. One week after ovulation, the corpus luteum has reached the peak of its activity, and looks as it does in Figure 2–20. There is a pronounced vacuolization of the theca and granulosa cells, which now

Follicular fluid   Granulosa

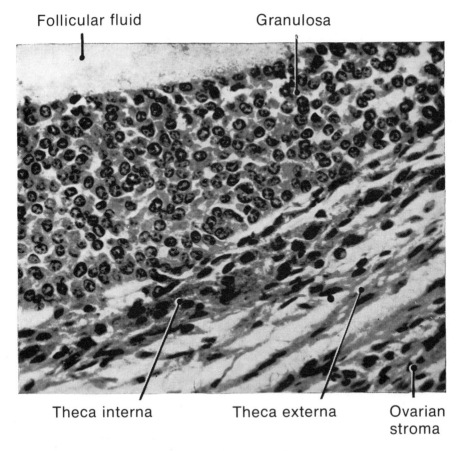

Theca interna   Theca externa   Ovarian
                                                      stroma

FIGURE 2–18.   Section through the wall of a mature Graafian follicle. (From Hellman and Pritchard, Williams' Obstetrics, 14th Ed. Appleton-Century-Crofts, New York, 1971.)

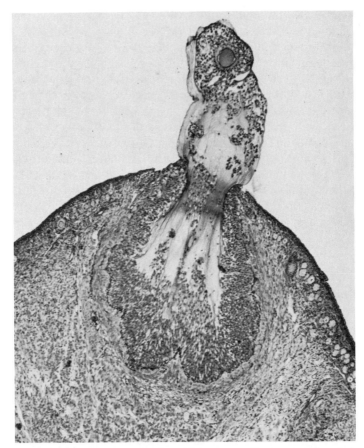

FIGURE 2–19.   Photograph of a rabbit follicle taken at the moment of ovulation. (Courtesy of Richard J. Blandau.)

resemble each other closely, and histochemical reactions give evidence of intensive hormonal activity. Many venules and collecting veins have appeared.

Now the corpus luteum begins to regress, and an increased quantity of connective tissue grows into its center. There is a progressive vacuolization of the granulosa cells, and by the time of the menstrual period, extensive degeneration has occurred. Eventually, the corpus luteum regresses and ends up as a tortuous white scar, the corpus albicans.

Uterine Tube

Under the peritoneal coating of the tube is a thin muscular layer with an inner circular and outer longitudinal coat, and then a mucosa which consists of a layer of columnar cells, of which some are ciliated and some secretory.

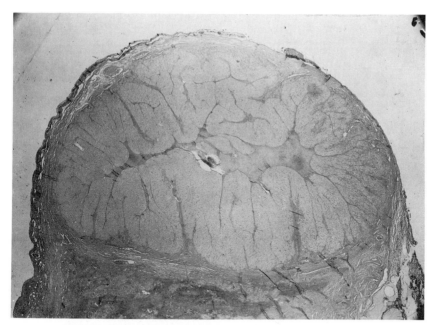

FIGURE 2–20.   Low-power view of mature human corpus luteum.

The mucosa is thrown up into progressively more complicated folds as the lateral ostium is approached (Figure 2–21).

The fallopian tube may be divided into four rather characteristic segments. (1) The infundibulum is the funnel-shaped fimbriated ostium which lies near the ovary. Around the time of ovulation, the fimbria (Fig. 2–22) go through beckoning motions which may entice the ovum into the mouth (this has been observed through a culdoscope). If the fimbria are cut off or diseased, relative infertility results. (2) The ampulla occupies the distal and middle portions of the tube, and it is here that the sperm meets the egg. The fertilized egg may implant here upon occasion, resulting in a tubal pregnancy, or it can implant at any point in the tube or even in the ovary or abdominal cavity. (3) Between the ampulla and the uterus is the isthmus, a small, firm portion which feels like — and indeed has been confused with — the round ligament during tubal sterilization. (4) The interstitial part traverses the myometrium, and it is exceedingly small in caliber, the lumen measuring less than a millimeter. The ovum cannot traverse this tiny tunnel unless it sheds its crown of granulosa cells.

The motions of the mucosal cilia and the peristaltic movements of the muscular coat create a current which slowly carries the egg toward the uterine cavity. The muscular activity is influenced by estrogen and by prostaglandins, and reaches its peak about the time of ovulation.

The chief enemy of the tubal mucosa is gonorrhea, a surface-spreading infection producing a purulent exudate which fastens the mucosal surfaces

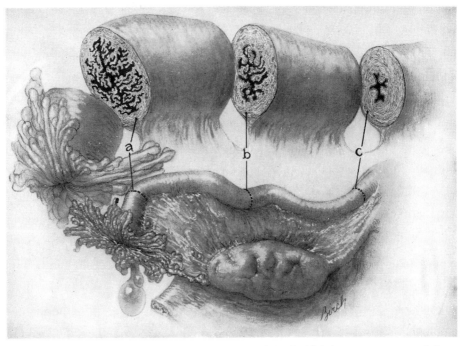

FIGURE 2–21.   The fallopian tube in cross-section, showing the gross structure of the epithelium in several portions: (a) infundibular, (b) ampullar and (c) isthmic. (From Hellman and Pritchard, Williams' Obstetrics, 14th Ed. Appleton-Century-Crofts, New York, 1971.)

FIGURE 2–22.   Scanning electron micrograph of the fimbria of the fallopian tube of a 50-year-old woman. Magnification × 10,500.

to each other, and often seals off the fimbriated end to produce a pyosalpinx. This is a bilateral process that results in tubal occlusion and sterility if it is not treated early and intensively with antibiotic drugs (see Chapter 17).

The patency of a tube may be determined by the Rubin test, a simple office procedure in which carbon dioxide gas is introduced into the uterus and the pressures measured, or by injecting a radiopaque fluid and making serial x-ray studies.

Myometrium

The bulk of the uterus consists of interlacing bundles of smooth muscle cells. Electron microscopy reveals light cells with a paucity of filaments and dark cells with large numbers of filaments, presumably actin, arranged obliquely to the long axis. In all probability, these are different functional states which depend upon the state of contraction and hormonal influences. Under the influence of estrogen, light cells are more numerous and the paucity of filaments may be the result of a state of depolymerization of the filaments. Contraction of the cell is accompanied by a loss of intracellular fluid, so that the intracellular proteins are packed more closely together, giving the dark appearance. Under the influence of progesterone, the dark cells with prominent filaments predominate. This is consistent with the higher resting tonus of the myometrium under progesterone influence. In both myometrial and endometrial cells, estrogen results in an increased cytoplasmic content of free ribosomes and endoplasmic reticulum, reflecting an increased production of RNA.

Endometrium

The uterine mucous membrane lies directly upon the myometrium, and varies in thickness from 1 to 6 mm., depending upon the stage of the menstrual cycle. The cyclic changes of the endometrium are described in Chapter 3. A low-power electron micrograph of the epithelial surface in the proliferative, estrogen-dominated phase would reveal microvilli, together with evidence of secretion, on the surface. Under the influence of progesterone, growth ceases and there is a decrease in the ultrastructural evidence of protein formation. Progesterone also causes the stromal cells to become large, pale, polyhedral cells filled with lipid and glycogen. When fully developed, they are called *decidual cells,* because they are shed during the menses or with the placenta at the time of delivery.

The endometrial glands reach down to the myometrium. They undergo striking changes during the ovarian cycle. On occasion they may penetrate the muscle deeply, forming adenomatous nodules, a condition known as *adenomyosis uteri.* Bits of endometrium may implant or "metastasize" to any portion of the pelvis, leading to a condition known as *endometriosis.* The endometrium may also form polyps which persist from one cycle to the next and which may cause increased flow or bleeding between periods (see Chapter 17).

Vagina

The vaginal mucosa consists of a stratified squamous epithelium which, under the influence of estrogen, has numerous layers. On the surface are large, flat cells with pyknotic nuclei, and at the base near the basement membrane are small, round basal cells with large vesicular nuclei. In between are parabasal and intermediate cells. Ascertainment of the relative numbers of these types found at any one time in a vaginal smear gives a reasonably good estimate of the estrogen status of the woman. After the menopause, if the women is not taking estrogen supplementation, the vaginal mucosa becomes thin, smooth and atrophic, and the surface cells which are shed are largely parabasal cells. If one should now treat such a vagina with a topical cream containing estrogen, one may observe a progressive shift to the intermediate and finally to the superficial cells.

The vaginal mucosa is rather firmly attached to a thick underlying musculofascial layer which changes from predominantly smooth muscle to mainly connective tissue cells in the deeper layers. As noted earlier, this fibromuscular tunic may be stretched or torn during childbirth, and it becomes atrophied and less vascular as the supply of estrogen decreases, so that such herniations as cystocele and rectocele are most common in multiparous, postmenopausal women.

In summary, a knowledge of how the female reproductive tract develops in embryonic life, of the gross and microscopic anatomy of the pelvic organs, and of how the various epithelia of the genital tract change with the ovulatory cycle and with the phases of a woman's life—this is the fundamental scientific base for understanding organic gynecologic diseases.

SUGGESTED SUPPLEMENTARY READING

Boving, B. G. Anatomy of Reproduction. *In* Greenhill, J. P., Obstetrics, 13th Ed., W. B. Saunders Company, Philadelphia, 1965, Chap. 1.
   All the "horizons" of embryonic development are presented in detail and beautifully illustrated. The chapter is particularly recommended for its descriptions of the ovary.
Patten, B. M. Human Embryology, 3rd Ed. Blakiston Division, McGraw-Hill Company. New York, 1968.
   Chapter 19 presents a thorough description of the embryological development of the urogenital system.
Sciarra, J. J. (Ed.). Gynecology and Obstetrics. Harper & Row, Publishers, Inc., New York, 1974.
   Chapters 1 to 5 are profusely illustrated descriptions of the female pelvis.

Chapter 3

THE FEMALE REPRODUCTIVE CYCLE: THE FUNCTIONAL BASIS OF GYNECOLOGY

HISTORICAL ASPECTS

It would be expected that a cyclic phenomenon such as menstruation should have produced a vast store of folklore, fancies and superstitions, mingled with a few facts, throughout the course of recorded history. Prior to the eighteenth century, it was believed that menstruation was a device of nature for the periodic excretion of accumulated poisons in women, under the control of the lunar cycle (hence the term *menses*), and in some way essential for the survival of an early embryo. By observation of various animals in estrus, it was erroneously concluded that the production of an egg occurred at the time of menstruation, and only as the result of a fruitful coitus. Many of these ancient beliefs are prevalent today among the people of underdeveloped countries and in the biologically uneducated populace of our own country.

The nineteenth century was marked by the discovery of the human ovum, the emerging concept of spontaneous ovulation and recognition of the fact that the uterus was under control of the ovaries. Following the description of the endometrial cycle by Hitschmann and Adler in 1907 and the correlation of these changes with the ovarian cycle by Schroeder in 1914, rapid progress was made by morphologists, culminating in Markee's meticulous observations of endometrial transplants in the anterior chambers of monkeys' eyes (1940). Indeed, it was the introduction of the rhesus monkey into the study of the sexual cycle by Corner in 1923 that paved the way for an experimental analysis of the primate cycle by numerous investigators.

The identification of the hormones responsible for the morphological changes and recognition of their precise roles awaited the development of

41

bioassay techniques. The estrous cycle of the rat, worked out by Long and Evans in 1921, the assay of estrogenic activity, described by E. Allen and Doisy in 1923, and the discovery of progesterone by Corner and W. Allen in 1929 laid the foundation for the discovery of the ovarian steroids and the gonadotropins. The period since 1935 has been marked by an exposé of intermediary steroid metabolism and the purification of the pituitary hormones. In the past decade, biochemists have become increasingly involved with the interaction of steroid hormones and intracellular components, and neurophysiologists have increasingly studied the hypothalamic control of the pituitary gland.

THE NEUROENDOCRINE CONTROL
OF PITUITARY SECRETIONS

The adenohypophysis, formerly regarded as the "master gland" which controlled other endocrine organs, should be regarded as the "servant" of the hypothalamus. The latter, in turn, is subject to many influences from higher nerve centers. Although no nerves connect the brain and the anterior pituitary, there are vascular connections by way of the portal system which carries neurosecretions from the area of the median eminence to the pituitary gland. The anatomic relations are shown schematically in Figure 3–1.

Final evidence for a neurovascular link was the isolation from hypothalamic tissue of several releasing factors, each capable of controlling the release of pituitary hormones. A growth hormone releasing factor (GRF) as well as a growth hormone inhibitory factor (GIF) are produced. The latter has now been isolated, its amino acid sequence determined and its name changed to somatostatin. It has a variety of effects in addition to its inhibition of GH release. Corticotropin releasing factor (CRF), the first of the hypothalamic hypophysiotropic substances to be identified as a distinct entity (though its chemical structure is still unknown), brings about the release of ACTH from the anterior pituitary. The first releasing factor to have its chemical structure determined was thyrotropin releasing factor (1970). In general, when the amino acid sequence of a hypothalamic factor is known, that substance is called a hormone. Thus the tripeptide composed of pyroglutamate, histidine and proline residues is referred to as thyrotropin releasing hormone (TRH). This hormone brings about the release of TSH and of prolactin from the anterior pituitary. Prolactin release is also regulated by a prolactin inhibitory factor (PIF) and, at least in birds, a prolactin releasing factor (PRF). The hypothalamus exerts an inhibitory influence on the secretion of prolactin under normal circumstances. Stimuli that induce a release of prolactin (suckling, for example) ultimately produce their effect by reducing the hypothalamic secretion of PIF. The secretion of MSH is also regulated by a releasing (MRF) and an inhibitory factor (MIF).

There are substances in extracts of the median eminence which can raise plasma concentrations of gonadotropins, induce release of FSH and LH from pituitary tissue in vitro, and rapidly lower the content of these hormones in the pituitary gland. It has been suggested that a single factor, gonadotropin releasing hormone, regulates the output of both gonadotropins. Porcine and ovine LH releasing hormone (LRH) and the peptide isolated from extracts of the median eminence show identical dose response curves in regard to release and synthesis of LH and FSH by pituitary tissue. As determined by assays both in vivo and in vitro, LRH is localized in the basal hypothalamus in a rather broad zone which extends fron the suprachiasmatic region to include the median eminence and pituitary stalk. It seems likely that the secretory neurons that elaborate LRH have their cell bodies in the suprachiasmatic region and axons which project to the median eminence and pituitary stalk, terminating there in juxtaposition with the hypophyseal portal vessels (Figure 3–1). Other neurons elaborating LRH have cell bodies in the arcuate nucleus, immediately overlying the median eminence. At present, evidence supports the concept that the arcuate neurons regulate the tonic discharge of LH, while the suprachiasmatic neurons regulate the burst of LH which triggers ovulation.

There are two ways in which the central nervous system controls gonadotropin secretion. One is through the hypothalamic neurosecretions, and the other involves the releasing factor regulating mechanisms, which include all of those portions of the brain that modify the activity of the median eminence cells or determine their threshold of sensitivity. These

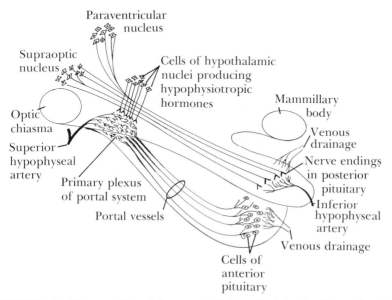

FIGURE 3–1. Diagram of hypothalamus and pituitary gland showing the portal system of the vasculature. (From Villee, D. B. Human Endocrinology. W. B. Saunders Co., Philadelphia, 1975.)

brain centers are concentrated in the limbic system, the thalamo-hypothalamic border and the anterior hypothalamus. These areas are sensitive to the action of the ovarian steroids (external feedback system) and probably to the pituitary gonadotropins as well (internal feedback system). For the most part, the higher centers exert an inhibitory action upon the hypothalamus, possibly by raising the neural threshold of sensitivity. It is believed, for example, that the amygdala exerts an inhibitory influence upon the hypothalamic centers which release LRH, and that it might be the progressive decline of this inhibition which initiates puberty. The amygdala also monitors emotional stimuli. Thus a stressful situation may suppress LRH, so that luteinizing hormone is not released, ovulation does not occur and amenorrhea results.

Positive feedback results when a steroid, an estrogen or a progestin, stimulates the release of a gonadotropin, such as the luteinizing hormone (LH). Negative feedback results when a hormone, such as a large amount of progesterone, prevents the release of a gonadotropin. A steroid hormone may have either effect, depending upon its concentration in the blood. For example, a small rising concentration of estradiol may trigger the elaboration of LRH, whereas a continuous high concentration will inhibit or alter the neurosecretion of the LH releasing factor. The latter effect is the basis for the use of estrogens in the sequential oral contraceptives.

EFFECT OF GONADAL HORMONES UPON MATURATION OF THE FEMALE FORE-BRAIN

There is a critical period during which the steroid-sensitive center of the brain which controls the ovarian cycle is not fully differentiated. In the rat, this includes the first 10 days of postnatal life, but in the rhesus monkey (and therefore most likely in the human subject), maturation occurs in the late prenatal period. If the brain of a female rat is exposed to a single dose of androgen during this undifferentiated period, the animal becomes permanently sterile and remains in constant estrus. Both FSH and LH are secreted in adequate amounts, but there is no inherent cyclic rise in LH secretion. The ovaries become polycystic and secrete estrogens at a constant rate. Similar sterilization results from administration of estrogens during the critical period. In many respects this resembles a disorder in human females who have amenorrhea associated with a polycystic ovary syndrome.

Barraclough believes that this steroid-sensitive center is in the preoptic suprachiasmatic area, as shown in Figure 3–2. When this area is exposed to androgen, as in a male fetus, it becomes incapable of cyclic responses to endogenous ovarian hormones, and the continuous tonic pattern of FSH and LH release is the result. As little as 10 μg. of testosterone injected into female rats two to five days after birth will convert a "female brain" into a "male brain" in 70 per cent of the animals. These neuroendocrine events

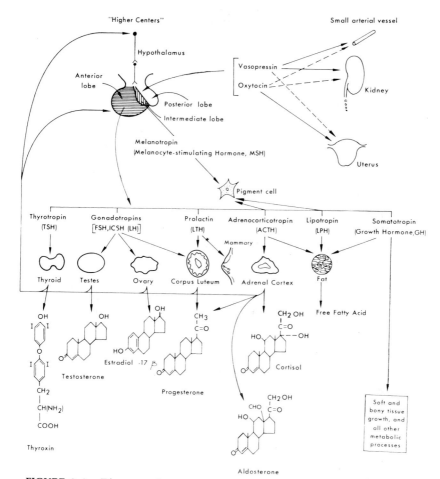

FIGURE 3-2. Diagrammatic summary of biological properties of pituitary hormones. (From Li, C. H. Proceedings of the American Philosophical Society, *116*:366, 1972.)

and their anatomic location have important implications for the normal ovarian cycle.

Studies of the sexual dimorphism of the rat hypothalamus, which indicate male and female behavioral patterns as well as cyclic versus non-cyclic release of hormones, cannot be applied directly to primates. Sexual differentiation of the central nervous system mechanism that mediates gonadotropin secretion in primates has not been confirmed. Indeed, there is some evidence that the female pattern is not eliminated by exposure to androgens in early life. Women with congenital adrenocortical hyperplasia, exposed to large amounts of androgen in utero, subsequently develop normal ovulatory cycles if treated adequately with cortisol. Both men and male monkeys respond to an acute rise in serum estradiol concentration with a surge of LH, indicating that the female pattern remains intact in the male primate. The low concentrations of estrogen and the high concentrations of androgen in the normal adult male produce the male non-cyclic pattern of gonadotropin release.

THE ADENOHYPOPHYSIS

The adenohypophysis is made up of an anterior and an intermediate lobe. It is derived from the primitive oropharynx as an evagination, called Rathke's pouch. It becomes separated from the oral cavity during early development and ultimately lies in apposition to the neural lobe of the pituitary (Fig. 3–1). Often the adenohypophysis is referred to as the anterior pituitary, since the intermediate lobe in the human is poorly defined. This endocrine gland produces the following ten different hormones: (1) Thyrotropin (TSH), (2) follicle-stimulating hormone (FSH), (3) luteinizing hormone (LH), (4) prolactin, (5) adrenocorticotropin (ACTH), (6) α-lipotropin (α-LPH), (7) β-lipotropin (β-LPH), (8) growth hormone (GH), (9) and (10) melanocyte-stimulating hormones (α-MSH and β-MSH).

Many different cell types exist in the adenohypophysis, and probably there is one cell type for each of the known hormones. The difficulty is that cell types are usually classified on the basis of their histochemical staining properties, and classifications vary according to the technique utilized. One investigator will select a fraternity of cells of one color, assign a Greek letter to that group and assume that they are all engaged in the same occupation. Another morphologist with a different technique will redistribute the same cells into two or more fraternities with different functional significance. Based upon the cellular affinity for dyes, there are three major groups: first, there are the chromophobes (about 24 per cent). These are believed to be the precursors of the chromophils, which in turn may be divided into the other two groups, the acidophil (37 per cent) and the basophil (39 per cent) cell groups. Further subdivisions may depend upon special histochemical reactions, electron microscopy, fluorescent antibody techniques, response of the cells to exogenous hormone administration or to the extirpation of endocrine glands, or the association of cell types with disease states. Based upon all of these considerations, Purves believes that specific functions may be ascribed to specific cells, each producing its own protein hormone.

FSH is a glycoprotein which promotes ovarian follicular growth; it must act before LH can act. Highly purified FSH alone will not cause the follicles to secrete estrogen in hypophysectomized animals; its primary function appears to be to instigate the morphological development of the entire follicle. In all likelihood, FSH is secreted continuously in both males and females, with a minor rise early in the follicular phase and a second, more sharply defined peak just prior to ovulation.

LH is also a glycoprotein which acts in concert with FSH and has multiple functions. It is responsible for stimulating estrogen secretion and, in higher concentration, for ovulation itself. It is also responsible for luteinization of the ruptured follicle. Whether it is necessary for normal proges-terone secretion by the corpus luteum is still controversial.

Both FSH and LH contain two subunits: an α subunit and a β subunit. Thus they belong to the family of glycoprotein hormones which are made up of a common α subunit (identical, or almost so in these hormones) and a hormone-specific β subunit. It is the latter that confers the unique biological properties to the hormone. TSH and hCG also belong to this family of gly-coprotein hormones.

Prolactin, another hormone of the anterior pituitary of special impor-tance in reproductive physiology, consists of a single polypeptide chain of 198 amino acids. Like GH it affects many tissues, but its primary function in mammals is undoubtedly the regulation of lactation (p. 47). It is capable of initiating and sustaining lactation, but the mammary gland must first be "prepared" by the actions of insulin, estrogens, progestins, GH and corti-costeroids. The secretion of prolactin by the anterior pituitary is stimulated by estrogens and TRH, and is inhibited by PIF.

Puberty

Strictly speaking, puberty is that period of time at which reproduction becomes possible, although the term is sometimes used to describe such prepubertal events as breast development and the growth of pubic and axil-lary hair. The term *menarche* refers to the onset of uterine bleeding. It is possible for ovulation to precede the first menses, but this is rare. The first several cycles are usually anovulatory, and full maturity or puberty, as judged by fertility, is ordinarily not attained until a year or more after the menarche.

The ovaries of very young girls are fully prepared to function if they receive the appropriate gonadotropic stimuli. Precocious puberty, the onset of ovulatory cycles considerably before the age of 10, may be idiopathic (or constitutional) or may be secondary to a variety of central nervous system lesions. Illustrative of constitutional sexual precocity, which occurs most often in females, is the well-known case of Lina Medina, who began menses at eight months and gave birth to a child by cesarean section at the age of five and a half years.

In the rat, all reproductive organs are prepared to function 25 days after birth, but puberty does not occur until 15 days later, when, it is believed, the hypothalamic neuroendocrine mechanism controlling the release of FRF and LRF has matured. A current theory states that this maturation results from a steady decline of the inhibitory influences emanating from the amyg-daloid nuclear complex, although there are a few who believe that the inhibition may result from biogenic amines produced by the pineal gland.

It was believed until recently that pituitary gonadotropins were absent from the blood and urine of prepubertal children. This was due to the use of bioassay methods which were relatively insensitive. The employment of highly sensitive radioimmunoassays has revealed the presence of both FSH and LH in the sera of fetuses, infants and children of both sexes. In both the fetus and infant the concentration of FSH in the sera of females is considerably higher than that found in the sera of males. There is also a clear-cut sex difference in FSH secretion in fetal life. The concentration in females at 20 weeks gestation approaches the adult castrate level, whereas the concentration in males is low or undetectable. There is no sex difference in the concentration of circulating estradiol, but during the middle trimester the concentration of testosterone in the male fetus is nine times that of the female. There is a decline in circulating gonadotropins near term, but following birth and the ensuing rapid decrease in the concentration of hCG and estrogens in the serum, there is a rise in the concentration of pituitary gonadotropins. In the male infant this is accompanied by increased testosterone concentrations in the blood. By about 4 months of age the gonadotropin levels return to the normal childhood range, and testosterone levels follow soon thereafter. This postnatal surge of gonadotropin secretion in both sexes is most likely a result of removal of feedback inhibition by placental steroids. The female, but not the male, shows a rhythmic pattern of FSH release in the first two years of life and later again during adolescence. Thus it is not the onset of pituitary activity which causes puberty, but some other combination of events. Some of the current hypotheses are as follows: (1) The immature hypothalamus is highly sensitive to the negative feedback effect of gonadal steroids, and puberty results from a gradual reduction of that threshold. This progressive decrease in the sensitivity of the LRH center to estrogen has actually been demonstrated in children. (2) As the hypothalamus matures, it becomes less sensitive to inhibitory influences from the higher centers in the brain. (3) The production of gonadal steroids gradually increases, under the low levels of LH and FSH prevailing before puberty, and eventually reaches levels which trigger the release of LRH to produce ovulation (positive feedback). (4) There may exist gonadotropin-inhibiting factors in the hypothalamus, perhaps originating in the pineal body, during the prepubertal period.

During the years preceding adolescence the quantities of FSH and LH are low, and therefore the amount of estrogen produced under their influence is small. In some children this small amount of estrogen is sufficient to stimulate breast development. This hypersensitivity of the breast to normal amounts of estrogen results in premature thelarche without other evidence of sexual maturation. The adrenals secrete increasing amounts of androgens in adolescence. These androgens induce growth of pubic and axillary hair. If adrenal androgens are produced in large amounts, prior to adolescence, premature adrenarche can occur. In these children pubic and axillary hair appears, but no other manifestation of sexual maturation is evident. It is the properly timed release of gonadal and adrenal hormones that brings about growth of the breasts, altered fat distribution, appearance of sexual hair, increased linear growth rate and maturation of bones. Finally, as enough estrogen is secreted by the theca interna cells of the ovary, a proliferative endometrium is produced from which bleeding occurs when the menstrual cycles are established. The rising estradiol concentrations have a positive feedback effect on the hypothalamus and trigger it to produce LRH. This stimulates the pituitary gland to release, rather abruptly and for a few days only, a large amount of luteinizing hormone which is the proximate cause of ovulation.

Endocrine Events in the Normal Cycle

Figure 3–3 illustrates the plasma concentrations of FSH, LH, and estradiol during a typical ovulatory cycle in women. The concentrations of FSH and LH tend to be higher in the preovulatory phase than in the corpus luteum phase, and there is a marked surge of LH at midcycle which

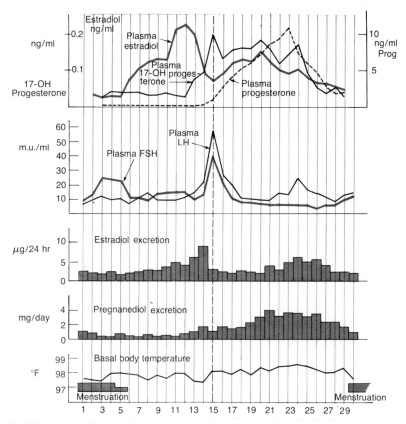

FIGURE 3–3. Diagram illustrating the concentrations of gonadotropins, estrogens and progestins in the plasma during a single human menstrual cycle. The urinary excretion of estradiol and pregnanediol and the changes in basal body temperature are also shown. (From Villee, C. A. Biology, 6th Ed. W. B. Saunders Co., Philadelphia, 1972.)

precedes the event of ovulation by about 24 to 36 hours. The smaller peak of FSH coincides with the LH peak.

The sharp rise of plasma estradiol precedes the LH surge, and the LH peak (which is the proximate cause of ovulation) ordinarily occurs while the estradiol concentration is falling. It is generally believed that the rise (or fall or both) of estradiol concentration is the factor which triggers the cycling center in the hypothalamus to release LRH.

The concentrations of plasma progesterone are minimal or absent until after ovulation. The rise and fall of progesterone parallels the activity of the corpus luteum. The use of oral contraceptives reduces the FSH concentrations to undetectable levels, and abolishes the preovulatory LH surge, although variable and smaller LH peaks are frequently observed at various times in the cycle during the use of low estrogen-containing oral contraceptive agents.

Following ovulation there is a brief drop in estrogen secretion, followed by a second rise which corresponds to the life of the corpus luteum. The drop in estrogen following ovulation is sufficient to cause microscopic withdrawal bleeding in about 90 per cent of all cycles studied. In approximately 10 per cent of the cycles, the bleeding is sufficient to be macroscopic or visible, and this has been referred to as ovulatory bleeding. If the first estrogen peak is just prior to ovulation, it would be more appropriate to refer to these episodes as postovulatory bleeding.

THE OVARIAN CYCLE

During the preovulatory phase, from three to 30 primary follicles in each ovary begin to mature. The granulosa cells increase in number through mitotic activity and become multilayered, and the adjacent connective tissue cells similarly multiply and differentiate into the theca interna. Both types of cells elaborate estrogen, probably in the form of 17-β-estradiol. An increasing amount of follicular fluid creates an antrum which, with its eccentrically placed ovum, forms the graafian follicle. This initial phase is believed to be self-regulated by its own estrogen acting locally. At this point, one follicle ordinarily "gains competence" to respond to the available FSH concentration. The first meiotic maturation division of the ovum takes place. A few hours before rupture, the follicle has reached its maximal size of 10 to 20 mm. in diameter, and the most peripheral portion has thinned out. Cause of the actual rupture has not been elucidated, but the dehiscence is not due to the mechanical pressure of the follicular fluid nor to local proteolytic enzyme activity. It is remarkable that in man and the higher primates the quantities of gonadotropins released in 99 per cent of cycles are just sufficient for a single ovulation, multiple ovulations occurring in about 1 per cent of all cycles. Attempts to duplicate these quantities by the injection of mixtures of human FSH and LH into amenorrheic women have been fraught with difficulties, inasmuch as there is a small margin between the amounts which fail to produce ovulation and the amounts which result in multiple ovulation and multiple pregnancy.

THE CORPUS LUTEUM

The process of luteinization, which begins prior to ovulation, continues in the theca interna and the remains of the granulosa. For two or three days (the stage of follicle proliferation) there is a rapid growth of these cells, with abundant mitotic activity. From the periphery, connective tissue and capillaries grow inward to begin the stage of vascularization. By the eighth day after ovulation, the corpus luteum reaches its peak of functional activity, with secretion of both estrogen and progesterone. At this time the corpus luteum is bright orange-yellow in color, due to the accumulation of lipids, and it may easily be recognized grossly. It is at about this time that the fertilized ovum implants in the endometrium. If it does not, and if, therefore, there is no chorionic gonadotropin to "nourish" the corpus luteum and extend its life span, the corpus luteum begins to regress. The capillaries collapse and the cells show signs of fatty degeneration and atrophy. Coincident with these morphological changes is a steadily decreasing production of steroids, leading to the phenomenon of menstruation about two weeks after ovulation.

Precisely what sustains the corpus luteum and what causes its downfall in the human species is not known. The subject is complex because the mechanism of luteotropic support and luteolysis varies from species to species. For example, in the guinea pig, the cow, the pig and the sheep, a luteolytic substance emanates from the uterus, and hysterectomy following ovulation results in a marked prolongation of the life of the corpus luteum. This is not so in the monkey or in man. In the rat, it is known that prolactin is luteotropic and that the LH hormone is luteolytic. In some way, the balance between these two protein hormones determines the life of the corpus luteum, but this situation may be peculiar to the rat. In the human species there are far more hypotheses than data, but there is some evidence to suggest that the corpus luteum is autonomous and has an inherent life span of two weeks. This concept is supported by the experience of Vande Wiele et al. They treated a hypophysectomized amenorrheic patient with human FSH and human chorionic gonadotropin (hCG); ovulation was achieved and the life span of the corpus luteum was 14 days, despite the fact that the gonadotropins were not administered beyond the second day following ovulation. Although the patient was receiving growth hormone injections, it seems evident in this case that the corpus luteum was not under the influence of endogenous gonadotropic hormones.

This does not necessarily prove that LH or prolactin might not play some role in the function of the corpus luteum in the intact individual or that continuing and rising levels of estrogen and progesterone during the corpus luteum phase exert a negative feedback effect on the hypothalamus so that luteotropic support is withdrawn. However, immunochemical measurements of the concentrations of FSH and LH in the plasma indicate only minor drops in their levels during the corpus luteum phase of the cycle.

THE ROLE OF PROSTAGLANDINS IN REPRODUCTION

The prostaglandins, derivatives of long-chain polyunsaturated fatty acids, are involved in every aspect of reproduction from erection and ejaculation to sperm transport, ovulation, formation of the corpus luteum, uterine motility, parturition and milk ejection. The discovery of the prostaglandins is credited in large measure to the efforts of Professor U. S. VonEuler and his colleagues at Karolinska Institute in Stockholm. In 1934 VonEuler observed that human semen, or extracts of sheep vesicular glands, will lower arterial blood pressure when injected intravenously. The same extracts cause contraction of smooth muscle preparations from the uterus or intestine. The active principle was identified in 1963 by Bergstrom as a long-chain fatty acid with a pentane ring in the middle. The prostaglandins of greatest importance physiologically are members of the E and F α series, which differ in the presence of a ketone group or a hydroxy group at position 9 (Figure 3–4). The number of double bonds is indicated by the subscript numeral after the letter. Thus PGE_1 and PGF_1 have a single trans double bond between carbons 13 and 14. PGE_2 and PGF_2 have, in addition, a cis double bond between carbons 5 and 6.

The enzyme system that catalyzes the conversion of the precursor unsaturated fatty acids to prostaglandins, prostaglandin synthetase, is located in the microsomes. This enzyme is inhibited by the anti-inflammatory drugs aspirin and indomethacin. Arachidonic acid is converted by this enzyme into an intermediate, 9,11-peroxy 5,8,12,15 eicosatetraenoic acid. An endoperoxide isomerase can convert this intermediate to prostaglandin E_2, or an endoperoxide reductase can convert it to prostaglandin $F_{2\alpha}$. Prostaglandins were initially found in the seminal fluid and are produced in considerable quantities by the seminal vesicle, but it has now been shown that they are made by essentially all tissues. Different tissues have different relative concentrations of prostaglandin E and prostaglandin F. Prostaglandins are metabolized very rapidly and 99 per cent of the prostaglandin in blood is cleared by a single circulatory pass, mainly by enzymes in the lung and liver. The precursor arachidonic acid is incorporated into the phospholipids of membranes and must be hydrolyzed before prostaglandin synthetase can act.

The production of the prostaglandins in man was measured by Samuelson; in four normal males the production rate ranged from 109 to 226 μg per 24 hours, and in two normal females it ranged from 23 to 48 μg per 24 hours. These experiments suggest that there is a sex difference in the rate of production of prostaglandins and that only a small proportion of the precursor essential fatty acids consumed in the diet are converted to prostaglandins. The daily human dietary intake of essential fatty acids is on the order of 10 gms.

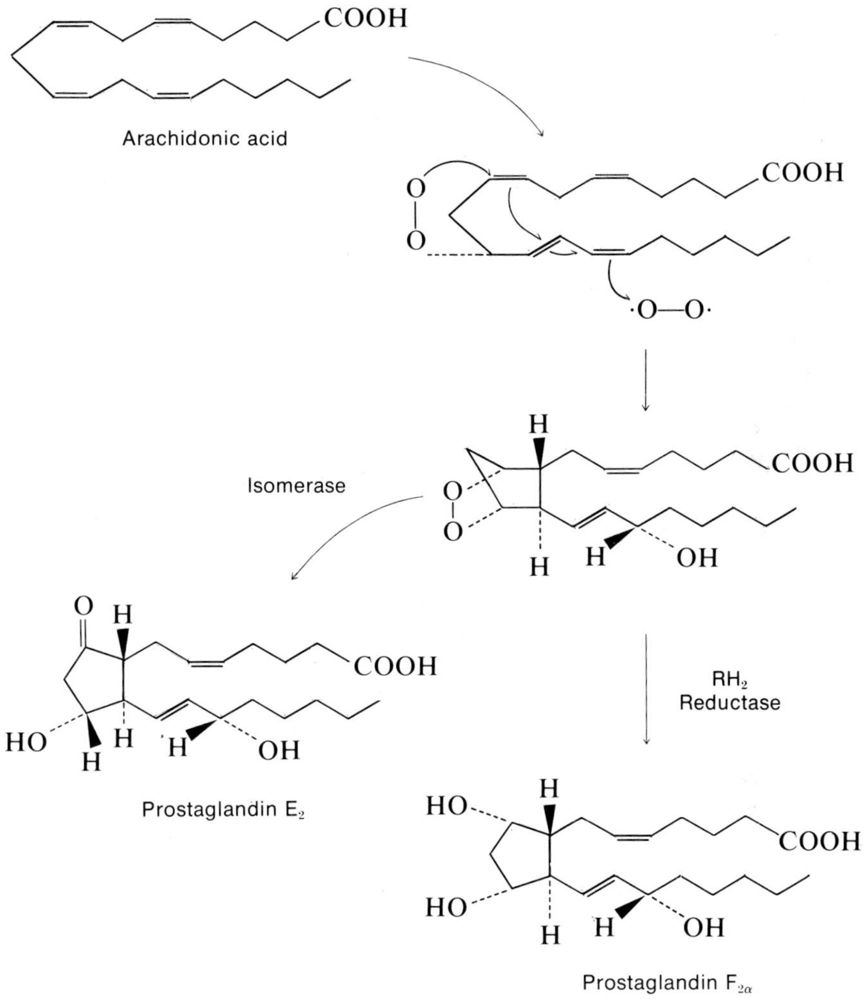

FIGURE 3-4. The biosynthesis of prostaglandins from arachidonic acid, a polyun-saturated fatty acid with a 20-carbon chain.

Experiments with labeled prostaglandins have shown that they are converted to metabolites and excreted in the urine. Prostaglandins are metabolized first by a 15-hydroxy prostaglandin dehydrogenase which forms the 15-keto group, then the 13,14 double bond undergoes reduction by prostaglandin reductase present in large amounts in the lung. This is followed by β oxidation and omega oxidation of the 15 keto dihydroprostaglandin to form shorter chain metabolites, which are then excreted in the urine. Some 95 per cent of the prostaglandin is removed by one circulation through the lungs and 80 per cent of the prostaglandin infused into the portal vein is removed by the liver; thus the body has efficient mechanisms for preventing prostaglandins from reaching the arterial circulation and resulting in decreased arterial pressure.

Plunkett reports that in several species the systemic injection of LH increases the concentration of $PGF_{2\alpha}$ in the ripening ovarian follicle, an event which may be blocked by indomethacin or by an antiserum to the $F_{2\alpha}$. When the LH-induced rise of prostaglandin is blocked, the ovum remains trapped within the follicle, although the corpus luteum functions normally. When $PGF_{2\alpha}$ is injected through a micropipette into the base of an ovarian follicle, the ovum is extruded into the ovarian stroma rather than through the surface of the follicle. When human tertiary follicles are placed in culture, the addition of LH stimulates the synthesis of $PGF_{2\alpha}$, probably within the granulosa cells. Thus a prostaglandin may play a key role in the process of ovulation, as well as in regulating steroid production by the corpus luteum.

The introduction of prostaglandins into the vagina or uterine lumen increases the motility of the uterus. The prostaglandins present in the seminal plasma may stimulate uterine motility at the time of intercourse and assist in transferring sperm up the uterus and oviduct to the site of fertilization of the egg. Prostaglandins decrease the gastric secretion of acid and pepsin and the total volume of gastric juice. The smooth muscle of the gastrointestinal tract is very sensitive to prostaglandins E and F, and responds with general contraction of the circular muscle. The oral administration of prostaglandin E leads to increased gastrointestinal motility, colic and diarrhea. The gastointestinal symptoms — vomiting, diarrhea and increased intestinal motility — are troubling side effects when prostaglandins are used intravenously to terminate pregnancy. Karim demonstrated in 1969 that prostaglandins can be used to induce labor, and they are now used to assist in the induction of labor (see Chapter 14). Intravenous prostaglandin $F_{2\alpha}$ or prostaglandin E_2 will induce abortion between the ninth and twenty-second weeks of pregnancy. The preferred method for middle trimester abortion is the direct intraamniotic injection of $PGF_{2\alpha}$ (see Chapter 10).

Unraveling the interrelations between prostaglandins and cyclic AMP has occupied the attention of endocrinologists for many years. Prostaglandin E_2 will mimic the effect of ACTH on the adrenal cortex and the effect of luteinizing hormone on the corpus luteum. Since cyclic AMP stimulates steroidogenesis it has been suggested the prostaglandins and ACTH exert their effects by a common pathway that involves the adenyl cyclase system.

The high concentrations of prostaglandins in human semen are apparently essential for normal fertility. The concentration of prostaglandin in the seminal fluid of infertile men was on the average 18 μg of PGE per ml. of serum, whereas in fertile men the average concentration was 55 μg per ml. The prostaglandin apparently does not alter the motility of the sperm, nor does it affect the oxygen uptake by the sperm or its metabolism of fructose. Prostaglandins have been tested as a possible once-a-month contraceptive administered as a single vaginal suppository at the time of the expected menstrual period. Its contraceptive action may involve an effect on the corpus luteum, causing it to regress, an effect on the myometrium, increasing uterine motility or a direct effect on the endometrium, leading to its regression and induced menstruation.

STEROIDOGENESIS: GENERAL CONSIDERATIONS

It is now abundantly clear that in all of the endocrine glands that synthesize steroids—the adrenal, the testis, the ovary and the placenta—the biosynthetic pathway is basically similar. The suggestion made by Butenandt and Rudzicka (1936) that steroids are derived from cholesterol has been shown to be correct by a great many subsequent experiments with a variety of tissues. The steroids and related compounds, such as the bile acids, are not synthesized on a template as nucleic acids and proteins are, but are assembled bit by bit in a sequence of enzymatic reactions. The final structure of the steroid is determined by the specificity of the enzymes that catalyze each step and not by the specificity of some template. These biosynthetic reactions require biologically useful energy—ATP—to drive them.

The classic experiments of Rudolf Schoenheimer in the early 1940's showed that cholesterol is synthesized from acetic acid and provided details of the incorporation of the carboxyl and methyl carbons of acetic acid into specific carbons in the cholesterol ring. Robinson (1934) suggested that sterols are produced by the cyclization of the 30-carbon unsaturated hydrocarbon, squalene. The stages in the process of sterol synthesis were visualized as acetate→isoprenoid→squalene→30-carbon sterol→27-carbon cholesterol.

Among the major metabolites of cholesterol are the 21-carbon, 19-carbon and 18-carbon steroid hormones. The 6-carbon side chain of cholesterol is removed by a reaction which begins when hydroxylations occur at carbon 20 and carbon 22 of the sterol nucleus. These reactions require NADPH and molecular oxygen. The 20α-hydroxylation of cholesterol is the rate-limiting reaction and appears to be the control point at which luteinizing hormone, LH, operates in the ovary and ACTH operates in the adrenal. These reactions convert cholesterol to $20\alpha,22R$-dihydrocholesterol, which subsequently undergoes cleavage to yield pregnenolone and isocaproic aldehyde. The isocaproic aldehyde is readily oxidized to the 6-carbon, branched chain isocaproic acid.

The steroid molecule consists of three fused cyclohexane rings, A, B, and C, and one cyclopentane ring, D. The carbons in the rings are numbered as shown in Figure 3–5. The basic ring structures for the steroids are the 21-carbon pregnane, the 19-carbon androstane, and the 18-carbon estrane rings. If both substituents on the two carbon atoms shared by two rings are on the same side of the plane of the rings, ring fusion is said to be *cis*. If the substituents at the carbon atoms shared by two rings are on opposite sides of the ring system, then the fusion is said to be *trans*. In all of the naturally occurring sterols and steroids, the ring fusion between rings B and C and between C and D is always the *trans* configuration. The fusion between ring A and B may either be *cis*, as in 5β-androstane, or *trans*, as in 5α-androstane. The spatial orientation of any substituent in the rings is referred to the methyl carbon 19 attached to carbon 10. This, the reference carbon for stereochemical relationships for all the steroids, is shown pro-

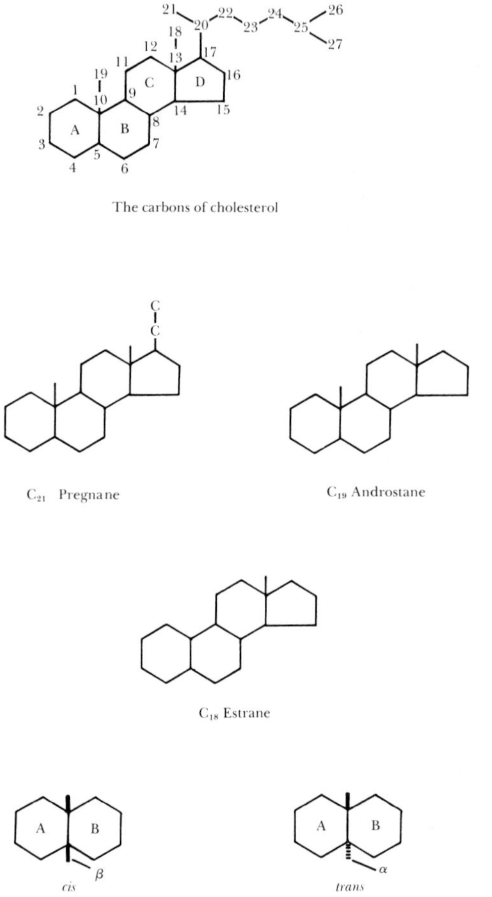

The carbons of cholesterol

C_{21} Pregnane

C_{19} Androstane

FIGURE 3–5. The numbering system of the carbons in the sterol molecule and the structures of the parent compounds, pregnane, androstane, and estrane.

C_{18} Estrane

cis

trans

jecting up from the relatively flat ring system, and by convention is termed *β* and designated as a solid line. Any other substituent that is oriented *cis* to the methyl group at carbon 10 — that is, that also projects up from the plane of the ring — is also termed *β*, and this orientation of the bond is indicated by a solid line. Those substituents that project down from the plane of the ring structure — that is, that are oriented *trans* to the methyl group at carbon 10 — are termed *α*, and the orientation of the bond is indicated by a dotted line.

A parent hydrocarbon is indicated by a name ending with the suffix *-ane*, e.g., pregnane. The presence of a double bond is indicated by replacing the suffix *-ane* with the suffix *-ene* and inserting before it the number of the carbon in which the double bond begins, e.g., pregn-4-ene. If there are several double bonds in a single molecule, this is indicated by the suffixes *-diene* and *-triene*, together with numbers indicating their location, e.g., estra-1,3,5(10)triene. Two substituents of importance in steroids are hydroxyl and ketone groups. A ketone group, $C = 0$, is designated by the suffix *-one* preceded by the number of the carbon atom to which the ketone group is attached. The suffixes *-dione, -trione* and so on are used to indicate compounds having more than one ketone group. To name a compound with a hydroxyl substituent, the suffix *-ol* is added, immediately preceded by the number of the carbon atom bearing the hydroxyl group, together with its configuration, that is, 17β-ol. The suffixes *-diol, -triol* and so on indicate compounds having two, three or more hydroxyl groups. If the compound has a substituent in addition to a hydroxyl group, the prefix *hydroxy-* is used, preceded by the number of the carbon atom bearing the hydroxyl group and its configuration, e.g., 3β-hydroxyandrost-4-ene-17-one.

There are a variety of enzymes that act upon the steroid nucleus and its side chain, and the particular steroid or steroids synthesized by any given gland, such as the ovary or testis, is a function of the kinds and amounts of the enzymes present. An enzyme that adds a hydroxyl group is called a hydroxylase, and steroid hydroxylases that add OH groups at carbons 2, 6, 7, 11, 15, 16, 17, 18, 19 and 21 have been described. Cortisol has hydroxyl groups at carbons 11, 17 and 21; estradiol has hydroxyl groups at positions 3 and 17. Another group of enzymes, termed dehydrogenases, catalyze the conversion of a hydroxyl group to a ketone group. An important one is the 3β-hydroxysteroid dehydrogenase, which converts pregnenolone to progesterone or dehydroepiandrosterone to androstenedione. 17β-Hydroxysteroid dehydrogenases convert testosterone to androstenedione or estradiol to estrone. There are 20α- and 20β-hydroxysteroid dehydrogenases that catalyze the conversion of progesterone to 20α-dihydroprogesterone and 20β-dihydroprogesterone, respectively. The reductases are enzymes that saturate double bonds such as the one between carbons 4 and 5. The saturated hydrocarbon that results may have its hydrogen group at position 5 lying below the plane of the ring structure and is termed 5α. Alternatively, the reductase may insert the hydrogen in a position above the plane of the ring structure and yield a 5β configuration. Enzymes termed desmolases cleave the carbon to carbon bonds to remove the side chains. Thus, the enzyme that removes the side chain of cholesterol is termed $20\alpha,22C_{27}$-desmolase. The $17\alpha,20C_{21}$-desmolase converts 17-hydroxyprogesterone to androstenedione or 17-hydroxypregnenolone to dehydroepiandrosterone and releases acetic acid as the other product. Finally, steroids may be conjugated with sulfate or glucuronic groups by sulfurylases and glucuronosyl transferases, respectively. The conjugated products are much more water-soluble than the free steroids, and are inactive metabolites for excretion.

STEROIDOGENESIS IN THE OVARY

The ovary synthesizes and secretes two classes of steroids, the 21-carbon progestins and the 18-carbon estrogens. The synthesis of progesterone involves a relatively short pathway, beginning with cholesterol (Figure 3–6) which may be synthesized in the ovary from acetyl CoA or may be obtained as preformed cholesterol from the blood supplying the ovary. Cholesterol is hydroxylated in positions 20α and 22R by reactions requiring NADPH and molecular oxygen. The resulting $20\alpha,22R$-dihydroxycholesterol is acted upon by the desmolase, which cleaves off a 6-carbon side chain as isocaproic aldehyde, and leaves pregnenolone, 3β-hydroxypregn-5-ene-20-one. The conversion of pregnenolone to progesterone involves two enzymes, an isomerase that changes the double bond at carbon 5 to one at carbon 4 and the 3β-hydroxysteroid dehydrogenase that converts the hydroxyl group at position 3 to a ketone group.

The pathway for the synthesis of estrogens diverges from that for the synthesis of progestins at pregnenolone. Pregnenolone undergoes hydroxylation at carbon 17 to yield 17-hydroxypregnenolone. This serves as substrate for the desmolase which removes carbons 20 and 21, yielding dehydroepiandrosterone and a 2-carbon fragment, acetic acid. Dehydroepiandrosterone is acted upon by the isomerase, 3β-hydroxysteroid dehydrogenase system, to yield androstenedione (androst-4-ene-3,17-dione). The enzyme 17-hydroxysteroid dehydrogenase catalyzes the interconversion of androstenedione and testosterone. Either one of these may undergo a process in which ring A is converted to a phenolic ring by a sequence of enzymatic reactions that has been termed *aromatization*. The first step in this process involves the hydroxylation of carbon 19 attached to carbon 10 to yield 19-hydroxytestosterone or 19-hydroxyandrostenedione. A third desmolase removes carbon 19 as formaldehyde and leaves an aromatic ring which will be estradiol if the starting material was testosterone and estrone if the starting material was androstenedione. The third classic estrogen, estriol, has an additional hydroxyl group at carbon 16. In the non-pregnant woman, 16-hydroxylation of the estrogen is carried out primarily by an enzyme located in the liver.

In the first portion of the menstrual cycle, the follicular phase, the cells of the theca are vascularized. They take in substrate and secrete primarily estrogens with only small amounts of progesterone. It is Short's hypothesis that the granulosa cells have no 17-hydroxylase or desmolase. Therefore, in the luteal phase, when the granulosa cells proliferate and become the vascularized corpus luteum, they are the major source of ovarian steroids. If they were to lack the 17-hydroxylase and desmolase system, as Short suggests, then the pregnenolone would not undergo 17-hydroxylation, but would be acted upon instead by the 3β-hydroxysteroid dehydrogenase isomerase system to form progesterone.

MECHANISM OF ACTION OF STEROID HORMONES

Although peptide and steroid hormones each react with specific receptors in their target organs, they have fundamentally different mechanisms of

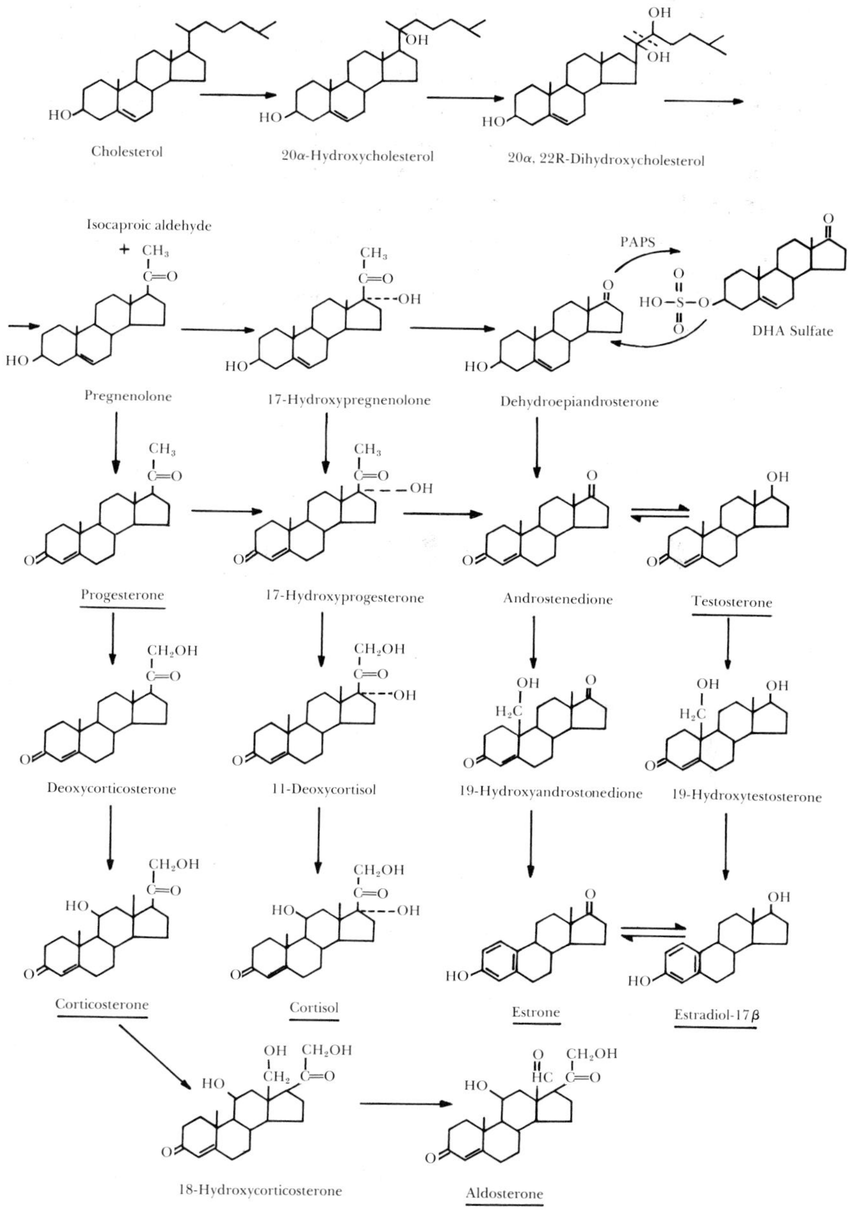

FIGURE 3–6. The synthesis from cholesterol of progestins, corticoids, androgens, and estrogens.

action within the cells of the target organ. Peptide hormones such as gonado-tropins interact with specific receptors located on the cell membrane and regulate the activity of the adenyl cyclase on the inner surface of the plasma membrane. The latter produces 3',5' cyclic adenosine monophosphate (cAMP), which activates a cascade of protein kinases and results in the modification of cellular structure and functions observed. Prostaglandins appear to play an intermediate role in the activation of adenyl cyclase. In contrast, steroid hormones enter the target cell and react with specific cytoplasmic protein receptors. These protein receptors are characterized by (1) high affinity for their ligand, (2) high chemical specificity for the steroid bound and (3) low capacity — that is, they can be saturated by a small amount of the steroid. The steroid-receptor complex undergoes a temperature sensitive transformation that changes its size, shape and properties; it then enters the nucleus, where it interacts with chromatin and results in an increased production of RNA (Figure 3–7). The heterogeneous nuclear RNA produced undergoes a processing with ATP in which a long tail of polyadenylic

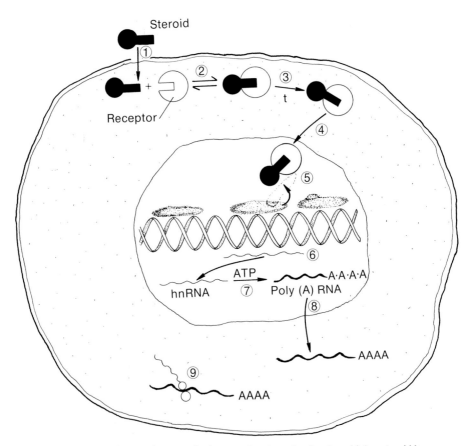

FIGURE 3–7. Successive steps in the postulated mechanism by which a steroid hormone stimulates the synthesis of specific proteins in its target cells. (1) The steroid enters the cell and (2) binds to a specific receptor protein. The steroid-receptor complex undergoes a temperature-dependent transformation process (3), enters the nucleus (4) and binds to the chromatin (5). This in some way facilitates the transcription of the DNA in that section of the chromatin (6). The resulting "heterogeneous nuclear RNA" undergoes processing with ATP (7), and a long polyadenylic acid tail is added. The poly (A) rich mRNA leaves the nucleus (8) and serves as the template for the synthesis on the ribosome (9) of the specific protein.

acid is added at the 3'-OH end. The resulting polyadenylic acid-rich messenger RNA passes to the ribosomes in the cytoplasm where the poly(A)-rich mRNA regulates the synthesis of the specific proteins for which it is coded.

The poly(A)-rich messenger RNA consists of triplet codons composed of sequences of the four nucleotides taken three at a time, each of which codes for one of the 20 amino acids. Each amino acid is activated by a specific enzyme and bound to a specific transfer RNA. These are smaller RNA molecules which have a specific anticodon, a sequence of three nucleotides which are complementary to the three nucleotides in the template RNA coding for that amino acid. The transfer RNA thus has two functions: to recognize and bind to a specific amino acid, and to bind to a specific triplet sequence of nucleotides (codon) in the messenger RNA.

When a hormone-receptor complex reacts with some specific section of chromatin and stimulates the transcription of the DNA at that point, the resulting poly(A)-rich messenger RNA passes from the nucleus to the cytoplasm, combines with ribosomes and leads to the synthesis of the specific protein. There is an abundance of evidence from many different experimental systems to indicate that the sex steroids bring about their effects in specific tissues by an effect on the genetic mechanism which stimulates the synthesis of specific forms of RNA.

Isolation and purification of the steroid receptors has been difficult because the proteins are unstable and have a tendency to aggregate. Estrogen binding receptors have been demonstrated in the uterus, vagina, mammary gland, pituitary and hypothalamus; experiments indicate there are some 100,000 receptor molecules per endometrial cell. Comparable specific receptors for testosterone or dihydrotestosterone have been found in the prostate and seminal vesicle, and for progesterone in the uterus and the chick oviduct. Each of these receptors is quite specific; thus the estrogen binding protein will bind both 17β estradiol and diethylstilbestrol, but it does not bind 17α estradiol. The binding of labeled 17β estradiol to the receptor is reduced by competition with unlabeled estradiol or unlabeled diethylstilbestrol but not by testosterone, progesterone and other steroids.

Experiments in the laboratories of O'Malley and Shimke have shown that estradiol in the chick oviduct stimulates the production of a specific protein, ovalbumin. It has been possible to recover the messenger RNA for ovalbumin and add it to a cell-free protein synthesizing system obtained from rabbit reticulocytes or wheat germ, thus demonstrating the production of ovalbumin. After the chick oviduct has been primed with estradiol, it will respond to administered progesterone with the synthesis of a second specific protein, avidin.

A target tissue, such as the endometrium, that responds both to estradiol and to progesterone would be expected to have receptors for both hormones, and this has been demonstrated experimentally. Indeed the amount of progesterone receptor seems to be regulated by the amount of estradiol available to the cell; the amounts of progesterone receptors in the uterus show considerable variation with the phases of the estrous or menstrual cycle, increasing and decreasing with the amount of estradiol circulating. Many of the effects of progesterone can be demonstrated only in cells that have previously been primed with estradiol. It was difficult to explain this estrogen dependence of progesterone action in terms of effects on the synthesis of RNA or protein, but it is adequately explained if the estrogen priming simply increases the number of receptor sites for progesterone up to an effective level.

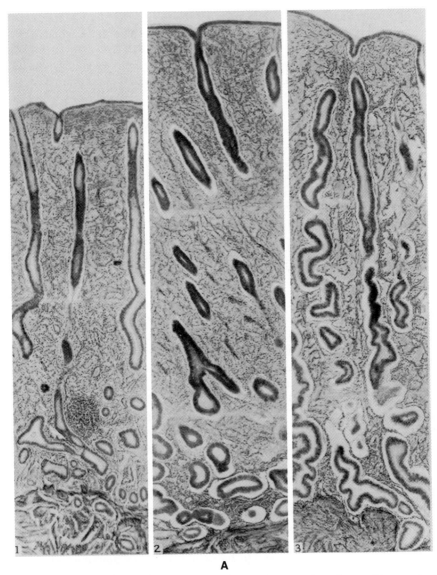

A

FIGURE 3–8.   Appearance of the human endometrium at various phases of the menstrual cycle. *A,* (1) Mid-follicular phase, (2) late follicular phase, and (3) early luteal stage.

The experiments of Segal showed that RNA extracted from the uterus of an estrogen-treated castrated rat would, when instilled into the uterus of another castrated rat, produce specific cytologic changes characteristic of those that occur in response to estradiol itself. Comparable studies by Niu and his colleagues showed that the activity of alkaline phosphatase in the uterus of the mouse was increased in response to RNA instilled into the

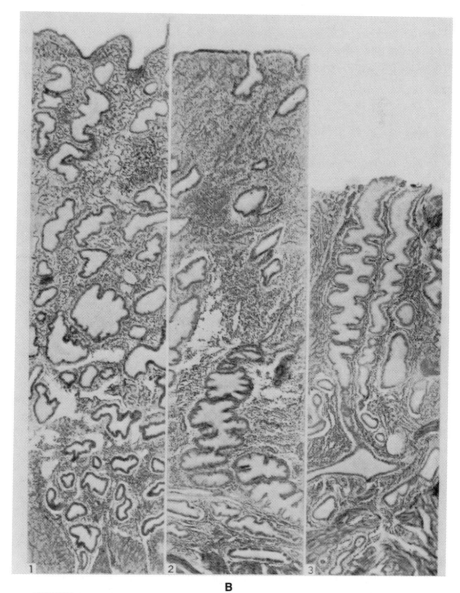

B

FIGURE 3–8. *Continued.*

B, (1) Mid-luteal stage, (2) late or premenstrual phase, and (3) menstrual phase. (From Papanicolaou, Traut, and Marchetti, The Epithelia of Woman's Reproductive Organs, Commonwealth Fund, 1948.)

lumen, just as it was in response to the injection of estradiol. Studies by Villee and his colleagues have shown that poly(A)-rich RNA extracted from the uterus of an estrogen-treated rat can increase the rate of protein synthesis and the synthesis of specific estrogen-dependent enzymes when instilled into the uterus of a control rat. Similarly, poly(A)-rich RNA extracted from the seminal vesicle of an androgen-treated rat will increase the rate of protein synthesis when instilled into the lumen of the seminal vesicle of a control rat. Thus, there is a considerable body of evidence supporting the contention that hormones, especially steroid sex hormones, act upon the genetic mechanism and lead to the synthesis of specific kinds of RNA that, in turn, stimulate the synthesis of specific proteins.

THE ENDOMETRIAL CYCLE

The normal endometrium is a mirror of the ovarian cycle, responding to the fluctuating concentrations of ovarian steroids in a rather precise manner. This cyclic response does not involve the full thickness of the endometrium. Three zones are described, although they are not sharply demarcated. The zona basalis is relatively thin; it interdigitates with myometrial cells, has a compact stroma and responds very little to the hormonal changes. Above this is the zona spongiosa, in which the endometrial glands, surrounded by a rather loose stroma, predominate. The zona compacta forms the surface surrounding the ostia of the glands, has a compact stroma, as its name implies, and in the luteal phase forms about a third of the total thickness. All of the compacta and variable portions of the spongiosa zones respond in a cyclic fashion and are shed during menstruation.

Variable terms have been used to describe the phases of the endometrial cycle. Utilizing a 28-day cycle, with day 1 representing the first day of menses and day 14 the time of ovulation, the phases may be described this way: The *menstrual phase* occupies the first four days, during which most of the functional zones of the endometrium are shed in fragments, occasionally as a cast. Days 4 to 7 constitute a *phase of repair* (or early proliferation), and days 7 to 14, the period of most rapid growth, comprise the *follicular phase* (or mid- and late proliferative phase). The remainder of the cycle, days 15 to 28, is referred to as the *luteal phase* (or secretory stage); the last few days are known as the *ischemic phase* (or stage of regression). Figure 3–8 illustrates the histologic appearance of the endometrium at various parts of the cycle.

By the use of endometrial biopsy, the day of the luteal phase can be dated with a reasonable degree of accuracy, using the criteria set forth by Noyes, Hertig and Rock (Figure 3–9). Such dating of the endometrium is of importance in assessing the adequacy of corpus luteal function, particularly in women with infertility problems. As examples, the microscopic appearance of the endometrium three and 11 days after ovulation are compared

DATING THE ENDOMETRIUM
APPROXIMATE RELATIONSHIP OF USEFUL MORPHOLOGICAL FACTORS

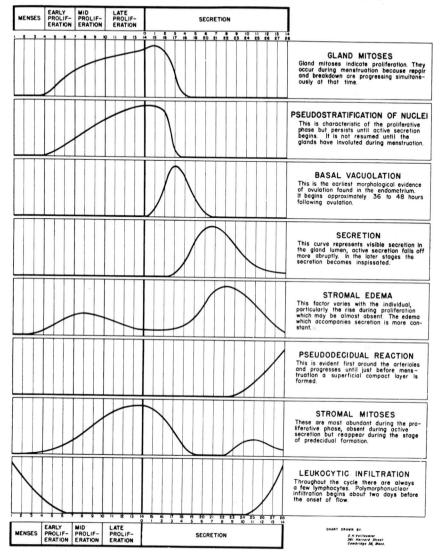

FIGURE 3–9. Criteria for dating the endometrium. Each curve represents the approximate quantitative change in each of eight factors considered most helpful. (From Noyes, Hertig, and Rock, Dating the endometrial biopsy. Fertil. Steril., *1*:3, 1950.)

in Figures 3–10 and 3–11, which illustrate some of the changes noted in the previous chart.

The key to the cause of menstruation lies in the activities of the spiral arteries which supply the functional zone of the endometrium. As demonstrated in Markee's ocular endometrial transplant studies in the monkey,

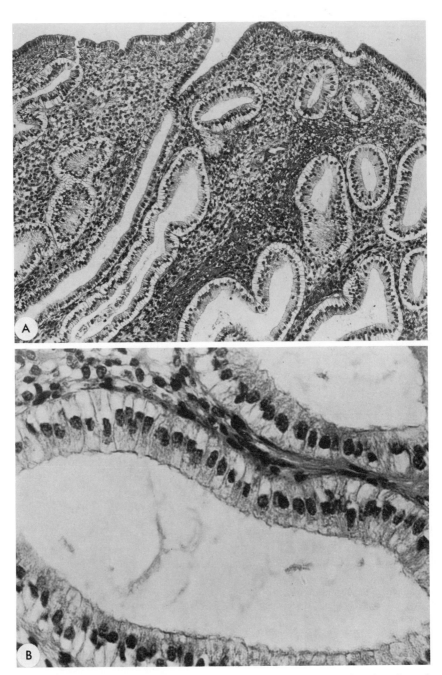

FIGURE 3–10. Low and high power photomicrographs of endometrium three days after ovulation. There is decreasing pseudostratification of the glandular epithelium, uniform basal vacuolization and beginning secretion. (From Noyes, Hertig, and Rock, Dating the endometrial biopsy. Fertil. Steril., *1*:3, 1950.)

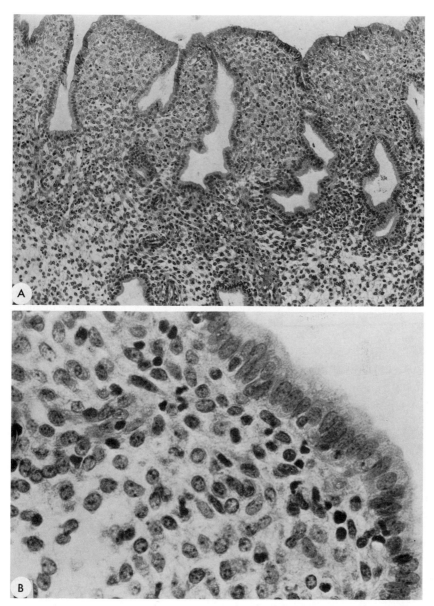

FIGURE 3-11. The endometrium 11 days after ovulation. Note predecidual development, stromal edema. basal location of epithelial nuclei and beginning leukocytic infiltration. (From Noyes, Hertig, and Rock, Dating the endometrial biopsy. Fertil. Steril., *1*:3, 1950.)

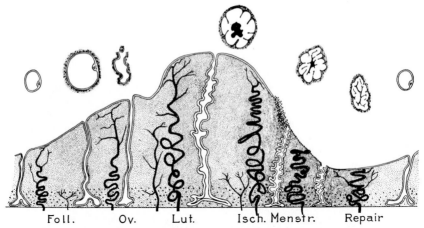

Foll.   Ov.   Lut.   Isch. Menstr.   Repair

FIGURE 3–12. Diagram indicating correlated changes in ovary and endometrium during a menstrual cycle. (From Bartlemez, Amer. J. Obstet. Gynecol., *74*:931, 1957.)

these vessels are highly sensitive to fluctuations in the concentration of ovarian steroid hormones. As the corpus luteum regresses and the rates of secretion of estrogen and progesterone fall, the spiral arteries respond by segmental intermittent constrictions, which obstruct the flow of blood. Inasmuch as the spiral vessels in the monkey and in man are "end arteries" without anastomoses, this leads to endometrial shrinkage, buckling of the arteries, and literally to crumbling and shedding of the functional zone and its vascular apparatus. During the phase of repair, a new capillary bed is formed from long precapillary channels which arise from the remaining arterial coils in the zona basalis and which are converted into new spiral arteries of the proliferating endometrium. Figure 3–12 is a correlative diagram based upon Bartelmez's observations in the monkey; these observations are equally applicable to the human.

It should be noted here that when ovulation does not occur, the continued ovarian secretion of estrogen results in an unremitting proliferation of the endometrium, which may ultimately result in marked endometrial hyperplasia. Under these circumstances, when estrogen concentrations do fall, the endometrium does not shed normally, but may erode in a patchy fashion, leading to excessive and prolonged (dysfunctional) bleeding.

The morphological changes in the endometrium are paralleled by important biochemical alterations. During the follicular phase there is a high content of ribonucleoproteins, which diminishes after ovulation as mitotic activity virtually ceases. Alkaline phosphatase activity during the growth period is replaced by acid phosphatase activity during the luteal phase. A week after ovulation, when the endometrium is prepared to receive the blastocyst, oxygen consumption, glycogen content, various dehydrogenase activities and lipid content all reach maximal values. In the menstrual phase, there is a sharp rise of fibrinolytic activity, which accounts for the rapid lysis of small blood clots as they are formed in the uterine cavity.

STATISTICAL ASPECTS OF THE "NORMAL" MENSTRUAL CYCLE

It has been said that the only thing regular about the human menstrual cycle is its irregularity; but for an understanding of what is meant by "abnormalities" of the reproductive cycle in women, it is necessary to describe those events which occur in about 90 per cent of all cycles.

In the United States, the average age of the menarche is 12.9 years, but the tenth to ninetieth percentiles include ages 11 through 16. Fluhmann believed that the age of onset is an index of the degree of civilization and well-being of the people of all countries. In 1900, the mean age of the menarche in a United States study of over 12,000 cases was 14.2 years. In Germany, the average age was 16.5 in 1870, and by 1950 this was reduced to 13.5 years. In this country, if menses have not occurred by the age of 16, we speak of delayed menarche; whereas if bleeding has not occurred by age 18, the diagnosis of primary amenorrhea is made.

The data on length of the menstrual cycle shown in Figure 3–13 are based upon the distribution of over 700 cycles in normal young California women. The length of 90 per cent of the cycles fell between 23 and 35 days, a range which may be considered, therefore, as including normal variations. Almost all of these variations are due to alterations in the duration of the follicular phase, inasmuch as studies based largely upon shifts of basal body temperature suggest that 90 per cent of luteal phases last 13 to 15 days. A luteal phase of less than 12 days is considered abnormal, and one which lasts longer than 16 days is almost diagnostic of pregnancy.

The mean duration of the menstrual flow is four and a half days, with a normal range of three to six days (94 per cent of cycles). The total blood loss is highly variable, but averages 50 ml. ± S.D. 30. When the measured blood loss regularly exceeds 80 ml., iron supplementation is required to

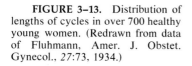

FIGURE 3–13. Distribution of lengths of cycles in over 700 healthy young women. (Redrawn from data of Fluhmann, Amer. J. Obstet. Gynecol., *27*:73, 1934.)

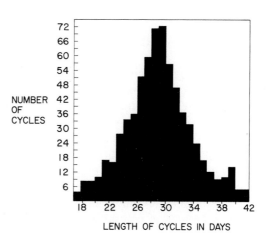

prevent secondary anemia; hence it is fair to conclude that menstrual blood loss in excess of this amount is abnormal.

In well over half of all cycles, the menstrual discharge does not clot. This is due to the fact that the blood has already clotted *in utero* and the fibrin has been lysed by the fibrinolytic enzyme. When, however, the blood escapes too rapidly at any given moment, clotting will occur in the vaginal canal.

THE DETECTION OF OVULATION

Proof that ovulation has occurred on a specific day rests upon such evidence as direct observation of ovulation by culdoscopy or laparoscopy, direct inspection or biopsy of the early corpus luteum, pregnancy resulting from an isolated artificial insemination, or the recovery of an ovum from the fallopian tube or of a dated conceptus from the uterus.

Symptoms which suggest that ovulation has occurred are the detection of postovulatory bleeding or the occurrence of sharp lower quadrant pains, the origin of which is not known but which in some cases may be due to slight follicular bleeding.

There are a number of indirect measurements which are based upon the production of progesterone and which therefore suggest that a corpus luteum is present. These include a rise in the basal body temperature (Figure 3–14), measurement of pregnanediol in the urine, dating the postovulatory endometrial biopsy, a loss of the fern pattern in dried cervical mucus, and serial studies of vaginal cytology.

The only method which would appear to anticipate ovulation is the serial measurement of LH in the plasma or urine, inasmuch as the rise begins one or two days prior to release of the ovum. There is no method, however, of predicting ovulation in such a way as to avoid pregnancy through abstinence, since spermatozoa are viable in the fallopian tubes for at least 72 hours.

SYMPTOMS OF THE MENSTRUAL CYCLE

The most common subjective disturbances which affect almost all women at some time are premenstrual tension and uterine cramps, or dysmenorrhea. The causes of these symptoms are not known, but they appear to be associated with the corpus luteum, because anovulatory cycles, whether they be spontaneous or induced by estrogen administration, are ordinarily asymptomatic.

The milder forms of premenstrual tension are evident as irritability, increased appetite, augmented but less purposeful activity, and fallacious faultfinding or flares of temper. More serious symptoms, observed rarely, include periods of emotional depression with inexplicable crying spells, an

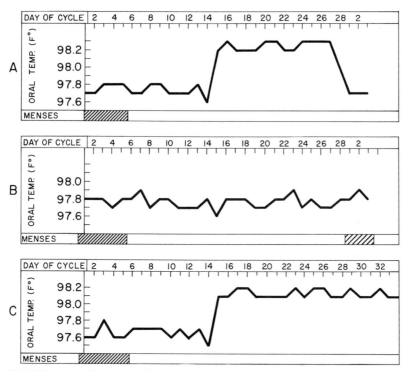

FIGURE 3–14. Typical oral basal body temperature curves in a normal cycle (A), an anovulatory cycle (B), and early pregnancy (C).

introspective indifference to environment or paranoid attitudes. The duration may vary from one day to two weeks. A variety of medications may give relief, but there does not seem to be any single regimen that is universally successful. The existence of a serious premenstrual tension syndrome in patients is unfortunately ignored all too often by physicians, but is of enormous importance in terms of the cost to industry and of increased marital strife, if not of the well-being of the woman herself.

Premenstrual edema is also common, and has been thought by some to be the cause of premenstrual tension. There may be an association, but diuretic therapy which successfully controls the edema does not relieve the premenstrual tension syndrome any more often than does a placebo.

Primary dysmenorrhea, that is, menstrual pain not secondary to organic disease, is most common in the young, sedentary nulligravida. It is the price of fruitless fertility inasmuch as it is essentially limited to ovulatory cycles and is ordinarily relieved after pregnancy. The onset of dysmenorrhea rarely coincides with the menarche, but rather suggests the time when ovulatory cycles have begun. It is possible that the pattern of sustained myometrial contractions induced by progesterone results in uterine ischemia and "uterine angina." It has been suggested that prostaglandins act as the mediator of dysmenorrhea. When the uterine vascular apparatus is more or less permanently increased by the occurrence of pregnancy and delivery, the organ may be less subject to impairment of blood flow.

Literally scores of remedies, medical and surgical, have been advocated over the centuries, all reporting good results in 60 to 90 per cent of cases. This suggests that a strong placebo effect prevails, which may be duplicated by hypnosis or psychotherapy, because dysmenorrhea is concerned to a large extent with the problem of pain perception. The simplest effective means of abolishing dysmenorrhea when it is severe or disabling is the employment of sequential oral contraceptives to inhibit ovulation.

EFFECTS OF ESTROGENS UPON THE BODY

The widespread effects which estrogens have upon various organs and tissues may be discussed for the sake of convenience under four general headings.

SPECIFIC TARGET ORGANS.   In the basal layer of the vaginal epithelium, the mitotic cycle is shortened by estrogen and true hyperplasia results. The mucosa becomes thickened, and rugae or folds develop, in contrast to the thin, smooth appearance of the postmenopausal atrophic vagina. Topical creams containing estrogen are utilized for treatment of the latter condition.

The glandular cells of the endocervix are stimulated to secrete a larger volume of clear mucus with low viscosity, which facilitates sperm migration. The increased content of water and sodium chloride at the time of ovulation is responsible for the delicate fern patterns which appear when the mucus is allowed to dry.

Estrogen is a growth stimulus for the endometrium and the myometrium, resulting in both cellular hypertrophy and hyperplasia. One of the earliest biochemical effects noted is an increased uptake of amino acids followed by an increased content of RNA. The myometrial cells show an increased concentration of actin and myosin and heightened excitation, and the uterine fundus undergoes cyclic clonic contractions.

The mammary ducts are stimulated to grow, resulting in enlargement of the breasts.

CENTRAL NERVOUS SYSTEM AND ADENOHYPOPHYSIS.   In addition to the effects upon steroid-sensitive centers of the hypothalamus discussed earlier, high concentrations of estrogen directly inhibit the adenohypophyseal synthesis of FSH, prolactin, and, to some extent, the synthesis of growth hormone. Estrogens are also used with partial success to inhibit the synthesis of growth hormone in the treatment of acromegaly.

The temperature-regulating centers, presumably those in the anterior hypothalamus, are affected in such a way that the basal body temperature is reduced. The nadir ordinarily occurs about the time of ovulation. Inasmuch as progesterone has the opposite effect, recording the temperature each morning before arising is widely utilized for estimating the "safe period" (i.e., the luteal phase) in the rhythm method of birth control and for the analysis of disordered menstrual patterns (see Chapter 17). The vasomotor centers are also affected in some obscure fashion, as will be noted in the description of the menopause.

METABOLIC EFFECTS. Both estrogens and androgens have anabolic effects on bone. The estrogens are thought to exert their effect by decreasing the responsiveness of bone to endogenous parathyroid hormone. This effect could occur as a result of direct inhibition of osteoclast function, thus favoring bone formation over bone resorption. In older women, or in young women some seven or more years after castration, osteoporosis becomes a common event, and when severe may result in vertebral collapse, the stoop of old age or the "dowager's hump."

Estrogens given to agonadal females increase the rate of linear growth but result in early epiphyseal closure. The estrogens also influence body configuration, resulting in narrower shoulders, wider hips, different areas of fat deposition, increased carrying angle in the arms, and the gynecoid shape of the pelvis which favors spontaneous birth.

Estrogenic hormones decrease plasma cholesterol and the cholesterol-phospholipid ratio, at the same time significantly increasing the α-lipoproteins and decreasing the β-lipoproteins. Testosterone has the reverse effects, a factor that allegedly favors the development of atherosclerosis and coronary heart disease in men. This is one of the reasons put forth for the low incidence of coronary heart disease in women prior to the menopause, and one of the arguments used to favor the indefinite use of oral estrogens in postmenopausal women. However, males given estrogen therapy for cancer of the prostate have increased mortality from cardiovascular disease.

There are systemic effects upon the capillaries, and one of these is a decreasing capillary strength with falling estrogen levels. These changes occur rather rapidly, so that in each menstrual cycle there is a brief decrease of capillary strength in the postovulatory phase and a longer decrease of capillary wall strength prior to and during the menses. On rare occasions this may lead to periodic nosebleeds, once called "vicarious menstruation."

Estrogen influences the hepatic cells to synthesize greater quantities of selected proteins, and among these are fibrinogen, ceruloplasmin, transcortin (which binds corticosteroids), and the thyroxin-binding protein, which results in an elevation of the protein bound iodine (PBI).

ANTIANDROGENIC EFFECTS. For specific androgenic target organs, estrogen exerts an inhibitory effect. For example, estrogen decreases the rate of prostatic growth, and this action is utilized for the prolongation of life of men with prostatic cancer. Libido is decreased in the male who ingests estrogen, leading to impotence. The activity of the sebaceous glands of the skin is inhibited, tending to counteract the effect of testosterone in producing acne.

EFFECTS OF PROGESTERONE UPON THE BODY

All of the morphological and biochemical changes which take place in the endometrium during the luteal phase are brought about by the addition of progesterone. This is a conditionally acting hormone which demands prior exposure of the tissue to estrogen. Indeed, the failure of the endometrium to bleed after the administration of progesterone, in the absence of pregnancy, is an indication of a paucity of estrogen secretion by the ovary.

Under the influence of progesterone, the endocervical glands secrete a thick mucus which will not 'form fern patterns and which impedes sperm migration, an important factor in the contraceptive effect of progestins. The myometrium becomes less excitable, and the motility pattern changes to one of a sustained tetanic type rather than a clonic type of contraction. In the breasts, progesterone causes alveolar development.

The influence of progestins upon the releasing factors of the hypothalamus has already been described. Other effects upon the brain include the elevation of basal body temperature and the stimulation of respiration, so that the carbon dioxide tension of the blood is reduced during the luteal phase as well as during pregnancy.

Contrary to former opinions, progesterone is natriuretic and protein catabolic and does not, therefore, account for premenstrual edema or for the protein anabolic state of pregnancy.

THE PERIMENOPAUSAL AND POSTMENOPAUSAL PERIODS OF LIFE

Woman is essentially the only creature who outlives her reproductive period. Ovarian senescence, with its cessation of ovulation and slow decline of estrogen secretion, is a gradual process which ordinarily takes place over a period of several years. This phase of life is referred to as the perimenopausal period, or climacteric, whereas the term menopause means the actual cessation of menses. Inasmuch as one never knows which is the last menstrual period until some six months of amenorrhea have ensued, the postmenopausal period of life may be arbitrarily defined as beginning six months after the last menstruation.

Siiteri and Macdonald have shown that essentially all of the estrogen produced by postmenopausal women is estrone or estrone sulfate, which arises from the peripheral extraglandular aromatization of plasma androstenedione of adrenocortical origin. The percentage of androstenedione converted to estrone may vary from 1 to 8 per cent, increasing with age and with the degree of obesity. Removal of both ovaries from postmenopausal women does not alter the total estrone production rate significantly. Some postmenopausal women convert a sufficient quantity of androstenedione to estrone; hence they do not demonstrate estrogen deficiency after the menopause, and therefore do not need replacement therapy. The majority of women, however, will become estrogen deficient within a year after cessation of menses. The simplest method of diagnosing such a deficiency is by the use of vaginal cytology.

The *usual age* for the menopause is between 47 and 52, and there is evidence that the mean age has been steadily increasing over the centuries in the highly developed countries. The pattern of menstrual cessation varies widely and may be abrupt, but more commonly consists of a series of irregularly spaced anovulatory cycles during which time episodes of dysfunctional (prolonged or excessive) bleeding are not uncommon. Any uterine bleeding occurring spontaneously after six months of amenorrhea at this age must be regarded as abnormal until proven otherwise.

With the decline of estrogen secretion there is an overproduction of FSH and LH, and both the plasma and the urinary concentrations of the gonadotropins are high. Urine collected from postmenopausal women is the source of the human gonadotropins currently used for the induction of ovulation. The aging ovaries, now essentially devoid of oocytes, no longer respond to gonadotropic stimuli.

The majority of perimenopausal women at some time complain of hot flashes and night sweats. These are thought to be due to an instability of the hypothalamic vasomotor centers, and might be regarded as withdrawal symptoms from years of addiction to endogenous estrogens. Women who are born without functioning ovaries or girls who are castrated before puberty because of ovarian neoplasms do not experience these symptoms unless they are placed on exogenous estrogen for a year or longer and are then abruptly withdrawn from the medication. The hot flashes and sweats are not due to the excessive gonadotropins, because they can be alleviated by quantities of estrogen which do not reduce the urinary excretion rates of FSH and LH; nor is there any correlation between the severity of symptoms and the plasma gonadotropin concentrations.

Symptoms other than the hot flashes and sudden sweating, such as nervousness, inability to concentrate, headaches or feelings of depression, may occur for a variety of reasons at any time in a woman's life and are not reliable indices of the menopausal syndrome. Any one of these symptoms, of course, might be due in part to a relative estrogen deficiency; if so, a therapeutic test is indicated, utilizing a moderate dose of oral estrogen, such as 0.5 mg. of diethylstilbestrol or 1.25 mg. of conjugated equine estrogens daily. Symptoms which are related wholly or in part to estrogen lack should improve within a week or two, and the dosage may then be cut in half to ascertain whether the improvement continues. When the uterus is present, oral estrogens should be discontinued for about five days each month (or for two days each week) to permit regression of endometrial growth. This is to prevent the development of endometrial hyperplasia which so often results in dysfunctional bleeding.

In the postmenopausal period of life, all of those effects estrogens have upon the body are slowly reversed. The majority of gynecologists believe that this represents a true endocrine deficiency which is deserving of replacement therapy for the remaining years of life. Essentially all physicians agree that if the uterus and ovaries are removed prior to the age of 45, replacement therapy should be instituted promptly and continued indefinitely. The controversy over "estrogens forever" for all women, which has waged for some

40 years, centers primarily about the growth effects exerted upon the myometrium and endometrium, the episodes of withdrawal bleeding which may require a curettage for diagnosis, or the somewhat tenuous belief of some that estrogen promotes the development of cancer of the endometrium or breast.

A large part of the bleeding problem results from overdosage. Many physicians do not realize that after menopausal symptoms are controlled by moderate cyclic dosages, the beneficial metabolic effects of estrogen may be maintained year after year by small amounts which rarely induce bleeding. Such amounts are the equivalent of 0.1 mg. of diethylstilbestrol or 0.3 mg. of conjugated equine estrogens, again given in a cyclic manner. *There is essentially no place for the parenteral administration of estrogenic hormones in the management of perimenopausal or postmenopausal women.*

If *any* substance necessary for normal bodily function is reduced in its concentration to the extent that health is impaired, this constitutes a *deficiency.* There is really no evidence to suggest that estrogen levels, as estimated by vaginal cytologic indices, must be kept as high in postmenopausal women as those observed during the reproductive period of life in order to preserve health. The amount of estrogen needed daily to prevent a deficiency, as defined above, is well below the usual amounts currently employed by most physicians.

The major contraindication to estrogen therapy is the history of treatment for endometrial or breast cancer within the prior five years, because the growth rates of these two forms of malignancy appear to be accelerated by estrogen administration. Theoretically, the frequency of endometrial carcinoma might be reduced if a periodic shedding of the endometrium were induced by progestin administration, and a number of gynecologists advocate the use of an oral progestin for five or six days once or twice a year in order to cause bleeding with endometrial shedding.

The proponents of estrogen administration "forever" — once estrogen deficiency has been demonstrated — believe that such therapy decelerates the aging process in skin and in the vascular system, prevents atrophic changes in the vagina, prevents the development of osteoporosis and even delays the onset of mental senility. Some of these claims are yet to be proven, but the advantages of treating estrogen deficiency states on a long-term basis appear to outweigh the disadvantages, provided that high dosages are avoided and some form of cyclic therapy is utilized when the uterus is present.

In summary, there are still some important gaps in our knowledge of the female reproductive cycle. The precise feedback mechanisms of various gonadal steroids upon the hypothalamus are imperfectly understood. The chemical structure of most of the releasing factors is yet to be elucidated. The cause of luteolysis in women is yet to be determined, and this is a most important bit of missing evidence, because it may hold the key to a more perfect means of contraception. Even the question of how a steroid hormone exerts its effect upon a cell is not fully known. Nevertheless, there have been tremendous strides during the past decade in our understanding of a woman's reproductive cycle, and that body of knowledge constitutes the essential scientific base for comprehending most of the functional disorders encountered in gynecology.

SUGGESTED SUPPLEMENTARY READING

Gold, J. J., Ed. Gynecologic Endocrinology, Harper & Row, Pubs., Inc., New York, 1975.
    An up-to-date text by multiple authors.
Grumbach, M. M., Grave, G. D. and Mayer, F. M. Control of the Onset of Puberty. John Wiley & Sons, Inc., New York, 1974.
Siiteri, P. K., and Macdonald, P. C. Role of extraglandular estrogen in human endocrinology. Chapter 28 in Handbook of Physiology, Endocrinology II, Part 1. Am. Physiol. Soc. Bethesda, Maryland, 1973.
Speroff, L., Glass, R. H. and Kase, N. G. Clinical Gynecologic Endocrinology and Infertility. Williams & Wilkins Co., Baltimore, 1974.
Villee, D. B. Human Endocrinology. W. B. Saunders Company, Philadelphia, 1975.
    A developmental approach from fetal life to senescence.
Williams, R. H. Textbook of Endocrinology. W. B. Saunders Company, Philadelphia, 1974.

Chapter 4

HUMAN SEXUALITY

Human reproduction is almost wholly a byproduct of human sexuality. Sexual drives, responses, customs and mores determine to a large extent the structure of society as well as the rate of growth and the composition of our populations. It is not always realized how completely an individual is involved in his or her sexual life from childhood to old age. Gender identification, or sexuality, is one of the most important determinants of human behavior, and its disorders are a major cause of crime, misery or illness. Many, if not most, aspects of sexual life are culturally determined, and the expressions of sex behavior or response are molded by personality, but they are also modified by hormones.

SEXUAL BEHAVIOR AND HORMONES

The sex of an individual is basically determined at conception, depending upon whether an X-bearing sperm or a Y-bearing sperm fertilizes the X-bearing egg (see Chapter 6). Ordinarily the XX or the XO embryo develops ovaries which are hormonally inactive during fetal life, whereas the XY embryo develops testes which secrete testosterone. It is, therefore, the presence of the Y chromosome which determines maleness, except in the XY androgen-insensitivity syndrome discussed in Chapter 2. In Chapter 3, it was noted that the administration of androgens to female mammals in the prenatal or early postnatal period could lead to the absence of ovarian cycles and to sterility. The early effects of gonadal hormones upon the midbrain also influence sexual behavior in the adult animal. If male rats are castrated before the fifth day of postnatal life, and are then treated with estrogen and progesterone at 120 days, they tend to exhibit feminine sexual behavior. On the other hand, female rats that are treated with androgens during the immediate postnatal period, and then given estrogen and progesterone as adults, exhibit the sexual behavior of males.

This work, resulting largely from the studies of Young and his coworkers, has been extended to the rhesus monkey by Goy and others. Pregnant females were injected with testosterone daily from the thirty-ninth to the sixty-ninth day of pregnancy with no treatment during the balance of the 168-day gestation. Genotypic female pseudohermaphrodites were born who in the next several years exhibited the aggressiveness, the rough-and-tumble play and the mounting behavior of males. In other words, in both lower mammals and higher primates, testosterone exerts a fundamental imprint on the organizational development of the midbrain, which determines to some extent whether the sexual behavior in adulthood will be masculine or feminine in character.

To some extent, the testosterone imprint upon the developing brain must operate in the human species. In the "testicular feminizing syndrome," all somatic cells are insensitive to androgens, and it would be presumed that the fetal testes would be unable to influence the developing brain. A group at Johns Hopkins found these individuals to be completely feminine in their behavior and attitudes. Conversely, a group of women with congenital adrenocortical hyperplasia, genetically and gonadally the antithesis of the androgen-insensitive group, were sexually dimorphic. Half showed homosexual inclinations and the majority reported sexual arousal through perceptual material, which is unusual in the normal human female. It would appear that gender identity is the result of a complex pattern of neurohormonal, genetic and psychosocial factors, and not simply the end result of postnatal environment.

After puberty, the maintenance of sexual drive and potency in men is dependent to a large extent upon androgen secretion. Castration of the adult male will, after a latent period of several months, result in impotence and a diminution of aggressive behavior. The sexual behavior of the human female, on the other hand, is relatively independent of ovarian secretions. The castration of women abolishes neither libido nor sexual response so long as a general well-being is maintained, and the administration of supraphysiological amounts of estrogen to women is just as likely to reduce as to increase the libido. Ovarian hormones, particularly progesterone, may affect a woman's psyche in other ways, but she has become quite independent of the profound influences that these steroids have upon sexual receptivity in lower mammals.

PREADOLESCENT DEVELOPMENT OF HETEROSEXUALITY

Children are by no means devoid of sexual interests or capabilities, and it is largely during the preadolescent period that heterosexuality (or homosexuality) develops. Many of Freud's concepts of infantile sexuality (the oral and anal stages), which were developed by retrospective psychoanalytic studies of adults, are doubted by modern psychiatrists. Nevertheless, the Freudian theories of Oedipal stages in later childhood have been largely confirmed.

According to Kinsey's samples, homosexual or heterosexual play has occurred in about 25 per cent of both sexes by the age 12. About three-fourths of boys experience orgasm through masturbation prior to puberty. This type of play, however, does not appear to be associated with the development of homosexuality in adult life. It is the psychosexual development, rather than the sexual experiences, which determine heterosexuality. According to Broderick, four factors are of particular importance in childhood: (1) the parent (or foster parent) of the same sex must not be so weak or so cruel as to prevent identification of the child with that parent, (2) the parent of the opposite sex must not be so seductive or so punishing as to make the child distrustful of the opposite sex, (3) the parent (or surrogate) cannot reject a child's phenotypic sex or try to teach him to behave like the opposite sex, and (4) there must be established early a positive concept that marriage is an eventual goal.

During later childhood, there are further steps leading to heterosexuality: development of an emotional attachment to some member of the opposite sex, professing to be "in love," then a preference for the companionship of the opposite sex, and finally dating. By the time of puberty, heterosexuality should be well established, even though some aspects of homosexual behavior may persist.

ADOLESCENT SEXUAL BEHAVIOR

The battle of the sexes begins with the onset of dating. The boy, soon approaching the peak of his sexual drive, feels an urgent need for gratification, and in recognition of this intense urge, society tends to condone his search for premarital intercourse. In contrast to this, the adolescent girl in Judeo-Christian cultures has been educated to repress her desires for genital contact, although not oral and bodily contact. Her sexual behavior is largely designed to attract and to retain a male of her choice, and the average teenage girl engages in coitus primarily when she feels that intercourse is a means of preventing rejection. That this event is encouraged by the prevalent practice of "going steady" is suggested by a study of unwed mothers in whom the time between onset of steady dating and first coitus is fairly constant. For other girls, promiscuity with multiple partners is a deliberate form of rebellion against parental authority or moralistic training, or is the result of severe anxieties, and it is among this group that pregnancies occur more frequently.

There is no good evidence that true promiscuity is more common among adolescent females now than in prior generations, but there is evidence that premarital intercourse with a single partner is steadily increasing, as are adolescent marriages. About one-half of those girls who marry while in high school are pregnant at the time, and it is this group which also has a substantially higher than average rate of divorce.

The problem of the adolescent male revolves around his need to preserve an image of masculinity in a culture which tends to equate this with sexual prowess. He believes, rightly or wrongly, that there is a sexual revo-

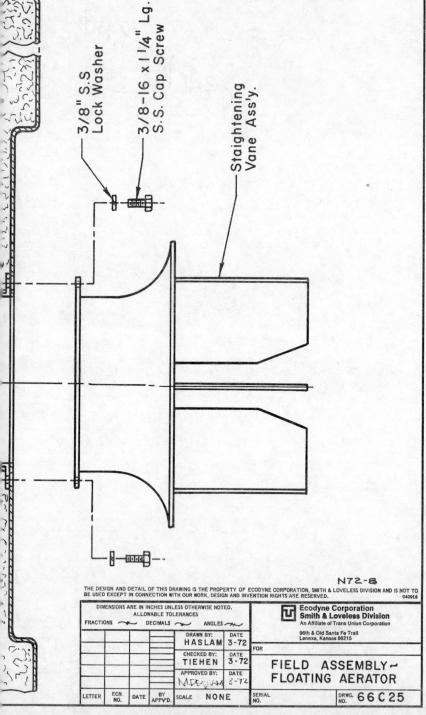

3/8" S.S Lock Washer

3/8-16 x 1 1/4" Lg. S.S. Cap Screw

Staightening Vane Ass'y.

N72-8

THE DESIGN AND DETAIL OF THIS DRAWING IS THE PROPERTY OF ECODYNE CORPORATION, SMITH & LOVELESS DIVISION AND IS NOT TO BE USED EXCEPT IN CONNECTION WITH OUR WORK, DESIGN AND INVENTION RIGHTS ARE RESERVED.   040916

DIMENSIONS ARE IN INCHES UNLESS OTHERWISE NOTED. ALLOWABLE TOLERANCES					Ecodyne Corporation Smith & Loveless Division An Affiliate of Trans Union Corporation
FRACTIONS	DECIMALS	ANGLES			96th & Old Santa Fe Trail Lenexa, Kansas 66215
		DRAWN BY: HASLAM	DATE 3-72	FOR	
		CHECKED BY: TIEHEN	DATE 3-72	FIELD ASSEMBLY~ FLOATING AERATOR	
		APPROVED BY: M.DEGMAN	DATE 3-72		
LETTER	ECN. NO.	DATE	BY APPV'D.	SCALE NONE	SERIAL NO.	DRWG. NO. 66C25

lution in which he should play some role, and that intercourse is the final proving ground for a convincing demonstration of his adequacy. Under such circumstances, an inadequate sexual performance may lead to panic or to severe depression.

The new morality assumes that coitus is related more to love than to marriage, but one problem with this philosophy is that, carried over into marriage, it becomes a justification for extramarital relations. It will be of interest to note whether the next generation returns to the "old" morality of relating sexual intercourse to marriage rather than regarding it simply as a means of heterosexual communication.

SEXUAL RESPONSE OF THE HUMAN MALE

Sexual stimulation of men may arise from a large variety of stimuli, verbal, tactile, visual, olfactory or lingual, and it is often accompanied by some degree of sexual fantasy. The nerve impulses causing compression of the venous channels of the penis, resulting in erection, originate in the S-2 and S-3 parasympathetic plexus. Continued stimulation of the glans penis causes a summation of efferent impulses which, upon reaching a specific threshold, results in a sudden contraction of the seminal vesicles and a propulsion of semen into the urethra. At this point, ejaculation becomes inevitable, and results from a reflex contraction of the bulbocavernosus and ischiocavernosus muscles, which causes a discharge of the semen in a series of spurts, accompanied by a general as well as a local pleasurable sensation referred to as the orgasm.

There is a considerable variation in the threshold at which the ejaculatory event occurs, both among individuals and in the same individual from time to time. When, for a variety of possible reasons, the threshold is low, premature ejaculation may occur. This situation, if persistent, is a form of impotence. Total impotence, the inability to achieve or sustain an erection, is often due to psychological causes, but may, of course, be secondary to organic disease.

The sexual drive of the human male reaches its peak between the ages of 15 and 19. During these years, or even earlier, some form of masturbation at some time is essentially universal. Thereafter, and quite unlike the human female, sexual drives diminish perceptibly, though slowly, and are moderately reduced by the forties and fifties. There is a steady increase of relative impotence after age 60, although the majority of men retain their sexual powers in old age. Accurate figures are hard to obtain, partly because the admission of impotence by men is quite traumatic to the ego.

SEXUAL RESPONSE OF THE HUMAN FEMALE

The anatomic and physiological correlates of the sexual response in women have been studied intensively by Masters and Johnson. For the sake of discussion, they have divided the response into four phases: excitement, plateau, orgasm and resolution. During sexual excitement, there is a moistening of the vaginal walls with droplets which may represent a transudate. At the same time, there is a lengthening and distension of the inner two thirds of the vagina. The nipples enlarge, and there may be a flushing of the skin, most notably on the chest, breasts and abdomen. During the plateau phase preceding the orgasm, there is a marked vasocongestion of the "sexual platform." The outer third of the vagina, including, of course, the erectile venous structures, becomes grossly distended with blood, so that the vaginal orifice is reduced in caliber. The Bartholin's glands may secrete a small amount of mucus at this time, but their contribution to vaginal lubrication is relatively minor. The clitoris enlarges and retracts to a more ventral position, actually away from the penile thrust. During orgasm, there are a series of muscular contractions around the outer third of the vagina, the only significant physical response confined purely to the orgasmic phase. The number, the intensity and duration of these contractions vary greatly with the intensity of the orgasm. During the phase of resolution, which may require 10 to 15 minutes, there is a progressive release of the vascular congestion and muscular tension.

Regardless of what the sexual stimulus may be, there is only one kind of orgasm in women. The Freudian concept of a transfer from a "clitoral orgasm" to a "vaginal orgasm" as a woman reaches sexual maturity is no longer tenable. To be sure, the achievement of a satisfactory sexual orgasm during coitus is to a large extent a learned response, and, unlike the inevitable ejaculatory response of the male, it is readily inhibited by emotional, physical and environmental influences.

Frigidity, the failure to achieve sexual arousal, is a relative term which should probably be avoided, since essentially any woman can be brought to orgasm by effective and appropriate stimulation. Whether she does reach the excitement and orgasmic phase depends upon her own psychosexual development, her state of well-being, the sexual attitudes of her cultural background, and the amorous efforts of her sexual partner. Under ideal circumstances, the sexual capacity of the human female is greater than that of men. There is no true refractory period following the orgasmic response, and with continued stimulation multiple orgasms are common. The peak of sexual drive is most commonly reached after the age of 25, and a high degree of libido and capacity for response may continue unabated into old age, assuming that she still has a potent partner.

In the development of western civilization, the suppression of women's innately high sexual drive has been a major preoccupation of every Judeo-Christian culture, and this may have been necessary for the development and protection of the family unit system. One Victorian writer stated that the sexual capacity of women was fortunately very low, for if it were not so, the "whole world would be a brothel." Queen Victoria herself advised her daughters that sex was a woman's burden to endure. But there has been a marked change in attitudes about sexual mores and activities during the past few decades, to the point where women are demanding equal orgasmic opportunities and an abolition of the double standard. Generations of cultural inhibitions, however, are no doubt still partially responsible for the fact that many wives only rarely achieve sexual satisfaction in marriage.

SEXUAL PROBLEMS IN WOMEN

The sexual complaints that women have may be grouped for convenience under two headings: anatomic factors, which interfere with or cause pain during coitus, and emotional factors, which suppress the libido or inhibit the achievement of orgasm during intercourse.

The problem of the virginal introitus and the rare cases of absence of the vagina were discussed in Chapter 2. Similar anatomic problems resulting in vaginal constriction may be iatrogenic in origin, resulting from combined anterior and posterior vaginal repairs or from overzealous closures of episiotomy incisions performed at childbirth. Vaginal atrophy occurring in postmenopausal women may result in excessive dryness or actual shrinkage. Vaginal infections or pelvic inflammatory diseases will cause dyspareunia, as will the development of endometriosis in the cul-de-sac. For the most part, these are readily recognized abnormalities for which there is appropriate therapy.

More common, and far more complex, are the emotional or psychological factors that interfere with libido or the sexual response. First of all, it is obvious that illness, excessive fatigue and a lack of well-being are not conducive to sexual interest or arousal. Both physical and nervous fatigue are important factors in the "tired-young-mother-cook-housewife syndrome."

Then too, there are situational or environmental influences which may inhibit the sexual response even though libido remains. Among these are distractions (a child crying, lack of total privacy, the presence of parents in an adjoining room, a bad odor) and preoccupation ("I have a million things on my mind"). Such factors are transient and often disappear while the husband and wife are alone on a vacation.

Finally, and far more serious, there are deep-rooted hostilities (which may result from marital conflicts), fears, depressions and anxieties. There must also be a reasonable adequacy of sexual stimulation, and, as some more sophisticated women state, there must be empathy. These are emotional factors that emerge during the taking of a sexual history. For the most part, they are disorders of love rather than disorders of sex, and may be symptomatic of a profoundly disturbed marriage.

Women who have never experienced an orgasm are preorgasmic, not frigid. They may be taught the orgasmic sensation through an appropriate masturbatory experience. Those who have already repeatedly achieved orgasm through masturbation, but who are unable to experience it with a sexual partner, are in another category. In most instances there are problems of communication with the partner, and these women should be counselled to be, not necessarily more aggressive, but more assertive as to what "turns them on" and what "turns them off." Any activity that gives pleasure and is rewarding should be considered acceptable, and the woman should be made to feel comfortable about such activity. Too much emphasis has been placed upon achieving orgasm at the "right time" during coitus. An orgasm is but the final refrain of a symphony; it is the entire symphony that is of importance.

Psychosomatic Manifestations of Sexual Frustrations in Women

The repetitive attainment of sexual excitement without orgasm and resolution may result in a sustained passive venous congestion of the pelvic organs, which in due time may actually lead to organic changes and chronic pelvic pain. The end result is sometimes referred to as the *congestion-fibrosis syndrome,* for which hysterectomy is all too frequently performed. Surgery, on the other hand, does not alleviate the underlying psychosexual problem, and quite typically, women who have chronic sexual frustrations undergo repeated surgical operations in futile attempts to alleviate their complaints. Chronic pelvic pain of long standing without demonstrable palpable changes in the pelvis is commonly encountered in medical practice, and is almost always due to psychiatric factors rather than to organic disease.

Chronic pelvic congestion may in addition lead to hypermenorrhea and to persistent excessive vaginal secretions. Sexual frustrations are also associated with multiple complaints, such as migraine headaches, constant fatigue (often a manifestation of depression), attacks of tachycardia, sweating and extreme nervousness. Sometimes these symptoms are falsely attributed to premature menopause. When the sexual basis of these rather nonspecific complaints is not recognized, women are frequently subjected to a never-ending sequence of drugs and hormones. The initial solution to the problem begins with an adequate sexual history, which must be undertaken whenever there are indications that there are major psychosomatic difficulties. Unfortunately, many physicians have not been instructed in the techniques of obtaining and evaluating a sexual history, or feel personally uncomfortable in discussing such subjects. It is important not only to ask the right questions, but to know what to do with the answers.

Taking a detailed sexual history is not indicated for every female patient, although some understanding of her emotional environment is an essential part of the history irrespective of the diagnosis. A sexual history is certainly needed when the primary complaint is infertility, dyspareunia chronic pelvic pain, chronic fatigue or nervousness, loss of libido, marital problems or episodic depressions, or whenever a long-standing complaint is believed to be of psychosomatic origin.

Taking a sexual history may begin in many ways, but always gently. Inquiries about her husband, if she is married, may begin with his state of health and emotional stability, leading up to the question of how sexually active he is and whether her relationships with him in that respect are satisfactory. If the patient is not married at this time, she may be asked whether she has been sexually active during the past year. Having introduced the subject, subsequent questioning must depend upon a perceptive interpretation of her responses, both verbal and non-verbal. The essential conclusion to reach is whether her sexual relationships are happy or unhappy for the most part or whether they are a potential source of emotional trauma.

PREMARITAL AND MARITAL COUNSELING

It is, perhaps, fortunate that many states have laws requiring some type of a premarital examination and medical certificate before a marriage license can be obtained. This gives the physician a unique opportunity to educate the patient with respect to sex when such education is needed, to detect congenital or organic pelvic disorders, to engage in genetic counseling when indicated, to advise dilatation of the virginal introitus when needed, to detect venereal disease, to advise on family planning, and to institute a plan for periodic examinations.

The prospective bride should be asked to return a few months after marriage should there be difficulties in sexual adjustments, marital conflicts or problems associated with contraception. Marriage counseling is much more productive in early years than in the later years of marriage when separation or divorce may be imminent because of long-standing conflicts. Ideally, both partners should be interviewed, separately as well as together, but the nature of present-day medical specialization is such that few physicians other than psychiatrists can (or will) give the necessary time. It is remarkable how often the simple question, "Tell me about your husband" is followed by such revealing answers as, "I have a wonderful husband. He leaves me alone." It is not suggested that all physicians practice psychiatry, but some insight into the sexual lives of their patients is vital to a full understanding of their health, their reactions to pregnancy and their reproductive efficiency.

The recent insatiable demand by the general public for sex therapy has led to the development of several thousand "treatment centers." Masters and Johnson believe that less than 100 of these are legitimate clinics administered by fully competent personnel. The current field of sexual therapy, according to Dr. Masters, is "dominated by an astounding assortment of incompetents, cultists, mystics, well-meaning dabblers and outright charlatans." This is a clear danger which must be recognized by all health personnel.

SEX EDUCATION

The psychosexual development of an individual occurs almost exclusively during childhood and adolescence. With prepubertal children, the long period of dependence compels parents to impart information about sexuality, verbally or non-verbally, to their children. How this is accomplished is of extreme importance. During adolescence, however, the lines of communication between parents and offspring are commonly clogged, and further knowledge about sex is more often acquired from peers or through sexual experiences, homosexual or heterosexual in nature. To a limited extent, information about reproduction is transmitted by way of the school or the church.

Sexuality plays such a momentous role in human existence that the subject deserves thorough and knowledgeable discussion with our youth. Too often, however, the material taught is the physiology of reproduction, not the impact of sexuality upon behavior. The entire question of appropriate sex education, of course, is mired in the morass of moralities. In the meantime, the incidence of venereal diseases is fast rising, the number of adolescent marriages (in almost half of which the brides are pregnant) exceeds half a million per year, an additional quarter of a million babies per year are born out of wedlock, and well over a million abortions are induced annually, all in the United States alone. Sex education *per se* will not alter these problems, except to the extent that they result from ignorance, nor will sex education increase promiscuity. It is society's attitudes about problems relating to sexual behavior, fertility control, abortion and sterilization, premarital and extramarital coitus, homosexuality and a host of related topics which will ultimately shape the psychosexual development of our youth. Indeed, education in the medical aspects of sexuality is still inadequate in our schools of medicine as well, although there has been some belated progress within the past decade.

The "new look" at sex education has been summarized by Kirkendall as follows: (1) All adults must recognize as fact that all young people are sexual beings with sexual needs. (2) Sex education must not be regarded as moral indoctrination. (3) The purpose of sex education is to help people understand sex in its relation to social patterns and attitudes. (4) More emphasis must be placed upon sex education for adults. (5) Instruction in sexuality must take into account the current freedom of choice and personal decision making. (6) The methodology of sex education must extend far beyond the public schools, and must involve mass media, professional schools, adult education programs, colleges and universities. (7) Finally, there must be a recognition that the final determinants of sexual behavior are not facts, but feelings of satisfaction and self-respect that an individual is able to develop about himself as a person.

SUGGESTED SUPPLEMENTARY READING

Barbach, L. For Yourself, the fulfillment of female sexuality. Doubleday and Company, Inc. New York, 1975.

This guide to orgasmic response, written by an experienced therapist, may be recommended for women with sexual problems.

Bellivean, F., and Richter, L. Understanding Human Sexual Inadequacy. Bantam Books, 1970.
This paperback summarizes in non-technical language the work of Masters and Johnson.

Green, R. Human Sexuality, a health practitioner's text. Williams and Wilkins Company, Baltimore, 1975.
Nineteen recognized authorities discuss all aspects of human sexuality.

Kaplan, H. S. The New Sex Therapy. Brunner-Mazel, New York, 1974.
This is an extensive 500-page book on the active treatment of sexual dysfunctions.

Masters, W. H., and Johnson, V. E. Human Sexual Response. Little, Brown and Co., Boston, 1966.
This book is devoted primarily to the anatomic and physiological aspects of sexual response.

Masters, W. H., and Johnson, V. E. Human Sexual Inadequacy. Little, Brown and Co., Boston, 1970.
Reports the continuing investigation of the Masters-Johnson group into the phenomenon of human sexuality.

Medical Aspects of Human Sexuality.
This monthly journal, which began publication in 1968, contains many fine articles, answers to questions, and personal viewpoints.

Chapter 5

FAMILY PLANNING AND POPULATION PROBLEMS

THE CONTROL OF FERTILITY

The primary purposes of contraception are to prevent the birth of children who are not wanted and to permit that spacing of births which is most conducive to the health of the mother. By decreasing the number of unwanted children, birth control is also an important step toward population control, but so long as the number of wanted children per family remains above two, the growth of populations is unremitting. Effective contraception, therefore, finds its greatest usefulness in the prevention of the overpopulation of families.

Cultural Background

In the records of various civilizations, it is clear that almost every culture has made attempts to limit the size of families. In some primitive tribes there has existed a long-standing taboo against coitus for a period of two years after the birth of a child, but this was largely associated with the practice of polygamy, which is fast disappearing. Among Eskimos, unwanted newborn and elderly infirm were abandoned on icefloes. Many African and American Indian tribes developed elaborate concoctions that were alleged to reduce fecundity, and among the Egyptians a variety of materials, including sponges, were placed in the vagina to prevent conception. Until the middle of the nineteenth century, however, abortion, infanticide, celibacy and late marriage were of greater importance from a demographic standpoint than contraception.

We have a legacy from the past that fertility is something which man can manipulate to correlate his family size with available material resources. On the other hand, in past centuries when underpopulation and excessively high death rates were prevalent, religious laws and hygienic rules were published to ensure a high fertility rate. Subsequently, the matter of deliberate contraception became moral and ethical issues that remain under debate. Only in 1965 was the last of many restrictive state laws prohibiting the use or the counseling of use of contraceptive devices overturned by the United States Supreme Court (Griswold vs. Connecticut).

The Measure of Effectiveness

The *theoretical effectiveness* of a contraceptive refers to its antifertility action under ideal conditions of use. This cannot be measured directly in large populations of people because of human errors, but is approximated in the most successful groups of users. In field trials, when any given method is offered to 100 women, the number of pregnancies which occur in a year is the expression of *use effectiveness*. Thus the "pregnancy rate per hundred woman-years" takes into consideration the duration of use, and should be the same for 50 women over a two-year period as for 200 women over a six-month period. The rate (R) is calculated by the following formula:

$$R = \frac{\text{total number of conceptions} \times 1200}{\text{total months of exposure}}$$

From the denominator are subtracted 10 months for each term pregnancy and 4 months for each abortion. With no contraception whatever, the pregnancy rate is about 80. With *any* kind of contraceptive effort, even those methods with the lowest theoretical effectiveness, the rate will fall below 40.

Using the coitally related methods discussed below, the reported pregnancy rates vary from six to 35 pregnancies per 100 woman-years, depending upon the population studied. Rates for the intrauterine devices range from two to four, but one must note in addition a cumulative expulsion or removal rate of about 25 per cent during the first year of use. The oral contraceptives have a use effectiveness ranging from zero to two, with a theoretical effectiveness approaching zero (Table 5–1).

Table 5–1.   THEORETICAL AND ACTUAL USE-FAILURE RATES (PREGNANCIES PER 100 WOMAN-YEARS)*

	"Perfect Use" Rate	"Actual Use" Rate
No contraception	80	80
Withdrawal	15	20–25
Calendar rhythm	15	35
Foam	3	30
Condom	3	15–20
Diaphragm	3	15–20
IUD	1–4	3–6
Low dose progestin	1–4	5–10
Oral contraception (combined)	less than 0.5	2–5

*Adapted from Contraceptive Technology, 1974–75

COITALLY RELATED CONTRACEPTIVE MEASURES

Prior to 1960, the modes of contraception most often advocated by physicians were the use of condoms by the male and the use of the vaginal diaphragm with contraceptive jelly by the female. The condom has an additional advantage of preventing the spread of venereal disease, and some authorities attribute a large part of the recent rise in the venereal disease rate to the progressive abandonment of condoms as a means of contraception. Both methods have a high degree of theoretical effectiveness, but the use effectiveness (that is, the overall acceptability and use in population groups) is relatively low.

The use of a vaginal diaphragm and jelly is still a relatively reliable and harmless means of birth control, and is considered by many physicians as the next choice when oral contraception is contraindicated. The device is coated with a contraceptive jelly and inserted prior to coitus, so that the cervix is covered, thus creating a detour for spermatozoa, which cannot traverse the distance without encountering a spermicidal compound. The diaphragm must be left in place for six or eight hours (preferably not longer than twelve), and with good care will last for two or three years. Its regular employment requires a high degree of motivation and persistence.

The precoital insertion of sponge and foam powder, jelly, cream, or a contraceptive foam are reasonably reliable methods also, but suffer from a reduced rate of acceptability when compared with non-coitally related methods.

The practice of withdrawal, or coitus interruptus, is a reasonably effective method for some, but not all, males and is widely practiced in some countries. Reliable estimates of its use effectiveness are difficult to obtain. Periodic abstinence, the "calendar rhythm" method, has limited effectiveness because of the unpredictability of ovulation, the fact that spermatozoa may retain their viability for three or four days in the fallopian tubes, and the biologic variations in the length of menstrual cycles. The "temperature rhythm" method, that is, restricting coitus to the corpus luteum phase of the cycle as determined by daily basal body temperatures, is highly effective but again requires a high degree of motivation.

NON-COITALLY RELATED CONTRACEPTIVE MEASURES

Oral Contraception

As long ago as 1942, Sturgis and Meigs showed that ovulation could be readily inhibited by oral estrogens, and they advocated the method for the control of severe dysmenorrhea. Continuous administration of large doses was used for the total suppression of menses for women with endometriosis, but little thought was given to the employment of estrogens for contraception, probably because their continued use without the addition of progesterone frequently leads to endometrial hyperplasia with irregular bleeding and prolonged hypermenorrhea.

The introduction of orally effective synthetic progestins permitted Pincus and his coworkers to utilize a combination of estrogen and progestin in field trials, which were reported in 1958. The ensuing decade witnessed an amazing acceptance of the method by an estimated nine million American women, together with an ever-increasing debate about the safety of the hormones over a prolonged period of time. Never in the history of medicine has a drug been used so extensively by so many for purposes other than the alleviation of symptoms or disease. This has placed a unique burden upon such policy-making groups as the Food and Drug Administration, and upon the judgment of the individual physician who must balance the potential side effects of the Pill against the risks of an unwanted pregnancy for any one individual.

MECHANISM OF ACTION

When 50 μg. of ethinylestradiol or 80 μg. of mestranol are administered daily, the mid-cycle peak of LH, as measured in serum by radioimmune assays, is abolished. The same effect has been observed when 2.5 mg. of norethindrone acetate alone is administered daily, but when lower dosages of a progestin are used, for example 0.5 mg. of chlormadinone acetate daily, the results are irregular, and ovulation frequently occurs, even though conception rarely takes place. Thus it is assumed that both estrogen and progestin act on the hypothalamus to prevent the cyclic surge of the LH releasing factor. To what extent combinations of estrogen and progestin exert a direct effect upon the ovaries of women has not been determined.

THE EFFECT OF PROGESTINS ALONE

In addition to their effect upon the hypothalamus, progestins have other antifertility actions. The endometrium is rapidly matured and eventually "exhausted," so that it is out of phase with the ovulatory cycle. The cervical mucus is thickened and rendered hostile to spermatozoa. Tubal motility is reduced, possibly decreasing the rate of ovum transport. Thus daily "mini-doses" of a progestin, such as 0.5 mg. of chlormadinone acetate, may prevent pregnancy without inhibiting ovulation. In 10 women receiving this dosage, laparotomy was performed 12 to 20 hours after coitus, and no spermatozoa could be recovered from the oviducts. A field trial in Mexico involving over 15,000 cycles revealed a failure rate of 3.7 pregnancies per 100 woman-years, but since only one in three of the women followed the regimen precisely, the theoretical effectiveness has not been estimated. A smaller field trial in Chile yielded a failure rate of 11. Inasmuch as most of the potentially serious side effects of oral contraceptives are believed to be due to the estrogen, the minidose progestin regimen offers interesting possibilities of fairly high effectiveness, high safety and low cost. Its chief drawback at present is the unpredictability of the bleeding pattern and its relative fallibility when compared with the estrogen-containing compounds.

COMBINED vs. SEQUENTIAL REGIMENS

A large number of estrogen-progestin combinations are available, and it must not be assumed that one product is the equivalent of another. The type of synthetic progestin contained may cause various side effects (possibly acne, amenorrhea, dysmenorrhea and depression), whereas other side effects may be related to the variable quantities of estrogen.

In the sequential regimen, estrogens alone are used for the first 15 days of the pill cycle, followed by five days of estrogen plus a progestin. Although the use effectiveness is very high, most gynecologists believe that forgetting to take the pill for one or two days is far riskier on the sequential regimen than it is with the combined hormones, because of the added antifertility actions of the progestin. In a Swiss study, the pregnancy rate with sequentials was 1.2, and with the combined forms it was 0.4.

The higher the estrogen content, the longer the lag time between cessation of medication and onset of bleeding, the greater the amount of flow, and the lower the incidence of early break-through bleeding.

SIDE EFFECTS

The typical pharmaceutical brochure which accompanies any sample package of oral contraceptives contains a list of over 60 side effects which have been reported, some serious, some trivial. Nevertheless, a few opponents of the oral contraceptives, by stressing these potential harms in the lay press without reference to the benefits, have created serious anxieties among the consumers. When it is realized that some 50 million women are taking the Pill, and that most of them, like women not taking the Pill, develop some kind of symptoms during the course of a year, the problem of relating cause and effect becomes enormous.

Let us consider one reported side effect, that of depression, to illustrate the difficulties of interpretation. The consensus in the medical literature is that whereas between 30 and 60 per cent of women "feel better" and enjoy a happier sex life, and a similar number claim no change in the way they feel, between 5 and 10 per cent of those placed on oral contraception become depressed. Psychological evaluation suggests that this last group contains a large number of women who reject sexual activity and receive no enjoyment therefrom, or who have a subconscious desire for children, or in whom there are religious conflicts. The fact that a similar number of women become depressed following the placement of an intrauterine device or after tubal sterilization suggests that in many cases psychological factors rather than the hormonal constituents of the Pill are the underlying causes.

Among the "nuisance" side effects, nausea (particularly in the first pill cycle), fluid retention, increased skin pigmentation, mastalgia, irregular spotting and break-through bleeding are fairly common and are related to the estrogen dosage. Other alleged effects, such as changes in libido, nervousness, alterations of appetite, vertigo, headaches, backaches, and a host of other symptoms are just as likely to be unrelated to the hormones.

Of much greater concern is the apparent increase of thromboembolic disease in users of oral contraceptives. Laboratory studies suggest that the coagulation accelerator factors VII, VIII, IX and X are increased among Pill users, as they are in pregnancy. In a British study, the mortality rate from thromboembolic disease in women between 20 and 34 years of age was found to be 0.2 per 100,000 in non-users of the Pill and 1.5 per 100,000 in users of oral contraceptives. Between the ages of 35 to 44, the comparable rates were 0.5 and 3.9, respectively. This would suggest that the risk of death from pulmonary embolism, even though extremely small, is about eight times as great for those taking oral contraceptives. In another British study, it was found that about one in every 2000 Pill users will be admitted to the hospital each year for non-fatal venous thromboembolism, compared to one in 20,000 women not on oral contraceptives. Data collected in the United States suggest a minimal increase in the death rate from pulmonary embolism, but the most recent studies from this country indicate that among women admitted to hospitals for thromboembolic disease, those on oral contraceptives outnumber those not on the Pill by 4.4 to one. It is obvious that oral contraceptives containing estrogen are contraindicated for patients who have had any thromboembolic episode in the past. The development of migraine after institution of oral contraception is also believed to be an indication for discontinuance because of a possibly higher incidence of cerebral thrombosis or oculo-vascular disease in this group.

The Collaborative Group for the Study of Stroke in Young Women reported in 1975 that the relative risk of thrombotic stroke among oral contraceptive users is four times and the relative risk of hemorrhagic stroke is twice that of the rate among controls. Heavy smoking enhances the risk of thrombotic stroke among users of the Pill, as does the existence of migraine. Severe hypertension increases the relative risk of hemorrhagic stroke more than 20 times in both the users and non-users, and the Collaborative Group believes that oral contraceptives should not be used by women with any degree of high blood pressure. The absolute risk of thrombotic stroke per year in young women on the Pill has been estimated by Vessey to be 1 in 10,000.

Two British studies published in August, 1975, indicated that women from 30 to 44 years, and increasingly with age, are from 2.7 to 5.7 times more likely to have heart attacks if they are on oral contraceptives. Fatal heart attacks in women over 40 are four times more common among pill-users than among non-users.

The 1974 interim report from the Royal College of General Practitioners indicated an increased reporting of gallbladder disease, urinary tract infections and Raynaud's syndrome among oral contraceptive users.

Of interest is a cooperative report from the U.S., the U.K. and Sweden indicating that thromboembolic disease in oral contraceptive users is four to five times more common in women with blood types A, B and AB than in those with type O. Of greater practical importance is the observation that the rate of thromboembolic disease is related to the quantity of estrogen contained in the pills. The sequentials and those combined preparations containing 0.1 mg. of mestranol were associated with two to three times the risk of thromboembolism when compared with the combined preparations containing only 0.05 mg. This observation suggests that most, if not all, women should be started on or switched to an oral contraceptive containing no more than 0.05 mg. of estrogen. In the past two years a number of products have been introduced which contain only 0.02 to 0:035 mg. of estrogen and from 0.5 to 1.5 mg. of the progestin. With these lower estrogen amounts pregnancy rates are slightly higher and breakthrough bleeding occurs with greater frequency, but hopefully the rate of serious side effects may be further reduced.

Other metabolic changes associated with oral contraceptives include a decreased glucose tolerance, increased plasma triglycerides, increased plasma ceruloplasmin concentrations, increased protein binding of thyroxine (resulting, of course, in elevated PBI concentrations), increased hepatic retention of BSP and increased plasma concentrations of renin substrate. All of these changes are reversible upon discontinuance of the medication. The elevated renin substrate (angiotensinogen) is thought to be responsible for the occasional development of hypertension in susceptible subjects, and this too is reversible. Although it is rare, its occurrence makes it mandatory to include measurements of the blood pressure before the use of oral contraceptives and periodically while they are being taken. The effects of oral contraceptives upon 100 laboratory determinations have been summarized by Miale and Kent.

EFFECTS UPON SUBSEQUENT FERTILITY

Oral contraceptives neither cause permanent sterility nor enhance fertility after discontinuance. The interval between stopping the Pill and the occurrence of ovulation is similar to the postpartum return of cycles in non-lactating mothers, and is commonly one to three months. Among women who take the Pill for a year or less, half of the subsequent pregnancies occur within three months and 90 per cent occur within 12 months. When conception has occurred within two months or so after discontinuance of oral contraception and spontaneous abortion has occurred, Carr has noted a pronounced increase in the incidence of polyploidy in the embryonic tissues. Even though this is universally fatal to the developing embryo, the observation suggests that it might be wise to advocate the use of some coitally related form of contraception for one or two cycles before attempting pregnancy.

Following the discontinuance of the combined preparations, from 2 to 4 per cent of users may have prolonged amenorrhea, occasionally accompanied by galactorrhea. This event appears to be more common in women who give a history of marked menstrual irregularity prior to beginning oral contraception. When the amenorrhea persists for longer than six months or a year, it is distressing for those couples who are planning a pregnancy. Fortunately, ovulatory cycles resume spontaneously in many; if they do not, the administration of clomiphene is usually effective in restoring ovulation during the cycle in which the drug is administered.

BENEFICIAL EFFECTS

The largest prospective study of oral contraceptive users has been conducted by the Royal College of General Practitioners in Great Britain. This interim report in 1974 on 23,611 pill takers and 22,766 controls matched for age revealed a use effectiveness of 0.34 pregnancies per 100 woman-years. There was no increase in any form of malignancy, nor in diabetes or other endocrine disorder. In addition to adverse effects noted above, the following beneficial effects were observed: (1) a 75 per cent reduction in benign breast neoplasia; (2) a significant decrease in dysmenorrhea and premenstrual tension; (3) protection against development of ovarian cysts; (4) decreased incidence of acne; and (5) decreased incidence of iron deficiency anemia. They stated that "it could well be that the beneficial effects actually exceed adverse reactions." As with most elective procedures in medicine, the risks and benefits must be fully explained to each woman so that she may make an informed decision.

ANTIFERTILITY EFFECT OF POSTCOITAL ESTROGEN

When, after an isolated coitus at the time of the fertile period, very large doses of estrogen are given, for example, 25 mg. of diethylstilbestrol · twice daily for five days, implantation does not seem to occur. This may be caused by some effect upon the endometrium that renders it unsuitable for implantation or by an accelerated rate of ovum transport. The technique has limited usefulness, for it usually causes nausea and vomiting, but has been of value in selected cases of rape or known contraceptive failures during the fertile period.

The Intrauterine Device (IUD)

For many centuries, Arab camel owners inserted small stones into their animals' uteri to prevent pregnancy; during the same period, natives of Indonesia introduced filaments into the uteri of unmarried women as a contraceptive device. The earliest mention of such a technique in the medical literature, however, is a note by Richter of Germany in 1909 advocating the insertion of strands of silkworm gut into the uterine cavity for contraception. The method went unnoticed until Gräfenberg of Berlin reported upon his successful use of an intrauterine coil of silver wire in 1931. During the next 25 years, the method fell into disrepute because of its general condemnation by physicians who had had no experience with the method. In 1959, interest was re-awakened by the reports of low pregnancy rates and the absence of serious side effects with the intrauterine ring technique in Japan and Israel.

TYPES OF DEVICES

The past decade has witnessed the development of a number of different devices, a few of which are illustrated in Figure 5–1. By means of a Cooperative Statistical Program, the results of IUD usage are reported in a standardized manner from all countries and are expressed as the rate of events during the first year of use per 100 women. Typical results from some major

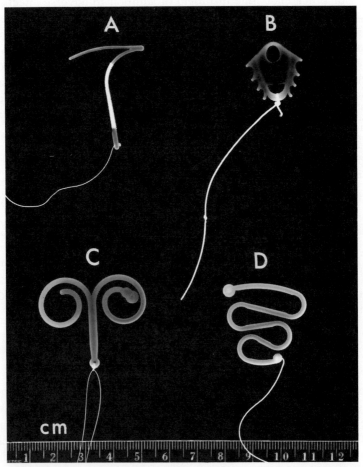

FIGURE 5–1. Four commonly used intrauterine devices. *A*, The Cu-7 (a slight modification of the Tatum copper-T). *B*, Dalkon Shield. *C*, Saf-T-Coil. *D*, Lippes Loop (size C).

studies for the Lippes Loop D, Dalkon Shield, Cu-7, Copper T, and a progesterone containing IUD are shown in Table 5–2. The lowest pregnancy and expulsion rates are associated with the Progestasert, although other reported series show higher expulsion and removal rates than indicated here.

The Dalkon Shield was developed because there is evidence that the contraceptive effectiveness is possibly proportional to the surface area of contact with the endometrium. Pregnancy occurring with the Dalkon Shield in place has been associated with an excess of deaths from septic abortion. Because of this, the device was removed from the market during 1975.

Expulsions and removals of IUDs for pain or bleeding problems or pelvic infections are more common in nulliparous than in parous women, and most gynecologists feel that an intrauterine device is best suited for women who have borne children. Uterine perforation may occur about once in two or three thousand insertions, and tubo-ovarian abscess has on rare occasions been a serious event.

MECHANISM OF ACTION

The mode of action in the human species is unknown. Experiments in monkeys suggest that the rate of ovum transport is accelerated, so that fertilized ova arrive in the uterine cavity prior to the optimal time for implantation. Others have suggested that the foreign body induces subtle ultramicroscopic changes that may alter the biochemical behavior of the endometrium. Workers in Turkey have demonstrated that the intrauterine

Table 5–2.   COMPARISON OF ONE-YEAR NET PREGNANCY, EXPULSION, REMOVAL AND CONTINUATION RATES FOR SELECTED IUDS*

Device	Author and Date	Pregnancy	Expulsion	Removals, Bleeding or Pain	Other	Continuation Rate
Lippes Loop D	Tietze 1970	2.7	9.5	11.7	6.6	77.4
Dalkon Shield	Snowdon 1973	3.8	3.9	4.6	7.0	82.4
Cu-7	Giber 1972	1.4	5.9	8.9	3.9	75.3
Copper T	Lewit 1973	2.2	8.3	6.4	6.7	76.4
Progestasert	Pharriss 1974	1.0	2.8	8.2	4.8	83.2

*Adapted from Population Reports, Series B, No. 2, January, 1975.

device promotes the accumulation of millions of macrophages, which they believe will attack a fertilized ovum by phagocytosis and will even consume spermatozoa.

EFFECTIVENESS

The IUD cannot compete with the oral contraceptives in terms of protection from pregnancy, but it is, of course, devoid of systemic side effects. The advantage of the method is the fact that it is the only readily reversible method of fertility control requiring only one decision and no sustained motivation on the part of one or both sexual partners. Trained personnel are required for the insertion, but repetition of supplies is not necessary and the long-term cost is low. The use effectiveness (two to four pregnancies per 100 woman-years) is relatively high, and the method does not require any degree of intelligence or education. In an attempt at some demographic control of populations, the method has been widely used in South Korea, Taiwan, India, Pakistan, Hong Kong and Singapore.

INDICATIONS AND CONTRAINDICATIONS

Intrauterine devices are indicated for women who want one, after all of the advantages and disadvantages have been explained. They are especially indicated for parous women who have had side effects with oral contraceptives in the past or in whom the oral contraceptives are contraindicated.

They are contraindicated for women who don't want one, or who are not willing to accept a 2 to 4 per cent risk of pregnancy. Additional contraindications include: active pelvic infection; an abnormally shaped uterine cavity; severe hypermenorrhea or dysmenorrhea; severe anemia; and the current use of anticoagulant or immunosuppressive drugs.

THE EFFICACY OF CONTRACEPTION

Tietze has classified the available methods of contraception into four groups, based upon their relative effectiveness:*

GROUP I: MOST EFFECTIVE
Combined oral estrogen and progestin
Sequential oral estrogen and progestin
"Temperature" rhythm method
GROUP II: HIGHLY EFFECTIVE
Intrauterine devices
Diaphragm and jelly or cream
Condom

---

*Tietze, C. In Calderone, M. S. (Ed.). Manual of Family Planning and Contraceptive Practice, 2nd Ed. Williams & Wilkins Co., Baltimore, 1970, pp. 268–275.

Group III: Less effective
              Vaginal foams
              Vaginal jellies or creams
              "Calendar" rhythm method
              Coitus interruptus
Group IV: Least effective
              Postcoital douche
              Prolonged breast feeding

In summary, there is at this time no method of birth control for sexually active women that is completely devoid of risk, either in terms of failure or in terms of serious side effects. Surgical sterilization of the female, irrespective of the method used, carries a risk of mortality which is not unlike the risk of therapeutic abortion or of pregnancy and delivery. Discounting venereal diseases for the moment, surgical sterilization of the male would appear to be the only method that would not subject women to some slight risk of death as the result of sexual intercourse. It is important for every physician who has the responsibility for the care of women to be familiar with all methods concerned with the prevention of unwanted pregnancy so that he may present an intelligent appraisal of the options and give an informed opinion of any risks involved.

POPULATION PROBLEMS

The Population Explosion

A glance at Figure 5–2 will suffice to show what has happened to the world's population since 2000 B.C. and what will happen by A.D. 2000,

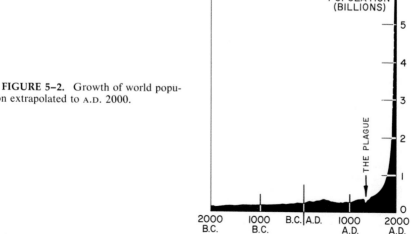

FIGURE 5–2. Growth of world population extrapolated to A.D. 2000.

which is less than 25 years away. The modern explosion of populations has been due entirely to the control of death. Every year, new information seems to require an enlargement of even the most pessimistic population forecasts.

The father of demography, Thomas Malthus, in the first edition of his Essay in 1798, wrote as follows: "I think I may fairly make two postulata: First, that food is necessary to the existence of man. Second, that the passion between the sexes is necessary and will remain nearly in its present state. Assuming then my postulata as granted, I say that the power of population is indefinitely greater than the power in the earth to produce subsistence for man. Population, when unchecked, increases in a geometrical ratio. Subsistence increases only in an arithmetic ratio."

Many writers in subsequent generations poured ridicule upon the Malthusian doctrine. Some of this resulted from a too-literal interpretation of his "arithmetic ratio" as applied to food production. Karl Marx in particular poured abuse upon Malthus, claiming that his essay was a shameless pack of lies, and that any problems of population growth were due entirely to capitalism and would not occur under a communist system. Until very recently, this was the official doctrine of communist countries.

The true facts, of course, are revealed in data such as those shown in Table 5–3. The world population, currently nearing 4 billion, will probably

Table 5–3. WORLD POPULATION DATA, 1975*

	Africa	Asia	North America	Latin America	Australia New Zealand	USSR	Europe	World
Population, mid-1975 (in millions)	401	2255	237	324	16.8	255	473	3967
Current growth rate, (% annually)	2.6	2.1	0.9	2.7	1.7	1.0	0.6	1.9
No. years to double population	27	33	77	26	40	69	116	36
Birth rate (per 1000 population)	46.3	34.9	16.5	36.9	21.5	17.8	16.1	31.5
Death rate (per 1000 population)	19.8	13.6	9.3	9.2	8.1	7.9	10.4	12.8
Infant mortality rate (per 1000 live births)	156	102	18	79	17	26	24	98
Life expectancy at birth (in years)	45	54	71	61	72	70	71	55
Per cent under the age of 15	44	38	25	42	28.5	36	24	36
Per capita gross national product (in $U.S.)	240	270	5480	650	2800	1400	2380	940

*Population Reference Bureau, Inc., Washington, D. C.

exceed 6 billion in the next 25 years. Some areas of the world, such as Europe, will be affected but little, for its annual rate of growth is only 0.6 per cent and its death rate has long been low. In contrast to this, Latin America, with its growth rate of 2.7 per cent, will double its population in 26 years. The average figures for large areas of the world do not, of course, indicate the imminent disasters which certain individual nations face. India, with its current overpopulation of 620 million, is perpetually on the verge of famine, yet will add over 16 million people in the next 12 months. El Salvador, with its rate of increase of 3.7 per cent, will double its population in 19 years. By way of contrast, East Germany, with its current negative growth rate, should slowly diminish its population size in the future.

Geometric projections of current trends may, of course, lead to some ridiculous conclusions, such as the "one square foot of space available for each human being," or an "outweighing of our planet by the mass of human bodies." The rate of growth of California (including immigration) exceeds that of the United States to such an extent that simple arithmetic predicts the year when every person in America will live in California! As in mammalian populations, excessive overcrowding may lead to wars of survival, to famine, to psychiatric disorders and other diseases secondary to stress, to migration, to reduced fecundity and to many other events that are unpredictable. To the human species, population control through the increase of death rates, whether by nuclear explosions, disease or famine, is unacceptable; the only possible alternative is the control of births.

The Population Implosion

Of considerable significance is the fact that people in most countries are progressively migrating from rural to urban areas. In the United States, 70 per cent of our population lives on 1 per cent of the land. Of the 100 million people to be added to this country in the next 30 years or so, it is estimated that 80 million will reside in the major metropolitan areas. The socioeconomic consequences are frightening, and indeed already confront us. It now costs the financially beleaguered city of New York over $20,000 to provide commuting facilities for every new individual moving to the suburbs, and the welfare costs required to care for many of those who pour in from poor rural areas is measured in the billions of dollars. Shortage of water, pollution of the air, public health problems, housing shortages, the concentration of poor people in ghettos, difficulties in transportation, the control of crime and riots, and a host of other serious consequences of excessive urbanization are here now and will inevitably increase.

The same problem exists in other countries where industrialization is occurring. In India, between 100 and 200 million people will migrate from rural to urban areas during the next 30 years, and it is estimated that the population of Calcutta will reach 66 million by the end of this century. In the past, migration to unpopulated areas of the earth solved these problems to a large extent, but with the exception of Siberia, Australia and very few other parts of our globe, this will no longer provide a solution. Migration to outer space remains in the realm of science fiction. There is no acceptable solution except the control of births.

Methods of Population Control

The control of births within any given population is quite a different problem from that of birth control within a given family. For example, the adoption of the vaginal diaphragm and jelly method by an American housewife may be, and has been, entirely satisfactory for her, but attempts to apply this to a large segment of the population have met with dismal failure. Let us assume that, as an experiment, 1000 women in Malaysia are fitted with a diaphragm, and that at the end of a year 300 are satisfied and 700 are dissatisfied, either because the method failed or because it was personally unacceptable to husband or wife. Each of the 1000 women tells 10 friends about her experience, so that 7000 of the next 10,000 women approached reject the method. By the end of two years, the experiment has failed. For demographic utility, therefore, any technique recommended must have a much higher degree of use effectiveness. The only methods to date which fall in such a category are (1) contraception with oral steroids or the IUD, (2) legal abortion, and (3) sterilization of the male or female. The factors which affect the demographic efficacy of the many ways of controlling births are compared in Table 5–4.

The Demographic Acceptance of Contraception

There are many reasons why the individual consumer rejects contraception, and there exist numerous anti-birth control attitudes on the part of the dispenser.

Some of the anti-birth control attitudes of groups of countries have been summarized by Guttmacher as follows: (1) Theological dogma. The reaffirmation by the Pope in 1968 that all methods of birth control other than periodic abstinence are illicit will no doubt retard the spread of family planning centers in some countries. Contrary to popular opinion, however, it is not the predominant religion of a country that determines its birth rate, but rather the degree of poverty and illiteracy. For example, in the 11 predominantly Roman Catholic countries of Europe, the birth rate of 18 per thousand is essentially identical with the rate in the 15 predominantly non–Roman Catholic countries. In Latin American countries, the high rates of growth are due primarily to the more recent control of death rates.

(2) "Birth control is a form of genocide aimed at minority groups." This dogma has unfortunately arisen among some minority groups in the United States and some underdeveloped countries, which claim that their

Table 5–4. FACTORS AFFECTING DEMOGRAPHIC EFFICACY OF DIFFERENT METHODS*

	Rhythm	Foam, Jellies	Dia-phragm	Condom	IUD	Oral Contra-ception	Legal Abort.	Female Steriliz.	Male Steriliz.
Effectiveness	+	+	++	++	+++	++++	++++	++++	++++
Limitations									
Personal and cultural	0	+	+	+	0	0	++	++	++
Need for special medical facilities	0	0	+	0	++	+	++	++	++
Need for repeat supplies	0	+	+	+	0	+	+	0	0
Need for high motivation	++	++	++	++	0	0	+	+	+
Need for higher education	+	+	++	+	0	0	0	0	0
Catholic objections	0	0	0	0	++	0	++	++	++
Unsuitable for nullipara	0	0	0	0	+	0	0	+	+
Unsuitable for low parity	0	+	+	+	0	0	0	+	++
High cost	0	0	0	0	+	+	+	++	+
Demographic utility	0	0	0	0	+	++	+	0	+

*(Adapted from Venning, G. R. *In* Advances in Reproductive Physiology. Academic Press, New York, 1966.)

numbers must rapidly increase in order to achieve political power. Counteracting such a false doctrine is a serious problem for politicians, sociologists and physicians alike.

(3) "Birth control, being more effective among the responsible, promotes overpopulation by the irresponsible segments of society." Responsibility, of course, is not genetically determined but is the result of economic and educational improvements.

(4) "A viable economy requires a rapidly increasing and large population." This is a fallacious economic doctrine that is prevalent in a number of the recently independent African States.

The most serious deterrent to the spread of family planning units in the developing nations, of course, is the lack of manpower in the health professions, the lack of money, and the more immediate need for the alleviation of death and disease.

Proposals Other than Voluntary Family Planning

There are many who feel that a massive extension of voluntary family planning cannot solve the population program, and Berelson has ably summarized and discussed 29 proposals or suggestions that have appeared in the literature. He has grouped these into eight categories:

1. Extensions of voluntary fertility control, such as the institutionalization of maternal care in developing countries or the liberalization of legal abortion.

2. The establishment of involuntary fertility control, such as the issuance of licenses for the bearing of children, the compulsory sterilization of men with three or more living children, required abortion for all illegitimate pregnancies, and so forth.

3. Intensified educational campaigns, world-wide and at all levels of education.

4. Incentive programs, such as rewards for not having children or for the initiation of contraception.

5. Tax and welfare benefits and penalties of many kinds.

6. Shifts in social and economic institutions in such ways that fertility of a population is lowered. These would include increases in the minimum age for marriage, the promotion of female participation in the labor force and a variety of other long-range social trends.

7. Approaches by way of political channels and organizations, such as the extension of national policies, insistence upon population control as the price of foreign aid, and the national promotion of zero growth in population.

8. Augmented research efforts, both in sociology and in improved contraceptive technology.

Berelson's appraisal of such proposals by specific criteria is shown in Table 5–5.

Table 5-5.   ILLUSTRATIVE APPRAISAL OF PROPOSALS BY CRITERIA*

	Scientific Readiness	Political Viability	Administrative Feasibility	Economic Capability	Ethical Acceptability	Presumed Effectiveness
Extension of voluntary fertility control	High	High on maternal care, moderate to low on abortion	Uncertain in near future	Maternal care too costly for local budget, abortion feasible	High for maternal care, low for abortion	Moderately high
Establishment of involuntary fertility control	Low	Low	Low	High	Low	High
Intensified educational campaigns	High	Moderate to high	High	Probably high	Generally high	Moderate
Incentive programs	High	Moderately low	Low	Low to moderate	Low to high	Uncertain
Tax and welfare benefits and penalties	High	Moderately low	Low	Low to moderate	Low to moderate	Uncertain
Shifts in social and economic institutions	High	Generally high, but low on some specifics	Low	Generally low	Generally high, but uneven	High, over long run
Political channels and organizations	High	Low	Low	Moderate	Moderately low	Uncertain
Augmented research efforts	Moderate	High	Moderate to to high	High	High	Uncertain
Family planning programs	Generally high, but could use improved technology	Moderate to high	Moderate to high	High	Generally high, but uneven on religious grounds	Moderately high

*(From Berelson, B. Beyond family planning. Publication number 38, Population Council, February, 1969.)

The year 1974 was designated as the World Population Year, and culminated in the Bucharest Conference in August, attended by representatives from 137 governments. For the first time in history, a World Population Plan of Action was adopted to encourage governments to improve their demographic positions. As might be expected, there were many controversies, political, religious and economic in nature. However, the final document stressed the need for socioeconomic development and for the reduction of desperate poverty in underdeveloped countries. Nevertheless, virtually every government emphasized the right of individuals to regulate their own fertility but at the same time to have universal access to both the information and the means to enable them to practice responsible parenthood. The conference went further by anticipating the United Nations' 1975 International Women's Year, and recommended "the full participation of women in the educational, social, economic and political life of their countries on an equal basis with men."

Some of the delegations from developed countries, including the United States, had hoped that there might be a general adoption of goals to reduce the rates of population growth in all countries to specified levels by 1985, but this was opposed by many countries. The final recommendation for the world population plan of action simply stated that "countries which consider their birth rates detrimental to their national purposes are invited to consider setting quantitative goals and implementing policies to achieve them by 1985." Nevertheless, the world conference should be considered a success if for no other reason than because of the intense focus of attention upon the social, economic and demographic problems facing the entire world.

Voluntary Sterilization

Satisfactory and safe methods for the sterilization of men and women have existed only since about 1920, and have not been officially advocated as a means of demographic control, except in India. A vast program of encouraging vasectomies was undertaken in India after 1960, and during the ensuing five years 1.3 million voluntary operations were performed upon men who had produced a specified number of children. To reduce its current birth rate of 41 per 1000 to as low as 25 per 1000 would necessitate over 5 million operations per year for the next 15 years, an almost insurmountable task.

Voluntary sterilization is now legal throughout the United States, although there are still some restrictive state laws pertaining to age, marital status, consent of spouse and waiting periods following request. For the most part all that is needed, except in the case of minors, is agreement between the patient and his or her physician. About one-quarter of married couples in the United States with the wife aged 30 or over have had a sterilizing procedure, about half of these have been by vasectomy.

Female sterilization may be accomplished by some type of procedure on the tubes by the vaginal or the abdominal route, by laparoscopy or by elective vaginal or abdominal hysterectomy. Many gynecologists consider that elective hysterectomy carries too high a risk of morbidity and mortality to justify its use for sterilization. In any event, female sterilization will not become a practical means of population control because of the high cost and the need for surgeons and hospital facilities.

Legal Abortion

A few countries, such as Japan, have adopted legal abortion as one means of population control, with a resulting decline in birth rates. Since the legalization of abortions in the United States, the number performed per year has steadily increased, but the influence on birth rate is difficult to ascertain because, in general, legal abortion has replaced illegal abortion with a concomitant fall in maternal mortality.

It would appear obvious that legal abortion performed by competent surgeons in approved hospitals is preferable to a self-induced or criminal abortion by untrained persons because of the frightful morbidity and appreciable maternal mortality associated with the latter. Apart from the moral issues involved, legal abortion alone can hardly become an effective means of population control because, like sterilization of the female, it involves a surgical operation, is costly, must be repeated in subsequent pregnancies, and is therefore inferior to contraception as a means of preventing unwanted births.

CONCLUSION

For most nations, the control of population growth is a matter of transcendent importance and utmost urgency. As stated by Bertrand Russell:

The desirable remedy does not lie in restoring the death rate to its former level. It does not lie in the promotion of new pestilence. Least of all does it lie in the vast destruction that a new war may bring. . . . During what remains of the present century, the world has to choose between two possible destinies. It can continue the reckless increase of population until war, more savage and dreadful than any yet known, sweeps away not only the excess but probably all except a miserable remnant. Or, if the other course is chosen, there can be progress, rapid progress, towards the extinction of poverty, the end of war, and the establishment of a harmonious family of nations.

That "other course" is obviously the control of births, and the most effective and acceptable means is through methods of contraception, with voluntary sterilization and legal abortion being only adjunctive measures.

But the voluntary control of births will not solve the problem so long as people marry young and desire large families. In the final analysis, social attitudes and structures must also change in such a way that each family need not feel it desirous to contribute more than two children to the world's population. To accomplish the latter is no mean task within a generation, but perhaps a first step would be to make sure that each individual has access to family planning services so that a freedom of choice will at least reduce the numbers of children who are unwanted.

SUGGESTED SUPPLEMENTARY READING

The following periodicals should be useful for those interested in contraception or demography:
    Advances in Planned Parenthood, Excerpta Medica Foundation, Amsterdam.
    Contraception; an international monthly journal, Volume 12, 1976.
    Demography. A quarterly journal. Vol. 13, 1976.
    Studies in Family Planning, Population Council, New York.
Berelson, B. Beyond family planning. Publication No. 38, Population Council, New York.
    A thoughtful analysis of the practicality of 29 proposals for limiting population growth.
Contraceptive Technology, 1974–75. Emory University School of Medicine, Atlanta, Georgia.
    This is the seventh edition of an 85-page handbook prepared for medical students and all others interested in contraception.
Intrauterine devices. Population Reports, Series B, No. 2, January, 1975.
    This entire series, published by the Department of Medical and Public Affairs of the George Washington University Medical Center, Washington, D. C., is one of the best and most complete sources of information on contraceptive methods.
Miale, J. B., and Kent, J. W. Am. J. Clin. Pathol., 57:80–88, 1972.
Peterson, William. Population, 2nd Ed., The Macmillan Co., New York, 1969.
    This is a standard textbook widely used by students of demography.

Chapter 6

HUMAN GENETICS

The basic principles of heredity were discovered by Gregor Mendel, a monk in Northern Moravia, and were reported in 1866. Mendel studied the inheritance of a number of characters in garden peas, from which he concluded that inheritance is governed by certain laws. He found that he could predict, with a high degree of accuracy, the types of offspring, and the relative frequencies thereof, resulting from the mating of two plants of known pedigree. Before Mendel's experiments, the various characters of the parents were believed to be "blended" in the offspring. Mendel's results showed that blending does not occur, but that the characters of the parents, although perhaps not expressed in the first generation of offspring, reappear unchanged in later generations. From his experiments, he inferred that inheritance is governed by "units" present in the cells of each individual. He concluded that there were two such units in the adult plant and that these units segregated in the formation of pollen or eggs so that there was only one such unit in the individual egg or sperm. This brilliant piece of reasoning was supported when the details of cell division and fertilization became known. His discovery received little or no recognition from the biologists and medical scientists of the day, and was neglected for more than 30 years.

Mendel's results led to the formulation of three principles, now called Mendel's Laws, which may be stated as follows:

1. THE LAW OF UNIT INHERITANCE. The inheritance of each kind of characteristic—in plants, such things as height, flower color and seed shape— is governed by unitary factors (now called genes) which exist in pairs in individuals.

2. THE LAW OF SEGREGATION. Characters, called genes, exist in pairs in each individual; in the formation of gametes, each gene separates or segregates from the other member of the pair of genes and passes into a different gamete. Thus each gamete has one and only one of each kind of gene.

3. THE LAW OF INDEPENDENT ASSORTMENT. The members of one pair of genes separate or segregate from each other independently of the members of other pairs of genes, and come to be assorted at random in the resulting gamete.

Three different investigators, de Vries in Holland, Correns in Germany, and von Tschermak in Austria, rediscovered Mendel's Laws in 1900, shortly after the details of mitosis, meiosis and fertilization were known. These three men, in searching through the literature, found Mendel's paper, and 16 years after his death the fundamental laws of inheritance were named after him. The universal nature of Mendelian inheritance was promptly recognized. The first example of Mendelian inheritance in man, alkaptonuria, was reported in 1902 by Garrod.

GENES AND CHROMOSOMES

Examination of a dividing cell under the phase contrast microscope or examining fixed and stained histologic sections under an ordinary microscope reveals elongate, dark-staining bodies called *chromosomes* within the nucleus. Each chromosome consists of a central thread, the chromonema, along which lies a series of bead-like structures, the chromomeres. At a fixed point along its length, each chromosome has a small, clear, circular zone, called a centromere, that may control the movement of the chromosome during cell division. Chromosomes are most visible at the time of cell division; at other times they appear as very long, thin, fine dark-staining strands called chromatin. Thus chromosomes are present as greatly extended but distinct physiological and structural entities between successive cell divisions, although they are not usually visible by light microscopy.

Mendel's unit factors of inheritance were christened *genes*, and are now defined as the hereditary factors that lie within a chromosome in a linear order. Each cell of every individual of a given species contains a characteristic number of chromosomes. Each cell in the body of every human being has 46 chromosomes. Many other species of animals and plants also have 46 chromosomes; hence, it is not the number of chromosomes that differentiates the various species. The chromosome numbers of most plants and animals lie between 10 and 50. Chromosomes generally exist in pairs; thus there are two of each kind in the somatic cells of higher plants and animals. The 46 chromosomes of man consist of two of each of 22 different kinds, plus the sex chromosomes X or Y (Fig. 6–1). The chromosomes differ in length and in shape, and in the presence of knobs or constrictions along their length. Generally the chromosomes vary enough morphologically that a cytologist can distinguish the different pairs. Patients with more or fewer chromosomes usually manifest some abnormality.

MITOSIS

The regularity of the process of somatic cell division ensures that each daughter cell will receive exactly the same number and same kind of chromosomes as the parent cell had. If a cell receives more or less than the proper

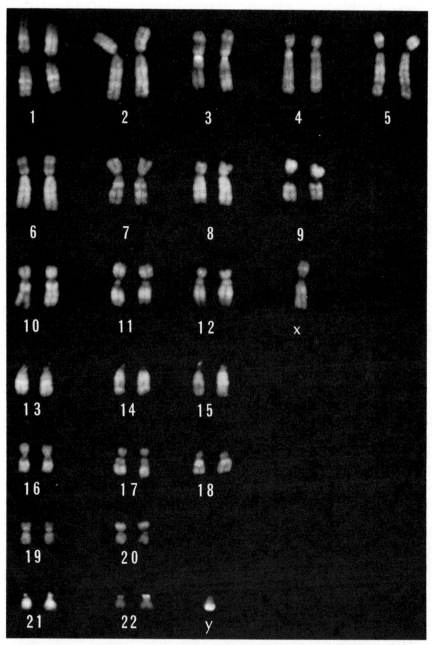

FIGURE 6–1. Normal human metaphase karyotype (from peripheral blood) stained with fluorescent quinacrine mustard, as described by Caspersson, Zech, Johansson, and Modest, Chromosoma, *30*:215 (1970), and as published by Caspersson, Zech, and Johansson in Experimental Cell Research, *62*:490 (1970). Magnification ×2500.

number of chromosomes by some malfunctioning of the process of cell division, it would have an unbalanced genetic complement and would probably show marked abnormalities. The term *mitosis* is applied to this regular division of the cell. It was once believed that each chromosome split longitudinally into two halves, but it is now clear that each original chromosome brings about the synthesis of an exact replica of itself. The new chromosome is manufactured during the S phase of the cell cycle, which precedes the initiation of the mitotic process. The old and new chromosomes (called chromatids until they separate) are identical in structure, and at first lie so close together that they appear to be one. As mitosis proceeds and the chromatids contract, the line of cleavage between them becomes visible. In each human cell undergoing mitosis, each of the 46 chromosomes has produced an exact replica of itself. As cell division proceeds, the 92 chromatids separate, and 46 go to one and 46 to the other daughter cell. The appropriate assortment of the chromosomes to the two daughter cells is regulated by a rather complex mechanism involving centrioles, centromeres, and a spindle (Fig. 6–2).

Although mitotic division is a continuous process, it has been divided, for descriptive purposes, into four stages: prophase, metaphase, anaphase and telophase. Between mitotic divisions the nucleus is said to be in the interphase or resting stage. The nucleus is "resting," however, only with respect to cell division, for during the interphase it is very active metabolically and synthesizes the DNA needed for chromosome replication.

PROPHASE

Prophase begins when the chromatin threads begin to condense and chromosomes appear as a tangled mass of thickening threads within the nucleus. The chromosomes grow thicker and shorter. Each chromosome has doubled during the previous interphase, and each part of the doubled chromosome is called a chromatid. The two chromatids are held together at the centromere, which remains single until the metaphase. The centriole, a small granular structure lying outside the nucleus, has divided during the interphase. The two daughter centrioles migrate to opposite sides of the cell, and from each extends a cluster of raylike filaments called an aster. Between the separating centrioles, a spindle appears, composed of protein threads with properties similar to those of the contractile proteins of muscle fibrils. The protein threads of the spindle are arranged like two cones placed together base to base, narrow at the poles near the centrioles and broad at the center or equator of the cell.

METAPHASE

When spindle formation is complete, the chromosomes, which have been getting shorter and thicker, line up across the equatorial plane of the spindle. This represents the beginning of the metaphase. At this time the

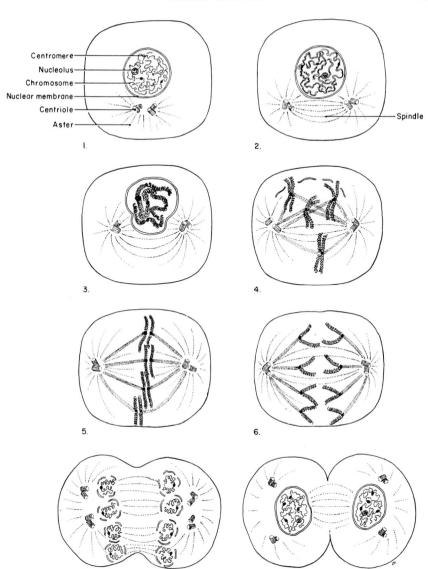

FIGURE 6-2. Diagram of mitosis in a cell of an animal with a diploid number of four (haploid number = 2). 1, Resting stage. 2, Early prophase: centriole divided. 3 and 4, Later prophase. 5, Metaphase. 6 and 7, Early and late anaphase. 8, Telophase: nuclear membrane has reappeared and cytoplasmic division has begun. (From Villee, C. Biology, 6th Ed. W. B. Saunders Co., Philadelphia, 1972.)

centromere divides and the two chromatids become completely separate daughter chromosomes. The centromeres of all of the chromosomes divide simultaneously under the control of some as yet unknown mechanism. The centromeres then begin to move apart, marking the beginning of anaphase. When human cells are observed under the microscope, prophase lasts 30 to 60 minutes, the metaphase a brief 2 to 6 minutes, and the anaphase some 3 to 15 minutes.

ANAPHASE

The mechanism by which chromosomes move to the poles is unknown. One theory suggests that the spindle fibers contract in the presence of ATP and pull the chromosomes to the poles. Spindle fibers isolated from cells about to divide can be induced to contract when ATP is added. According to another theory, the motive power is supplied by substances which get between the daughter chromosomes and absorb water and swell, pushing the chromosomes apart. The spindle fibers serve as guide-rails along which the chromosomes glide so that all of one set of daughter chromosomes are gathered at one pole and all of the other set are gathered at the other pole. Whatever force moves the chromosome to the pole is applied at the centromere, for a chromosome that lacks a centromere, perhaps as the result of exposure to x-radiation, does not move at all in mitosis.

TELOPHASE

When the chromosomes reach the poles, anaphase is completed and telophase begins. Telophase, lasting some 30 to 60 minutes, is the period in which the chromosomes elongate and return to the resting condition in which only chromatin threads are visible. A nuclear membrane forms around each daughter nucleus completing nuclear division (karyokinesis), and the division of the cell body (cytokinesis) follows.

The process of mitosis ensures the precise and equal distribution of chromosomes to each of two daughter nuclei. The chromosomes contain genetic information coded in their deoxyribonucleic acid (DNA), and a regular and orderly mitotic process ensures that this genetic information is precisely distributed to each daughter nucleus. Each cell has all of the genetic information for every characteristic of the organism. The problem of how the cells in the multicellular organism, all of which have the same genetic information, can end up with such widely different structural and functional characteristics is one of the key problems in present day biology.

MEIOSIS

The constancy of chromosome numbers in the cells of successive generations of organisms is ensured by the process of meiosis, which occurs during the formation of an egg or sperm. In meiosis a pair of cell divisions reduces the chromosome number by one-half; the gametes receive only one of each kind of chromosome (Fig. 6–3). When the two gametes, the egg and sperm, unite in fertilization, the fusion of their pronuclei reconstitutes the full number of chromosomes. To accomplish the precise distribution of

THE BASIC CONCEPTS OF CLASSIC GENETICS

Genetic research in the first four decades of this century established the concepts of transmission genetics, dealing with the mechanisms by which genetic information is transferred from one generation to the next. Research in the next three decades led to the clarification of the chemical and physical nature of the gene and the mechanism by which genes control the biochemical reactions of the cell to bring about the synthesis of specific proteins and other molecules. It was clear early in this century that the structure, function and chemical nature of an organism are determined by its specific pattern of genes acting in a specific environment. Since the action of any gene may be influenced by the activities of other genes in the same nucleus, it is clear that the term *environment* may actually include the genes. The physical basis of inheritance is the transmission from parent to offspring of genes, situated at specific points or loci on specific chromosomes. Each chromosome contains many, perhaps a thousand or more, genes arranged in a linear order. Since all the somatic cells of human beings contain two of each kind of chromosome, each cell contains two of each kind of gene. Each chromosome and each gene is duplicated from chemical substances present in the cell during the interphase preceding a mitotic division.

Genotype and Phenotype

Each gene may exist in two or more alternative states or alleles, which have different effects on the phenotype. Indeed, the inheritance of any trait can be studied only when two or more contrasting conditions exist. These contrasting conditions are inherited in such a way that an individual may have one or the other, but not both. In considering a cross between two individuals that differ in a single pair of genes, for example, brown eyes (BB) and blue eyes (bb), the operation of the Mendelian laws becomes evident. During meiosis in the male, the two bb genes separate so that each sperm has only one b gene (haploid). In meiosis in the female, the BB genes separate so that each ovum has only one B gene. The fertilization of this egg by a b sperm results in an offspring with the genetic formula Bb. An individual with two genes exactly alike—two genes for brown eyes, BB, or two for blue eyes, bb—is said to be homozygous for the trait. An individual with one B and one b gene is said to be hybrid or heterozygous. A heterozygous individual may resemble one parent or may exhibit a condition intermediate between the two parents. Usually one gene, termed the dominant gene, has a greater effect on the phenotype than the other, the recessive gene. The heterozygous individual may have a phenotype nearly identical to that of individuals homozygous for the dominant alleles. This phenomenon of dominance supplies part of the explanation as to why an individual may resemble one of his parents more than the other, despite the fact that both make equal contributions to his genetic constitution.

During meiosis in the gonads of heterozygous (Bb) individuals, the chromosome containing the B gene first synapses with and then separates from the chromosome containing the b gene. Each egg or sperm has either a B gene or a b gene but never both. Sperm (or eggs) containing B genes and those with b genes will be formed in equal numbers. With two types of eggs, B and b, and two types of sperm, B and b, four combinations are possible in fertilization and these will occur in equal numbers. There is neither any special attraction nor any repulsion between an egg and a sperm containing the same type of gene. The possible combinations of genes among the offspring can be determined by simple algebraic multiplication: $Bb \times Bb = BB + 2Bb + bb$. The ratio of genotypes, then, is 1:2:1. If, as is frequently the case, the heterozygous individuals (2Bb) resemble the homozygous dominant individuals (BB), the ratio of phenotypes will be 3:1 — three with the phenotype of the dominant gene and one with the phenotype of the recessive gene (Fig. 6–4).

It is important to realize at the outset that all genetic ratios are probability ratios. For example, if two individuals who are heterozygous for eye color mate and produce exactly four offspring, it is possible that there will be three with brown eyes and one with blue eyes, but all might have brown eyes or all might have blue eyes, although the latter possibility will occur very rarely. A simple application of the laws of probability will reveal just how frequently in a series of matings yielding exactly four offspring all four of the offspring would show the recessive trait. Since the probability that each offspring will have blue eyes is $\frac{1}{4}$, the probability that all four will have blue eyes is the product of the four separate probabilities: $\frac{1}{4} \times \frac{1}{4} \times \frac{1}{4} \times \frac{1}{4} = \frac{1}{256}$.

In any single mating with a small number of offspring, the ratio of brown-eyed to blue-eyed individuals can be almost anything. However, if one pools the data from many such matings, the ratio of individuals with brown and blue eyes will approach the theoretical 3:1 ratio. To state the probability in another way, in mating two heterozygous individuals (Bb), there are three chances out of four that any offspring will show the dominant trait and one chance out of four that he will show the recessive trait. Each mating is a sep-

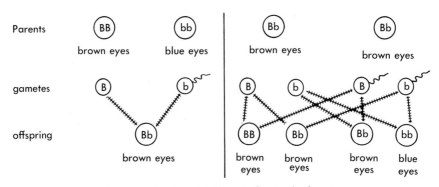

FIGURE 6–4.   The inheritance of eye color in man.

arate, independent event and its result is not affected by the results of previous matings. For example, if two heterozygous brown-eyed parents have had three brown-eyed children and are expecting a fourth child, this child has three chances in four of having brown eyes and only one chance in four of having blue eyes.

Since the distribution of one pair of chromosomes to the gametes has no effect on the distribution of other pairs of chromosomes, the inheritance of genes lying in different pairs of chromosomes is completely independent. This principle is used in calculating the probability of the simultaneous inheritance of two different traits.

Polygenic Inheritance

Although in certain instances the relationship between gene and trait is simple and clear—that is, each gene controls the appearance of a single trait—this relationship in other instances is quite complex. Several pairs of genes may interact to affect a single trait or one pair of genes may inhibit or reverse the effect of another pair of genes, or a given gene may produce different effects when the environment is changed in some way. In any event, the genes are inherited as units, but they may interact in a variety of ways to produce the trait.

Many human characteristics such as height, body form, intelligence, and skin color cannot be separated into distinct, alternative classes and are not inherited by a single pair of genes. Several, perhaps many, pairs of genes affect each characteristic. The term *polygenic inheritance* or *multiple factor inheritance* is used when two or more independent pairs of genes affect the same characteristic in the same way and in an additive fashion. Traits inherited by polygenes have a pattern of inheritance characterized by an F1 generation which is intermediate between the two parents and shows little variation and by an F2 generation which shows a wide variation between the two parental types. When many pairs of genes are involved and when this is coupled with effects of environment on the trait, the distribution of the trait results in a bell-shaped curve. Human beings, for example, cannot be separated into two classes of tall and short, or even into three classes of tall, medium and short. When the heights of 1083 adult men are measured and the number of people of each height is plotted against height in inches, the curve connecting the points is a bell-shaped curve of normal distribution (Fig. 6–5).

Multiple Alleles

When more than two alleles exist at a given locus, all of the alleles are termed multiple alleles. The classic example of inheritance by multiple alleles is the inheritance of the human blood types O, A, B and AB. Three

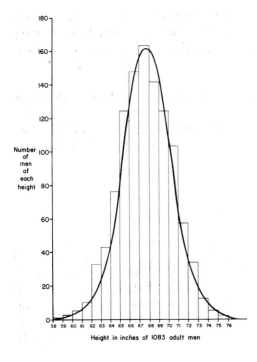

Height in inches of 1083 adult men

FIGURE 6–5. An example of a "normal curve," or curve of normal distribution; the heights of 1083 adult white males. The blocks indicate the actual number of men whose heights were within the unit range; for example, there were 163 men between 67 and 68 inches in height. The smooth curve is a normal curve based on the mean and standard deviation of the data. (From Villee, C. Biology, 6th Ed. W. B. Saunders Co., Philadelphia, 1972.)

alleles are known at this locus: gene A^a, which leads to the formation of agglutinogen A; gene A^b, which causes the formation of agglutinogen B; and gene a, which produces no agglutinogen (blood group O). The agglutinins, anti-A and anti-B, appear in those individuals lacking the corresponding antigens on their red cells. They are immunoglobulins, IgM, IgG and IgA, which appear after birth as a result of a variety of stimuli.

The difference between the A and B antigens resides in the terminal sugar in the oligosaccharide attached to the polypeptide chain. From a common precursor ("H substance") the N-acetyl galactosaminyl transferase coded by gene A^a produces antigen A by adding a terminal N-acetyl galactose, whereas the D-galactosyl transferase coded by the A^b gene produces antigen B by adding a terminal galactose residue using uridine diphospho-galactose as substrate. The a gene apparently codes for no enzyme. Neither A^a nor A^b is dominant over the other, and when both genes are present (A^aA^b), both agglutinogens are formed and the individual has blood group AB (Table 6–1). These blood groups, which are determined genetically, do not change during a person's lifetime, and blood tests can be helpful in settling cases of disputed parentage (Table 6–2). Although a blood test can never prove that a certain man is the father of a certain child, it might be able to prove that he could not be the father of a given child. More than a dozen other sets of blood types are inherited independently of the ABO blood types. These include the MN factors and a series of Rh alleles.

Table 6–1. THE ABO BLOOD GROUPS

Group	Genotype	Agglutinogen in Red Cell	Agglutinin in Plasma	Can Give Blood to	Can Receive Blood from
O	aa	none	anti-A, anti-B	O, A, B, AB	O
A	A^aA^a, A^aa	A	anti-B	A, AB	O, A
B	A^bA^b, A^ba	B	anti-A	B, AB	O, B
AB	A^aA^b	A, B	none	AB	O, A, B, AB

Linkage and Crossing Over

The chromosomes tend to be inherited as units—that is, they pair and segregate in meiosis as units—hence, all of the genes located in any given chromosome tend to be inherited together. During meiosis, while the chromosomes are undergoing synapsis, the homologous chromosomes may exchange segments, a process called *crossing over*. In this way, genes can be transferred from one chromosome to the other member of the homologous pair. The exchange of chromosome segments occurs at random along the length of the chromosome and several exchanges may occur at different points along the same chromosome at a single meiotic division. Since the exchange of segments occurs at random along the length of the chromosome, the greater the distance between the loci of any two genes on the chromosome, the greater the chance that an exchange of units will occur between them (Fig. 6–6). Thus, the distance between genes is measured in terms of cross-over units.

Genetic experiments with many different animals and plants have shown that the frequency of crossing over between different genes can be explained only on the assumption that the genes lie in a linear order on the chromosome. We can observe only the frequency of recombination among the offspring and not the actual frequency of crossing over. Two or more cross-overs may occur in a single chromosome; hence, the frequency of

Table 6–2. EXCLUSION OF PATERNITY BASED ON ABO BLOOD GROUPS

Child's Blood Group	Mother's Blood Group	Father Cannot Be
O	O, A, B	AB
A	O	O, B
A	A	–
A	B	O, B
B	B	–
B	A	O, A
B	O	O, A
AB	A	O, A
AB	B	O, B
AB	AB	O

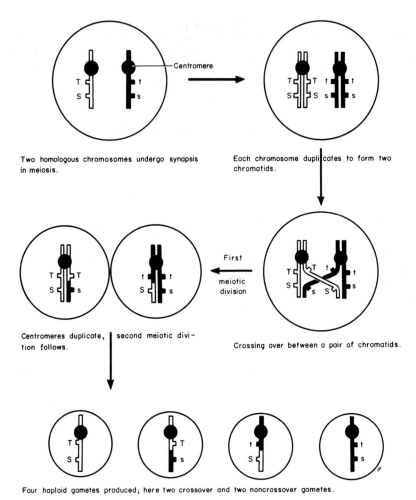

FIGURE 6-6. Diagram illustrating crossing over, the exchange of segments between chromatids of homologous chromosomes. Crossing over permits recombination of genes (e.g., tS and Ts); the farther apart genes are on a chromosome, the greater is the probability that crossing over between them will occur. (From Villee, C. Biology, 6th Ed. W. B. Saunders Co., Philadelphia, 1972.)

crossing over will be somewhat larger than the observed frequency of genic recombination. The simultaneous occurrence of two cross-overs between two particular genes will lead to the reconstitution of the original combination of genes, and this will be recorded as a *non–cross-over*.

By assembling the data on the frequency of crossing over between the various genes in a given chromosome, it is possible to establish chromosome maps. Fairly detailed chromosome maps have been drawn up for some organisms, and a beginning has been made toward mapping the X chromosome and certain other chromosomes in man (Fig. 6–7).

Sex Chromosomes

One exception to the general rule that all pairs of homologous chromosomes are identical in size and shape is provided by the sex chromosomes. The females of most species have two identical sex chromosomes called X chromosomes, whereas males have only one X chromosome and a smaller Y chromosome with which it undergoes synapsis during meiosis. Thus, human males have 22 pairs of ordinary chromosomes or autosomes plus one X and one Y chromosome, and human females have 22 pairs of autosomes plus two X chromosomes. In man and other mammals, maleness appears to be determined in large part by the presence of this Y chromosome. An individual with the XXY constitution is a nearly normal male in external appearance, although he may have underdeveloped gonads (Klinefelter's syndrome). An individual with one X but no Y chromosome has the appearance of an immature female (Turner's syndrome). A normal male, with one X and one Y chromosome, produces two kinds of sperm. Half of his sperm contain an X chromosome and half contain a Y chromosome. All eggs contain a single X chromosome. Fertilization of an X-bearing egg by an X-bearing sperm results in an XX (female) zygote. The fertilization of an X-bearing egg by a Y-bearing sperm results in an XY or male zygote. Since there are approximately equal numbers of X- and Y-bearing sperm, approximately equal numbers of each sex are born. There are some 106 male babies born for every 100 girl babies, and the ratio of males to females at conception is believed to be even higher. One possible explanation for this numerical difference is that the Y chromosome is somewhat smaller than the X. Therefore, a sperm containing a Y chromosome might be lighter, and be able to swim a little faster than an X-bearing sperm, and thus win the race to the egg slightly more than 50 per cent of the time.

Sex Chromatin

In 1949, Barr found that certain cells, but not others, show a chromatin spot at the periphery of the nucleus (Fig. 6–8). Further investigation revealed that cells showing this spot had come from females, whereas cells which did not show this spot came from males. With this characteristic, it is now possible to carry out "nuclear sexing" to determine whether an indi-

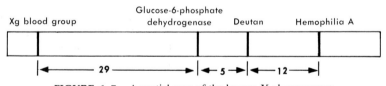

FIGURE 6–7. A partial map of the human X chromosome.

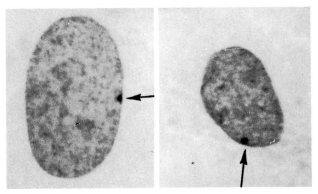

FIGURE 6–8. Sex chromatin in human fibroblasts cultured from skin of a female. The chromatin spot at the periphery of each nucleus is indicated by the arrow. (Feulgen, magnification ×2200. Courtesy Dr. Ursula Mittwoch, Galton Laboratory, University College, London. From Villee, C. Biology, 6th Ed. W. B. Saunders Co., Philadelphia, 1972.)

vidual is genetically female or male. Subsequent studies showed that this spot represents one of the two X chromosomes in the female, which becomes dense and dark-staining. The other X chromosome, like the autosomes, remains as a fully extended thread and is not evident by light microscopy during the interphase.

Mary Lyon has suggested that only one of the two X chromosomes of the female is active, and that the other is inactive. Which of the two becomes inactive in any given cell during early embryonic development is a matter of chance. The cells of a woman's body are of two kinds, in which one or the other X chromosome has become inactive. Since the two X chromosomes may have different genetic complements, the cells in a woman's body may differ in the effective genes present, so that she is a genetic mosaic. In cats, where there are several sex-linked genes for certain coat colors, a female heterozygous for such genes may show patches of one coat color in the midst of an area of the other color ("tortoise-shell"). The inactivation of one X chromosome apparently occurs early in embryonic development, and thereafter all of the progeny of that cell have the same inactive X chromosome.

Sex-linked Genes

Since the human X chromosome contains many genes, whereas the Y chromosome contains relatively few — principally the genes for maleness — the traits controlled by genes located in the X chromosome are termed sex-linked, because their inheritance is linked with the inheritance of sex. A male offspring receives a single X chromosome, and hence all of his genes for sex-linked characters, from his mother. A female receives one X from her mother and one X from her father. Males, having but one X chromosome, have only one of each kind of gene located in the X chromosome. Hemophilia and color-blindness are two of the more famous sex-linked traits in men. Hemophilia is a disease in which there is a deficiency in the formation of thromboplastin, due to a deficiency of the anti-hemophilic globulin. If a sex-linked gene is recessive and relatively rare — that is, present in the population in low frequency — the trait will appear much more frequently in males than in females, for only a single gene dose will produce the trait in males, whereas two doses are necessary in the female. Color-blindness, for example, affects about 4 per cent of all human males but less than 1 per cent of human females. Hemophilia is a rare trait in human males and was unknown in human females until 1951. How could you account for colorblindness in a woman with two normal visioned parents?

MOLECULAR BIOLOGY OF THE GENE

Since the rediscovery of Mendel's Laws in 1900, geneticists have been dissatisfied with the simple statement that a particular gene produces a given trait, such as blue or brown eyes. Attempts in the past three decades to discover the physical and chemical mechanisms involved in the transmission and expression of inherited traits have greatly increased our understanding of the chemical and physical nature of the genetic material and the mechanism whereby genes control the development and maintenance of the organism.

Our present working hypothesis states that genes are composed of deoxyribonucleic acid (DNA) and that they are located within a chromosome. Each gene contains information, coded in the form of a specific sequence of purine and pyrimidine nucleotides in a polynucleotide chain. The unit of genetic information is a group of three adjacent nucleotides, termed a *codon,* which ultimately specifies a single amino acid. Thus the genetic code is said to be a triplet code. To specify a peptide composed of 200 amino acids in a specific sequence, the DNA would have to have 600 nucleotides arranged in a specific sequence. The DNA molecule consists of two complementary chains of polynucleotides twisted about each other in a regular helix and joined by specific hydrogen bonds between specific pairs of purine and pyrimidine bases. Thus, adenine pairs with thymine and guanine pairs with cytosine. This complementary feature of the DNA molecule permits the specific replication of the DNA when the two strands separate, and each acts as a template for the formation of a new complementary strand.

The DNA of each gene has a sequence of nucleotide triplets that differs in some respects from that of every other gene. The information is transcribed from the DNA of the gene to a kind of RNA termed *messenger RNA.* This is synthesized in the nucleus and passes through pores in the nuclear membrane to the ribosomes in the endoplasmic reticulum. There, the messenger RNA combines with the ribosomes and serves as a template for the synthesis of an enzyme or other specific protein.

The first step in the synthesis of a peptide is the activation of an amino acid and its combination with a specific *transfer RNA*. Each amino acid is activated by a separate specific activating enzyme which catalyzes the reaction of the amino acid with ATP to form the amino acid–adenylic acid compound. Then the same enzyme catalyzes the transfer of the amino acid to the specific transfer RNA for that amino acid. The amino acid is attached to the ribose of the terminal adenylic acid at the end of the transfer RNA which contains cytidylic, cytidylic and adenylic acids (CCA). The transfer RNA acts as an adaptor to bring a specific amino acid into line in the growing polypeptide chain in the appropriate place. It is smaller than messenger or ribosomal RNA, and is able to function as an adaptor because it contains an anti-codon, a group of three nucleotides complementary to the triplet codon in messenger RNA. Thus there is at least one specific transfer RNA for each amino acid.

Transfer RNA's are polynucleotide chains of some 70 nucleotides. The chain is doubled back on itself so that some 25 bases on one limb are hydrogen-bonded and paired with an equal number on the other limb to form a double helix. These base-paired limbs are joined by a loop in the center of the chain which contains the triplet anti-codon (Fig. 6–9).

The amino acid–transfer RNA complex moves to the ribosome, which serves to provide the proper orientation of the amino acid–transfer RNA with the messenger RNA, so that the genetic information supplied by the messenger RNA can be read accurately. The messenger RNA appears to

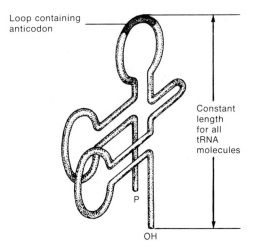

FIGURE 6–9. A diagram of the three-dimensional clover-leaf structure of transfer RNA. One loop contains the triplet anticodon which forms specific base pairs with the mRNA codon. The amino acid is attached to the terminal ribose at the 3′ OH end, which has the sequence CCA of nucleotides. Each transfer RNA also has guanylic acid, G, at the 5′ end (P). The pattern of folding permits a constant distance between anticodon and amino acid in all transfer RNA's examined. (From Villee, C., and Dethier, V. Biological Principles and Processes, 2nd Ed. W. B. Saunders Co., Philadelphia, 1976.)

move across the site on a ribosome at which protein synthesis occurs, thus bringing successive codons into position to combine with the appropriate amino acid–transfer RNA anti-codon (Fig. 6–10). The transfer of the amino acids to the growing peptide chain on a ribosome requires guanosine triphosphate, certain specific enzymes and glutathione. The peptides are synthesized by the sequential addition of amino acids, beginning at the N-terminal end of the peptide chain. Each ribosome appears to ride along the extended messenger RNA molecule, reading the message as it goes. As peptide synthesis proceeds, the growing polypeptide chain is not transferred from one ribosome to another, but remains attached to its original ribosome. After completing the reading of one molecule of messenger RNA and releasing the polypeptide that has been synthesized, the ribosome appears to drop off the end of one messenger RNA chain and find a new one (which may code for a different peptide chain) to read. All of the processes of gene replication, gene transcription and protein synthesis depend upon the formation of specific, though relatively weak, hydrogen bonds between specific base pairs (A to T, G to C). The specificity of these bonds insures the remarkable accuracy of the processes. Mistakes in base pairing occur less than once in 1000 times.

REGULATION OF PROTEIN SYNTHESIS

The system just outlined provides an explanation for the molecular mechanisms involved in the synthesis of a specific protein. Since the amount of any specific protein synthesized per cell may undergo wide variations, there must be some additional mechanism that controls how much of this particular enzyme is synthesized in a given cell at a given moment. The rate of synthesis of a given protein may be controlled in part by the genetic apparatus itself and in part by factors entering from the external environment. Most of the data relating to the control of protein synthesis is derived from experiments with microbial systems, especially *Escherichia coli*. From the length of the chromosome of *E. coli* and the estimate that an average gene contains about 1500 nucleotide pairs (codes for a polypeptide chain of 500 amino acids), you can calculate that the genes in *E. coli* will code for between 2000 and 4000 different polypeptides.

The number of different enzymes required by *E. coli* growing on glucose has been estimated to be about 800. When *E. coli* cells that have been growing on glucose are transferred to a medium that contains lactose as the sole carbon source, the cells respond by synthesizing the enzyme β-galactosidase, which is normally present in very small amounts. In the presence of lactose, the cells synthesize β-galactosidase until it comprises as much as 3 per cent of the total protein of the cell. This represents at least a thousand-fold increase over the amount present in cells growing on glucose. A substance – such as lactose – that elicits an increased amount of an enzyme is called an inducer, and enzymes that respond to inducers are termed inducible enzymes.

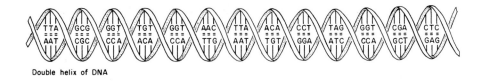

Double helix of DNA

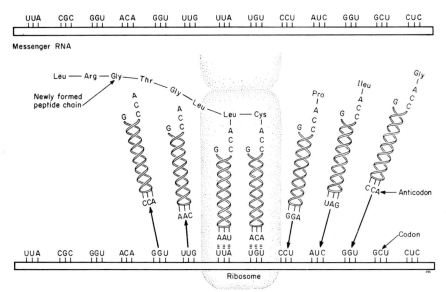

FIGURE 6–10. Diagram of the postulated mechanism of protein synthesis on the ribosome, illustrating the relationship between the triplet code of the DNA helix, the complementary triplet code of messenger RNA and the complementary triplet code (anticodon) of transfer RNA. Molecules of tRNA charged with specific amino acids are depicted coming from the right, assuming their proper place on mRNA at the ribosome, transferring the amino acid to the growing peptide chain, and then (left) leaving the ribosome to be recharged with amino acids for further reactions. The growing polypeptide chain remains attached to its original ribosome. (From Villee, C. Biology, 6th Ed. W. B. Saunders Co., Philadelphia, 1972.)

Conversely, cells of *E. coli* grown on a medium without amino acids contain the whole spectrum of enzymes required for the synthesis of all 20 or so amino acids needed for the assembly of protein molecules. When an amino acid is introduced into the incubation medium, there is a great decrease in the amount of the biosynthetic enzymes required for the production of that amino acid. Enzymes that are reduced in amount by the presence of the end product of a biosynthetic sequence are called repressible enzymes, and the small molecule that brings about the repression (the amino acid in this example) is called a co-repressor. Both the induction and the repression of enzymes are adaptive phenomena that should be of survival value to the bacterium.

Differences in the amount of a specific protein in the bacterium are generally the result of variations in the rate of synthesis of that enzyme, which are in turn due to the amount of messenger RNA for that enzyme present per cell. It has been postulated that the amount of effective messenger RNA template for a given enzyme is controlled by a special kind of protein called a repressor, which blocks the synthesis of messenger RNA. It has been further postulated that these repressors are coded for by special genes, termed regulatory genes. It has been postulated that repressors block the synthesis of specific proteins by combining with a specific site on the DNA and blocking its transcription to form messenger RNA. In 1968, the repressor for β-galactosidase was isolated and identified as a protein with a molecular weight of about 130,000.

If regulatory genes made repressors all the time, the synthesis of messenger RNA would always be inhibited. It has been postulated that repressors may exist in either active or inactive forms, depending on whether the repressor is combined with a specific small molecule, the inducer or the co-

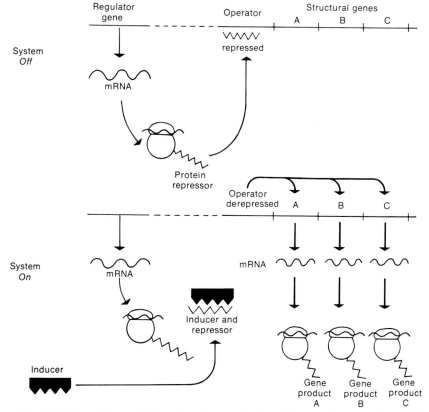

FIGURE 6–11. The regulation of genic transcription. Diagram of a means by which a regulator gene may produce a messenger RNA which codes for a protein repressor. *Above*, the repressor in turn inhibits an operator and thus prevents the transcription of the structural genes. *Below*, the messenger RNA produced by the regulator gene again produces a protein repressor, but this is combined with an inducer. The operator gene is thereby derepressed, permitting the transcription of structural genes A, B and C and the formation of the protein gene products A, B and C. (From Villee, C., and Dethier, V. Biological Principles and Processes, 2nd Ed. W. B. Saunders Co., Philadelphia, 1976.)

repressor. When an inducer is attached to a repressor, the repressor is inactivated (Fig. 6–11). The combination of lactose with the repressor blocks the repressor and permits the synthesis of the enzyme. The binding of repressors with their specific inducers or co-repressors is believed to be accomplished by weak bonds, such as hydrogen bonds.

In a number of systems, two or more enzymes have been shown to vary in amount in a coordinate fashion, suggesting that they are under the control of the same repressor system. It has been inferred that a single repressor may control the formation of the messenger RNA for these coordinately repressed enzymes. The genes with codes that are transcribed on a single messenger RNA molecule and that are under the control of a single repressor have been termed an operon. Originally it was believed that all the genes controlled by the same repressor must be in the same operon and lie closely adjacent in a chromosome, but more recently, examples have been found of widely separated genes that appear to be coordinately repressed by the same repressor. A further entity, the operator gene, has been postulated to account for the control of the operon. Operators are postulated to be adjacent to the genes in the operon and are the sites on the DNA to which active repressor molecules are bound, thereby inhibiting the synthesis of messenger RNA by the genes in the adjacent operon. In the absence of an active repressor, the genes in the operon are free to be transcribed, messenger RNA is formed and enzymes are produced on the ribosome. It is possible that the operator site in some way provides a code that directs the RNA polymerase to begin transcribing the DNA at that specific point at the end of the operon.

Relatively little is known about genetic regulation in human or other mammalian systems, but many investigators have suggested that some similar mechanism coordinates protein biosynthesis in the cells of higher organisms. A number of the enzymes in the liver are essentially absent before

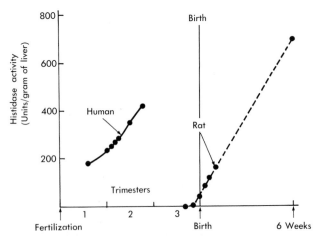

FIGURE 6–12. The development of histidase activity in the livers of rat and man. (From data of Cornell, N. W. and Villee, C. A. Comp. Biochem. Physiol., 27:603, 1968.)

birth and then appear suddenly and increase rapidly to the normal adult level of activity within a few days or a week after birth (Figure 6–12). The factors controlling this particular aspect of protein synthesis are unknown.

CONTROL OF ENZYME SYNTHESIS AND FUNCTION

The process of development in multicellular organisms must involve some very precise and efficient mechanisms to control the successive activation and perhaps inactivation of specific genes in specific tissues. The mechanism just outlined would account for the presence or absence of an enzyme, depending on whether the specific gene was being transcribed or not. The gene may, in addition, be able to control the amount of enzyme made per cell and also the time in the course of development at which the enzyme appears. In the human inherited disease, galactosemia, individuals with two normal genes—homozygous normal individuals—have approximately two times more of the specific enzyme, galactose-1-phosphate uridyl transferase, than do individuals with a single normal gene—the heterozygous carriers. Individuals with galactosemia, homozygous for the mutant allele, have very little, if any, of the enzyme. This suggests that there may be quantitative as well as qualitative relationships between a gene and its enzyme.

Control of Messenger RNA

The turning on and off of the synthesis of a specific protein (which might be termed "differentiation at the molecular level") could occur by a process operating at the genic DNA or at the messenger RNA, in the processes occurring on the ribosome during protein synthesis, or in some conformational change involved in the conversion of the linear peptide chain into the ultimate three-dimensional protein. Since mitosis ensures the regular and precise distribution of genes to daughter cells, cellular differentiation must involve some mechanism by which the genes in different cells and tissues undergo differential activation. A control system that would operate at the transcription process would perhaps be the most economical one biologically, for it would clearly be advantageous to the cell not to have its ribosomes encumbered with non-functional molecules of messenger RNA. It would be more economical for the cell to produce only those kinds of m-RNA required at the moment for the synthesis of a specific protein. Since the messenger RNA for each kind of protein has a finite half-life, ranging from a few minutes in microorganisms to 12 or 16 hours or more in mammals, it is clear that the continued synthesis of any protein requires the continued synthesis of its corresponding m-RNA. Each molecule of m-RNA can serve to direct the synthesis on the ribosome of many molecules of protein, but ultimately it is degraded and must be replaced. Without this, the cell would have no way of changing the kind of protein being synthesized in response to stimuli from inside or outside the cell.

Lethal Genes

Certain genes produce such a tremendous deviation from the normal development of an organism that it is unable to survive. The presence of such lethal genes can be detected only by certain upsets in the expected genetic ratios. For example, when two mice with yellow coats are bred, the offspring appear in the ratio of two yellow to one non-yellow instead of the expected 3:1 ratio. Analysis of the phenomenon showed that yellow is a dominant gene that results in death of the fetus at an early age when the fetus is homozygous for the yellow gene. Several genes are suspected of being lethal in human embryos. For example, the marriage of two individuals with brachyphalangy produced a child with no fingers or toes who died shortly after birth. Lethal genes appear to be mutants that cause a deficiency of some enzyme of such primary importance in intermediary metabolism that development of the organism is prevented.

Penetrance and Expressivity

It has been found that many genes do not produce the expected phenotypes in every instance, and the term *penetrance* has been established to describe this effect. Penetrance refers to the statistical regularity with which a gene produces its effect when present in the requisite homozygous or heterozygous state. The percentage of penetrance of a given gene may be changed by altering the physical conditions under which the organism develops.

A population of individuals with identical genotypes may show considerable variation in the phenotypic appearance of a given trait. For example, fruit flies that are homozygous for a recessive gene that produces shortening and scalloping of the wings may exhibit wide variation in the degree of shortening and scalloping. Such differences are termed variations in the *expressivity* of the gene. Such human traits as shortening of the digits or the presence of a sixth digit show considerable variations in their expressivity. There may be variations in the expression of a trait not only among different members of a family, but also between the right and left sides of a single individual. In view of the long and perhaps tenuous connection between a gene and the final production of its trait, it is easy to understand why the expression of a trait might vary from one individual to the next (expressivity) or why the mutant trait might fail to appear in a certain fraction of the individuals with the mutant genotype (penetrance).

DIFFERENTIATION OF TISSUES

The properties and constituents of each kind of cell or tissue are ultimately determined by the complement of enzymes present. Both qualitative and quantitative differences in the enzymic contents of cells in man and other organisms have been demonstrated. Mammalian liver cells have glucose-6-phosphatase, but mammalian muscle cells do not. Thus, mammalian muscle cells cannot convert glycogen or other precursors to free glucose, whereas the liver can. The enzymes that carry out a given reaction in different tissues may differ in their molecular size, in their amino acid composition and in their immunologic properties, and they may show different responses to hormones and other control mechanisms. The phosphorylases of liver and skeletal muscle differ in molecular size, immunologic properties and in their response to the hormone, glucagon.

The course of development may reveal (1) the sequential appearance of the same protein in different tissues, (2) the sequential appearance of specific proteins in a single tissue or (3) an increase in the amount of a particular protein in a single tissue. For example, the serum of the fetal rat at successive stages of development contains not only a greater total concentration of protein, but an increased number of immunologically distinct proteins. There are five serum proteins at 17 days of gestation, but 15 or more at 22 days of gestation.

A given enzyme may appear at different times in development in different tissues. Glucose-6-phosphatase appears at 10 weeks in human fetal liver, at 11 weeks in the lung and at 14 weeks in the kidney (Fig. 6–13). The activity of an enzyme, taken as a measure of the amount of the enzyme, may increase steadily in the course of development. For example, the rate of glycolysis in human fetal cerebral cortex increases approximately linearly with time during weeks 8 to 25 of gestation.

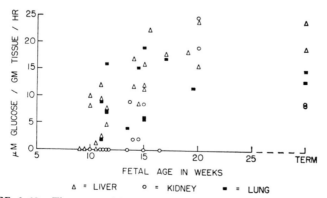

FIGURE 6–13. The sequential appearance of glucose-6-phosphatase in human fetal tissues, measured by the ability of the tissue to produce free glucose.

DEVELOPMENT OF ENZYMES IN THE FETUS AND PLACENTA

Certain biochemical characteristics of the cell appear to be established very early in the course of development, and some, of course, are present from the time of the fertilized egg. The several enzymes concerned with the metabolism of carbohydrates and fatty acids are so fundamental to the production of biologically useful energy needed for survival and growth that it would be difficult to imagine a cell lacking them. Characteristic changes have been observed in the activities of other enzymes at certain times in development. Several enzymes catalyzing the metabolism of amino acids undergo marked increases in activity and amount either just before or after the time of birth. Tyrosine transaminase and tryptophan 2,3-dioxygenase undergo a rapid increase from essentially zero activity to nearly the adult level within a few days after birth in the rabbit. This implies that the specific DNA's coding for these enzymes become activated—perhaps freed from histones or other protein repressors—and release a burst of messenger RNA's that code for the enzymes involved. In the course of development of the fetal adrenal, the enzymes concerned with 17-hydroxylation, 21-hydroxylation and 11-hydroxylation of the steroid nucleus appear in sequence from about the eighth to the fifteenth week.

It would be wrong to conclude that cellular differentiation depends solely on nuclear functions, for the experiments of Grobstein indicate that, at least in certain systems, extrinsic factors, together with complex cell interactions, may regulate the process of differentiation. Grobstein found that embryonic pancreatic epithelium will continue to undergo differentiation in organ culture only in the presence of mesenchyme cells. This requirement can be met not only by mouse pancreatic mesenchyme, but also by mesenchyme from a variety of other sources, even chick embryo mesenchyme or a chick embryo juice. The active principle of the juice is inactivated by treatment with trypsin, but not by ribonuclease or deoxyribonuclease, and can be sedimented by centrifugation at 100,000 times gravity. It thus appears to be a large molecule, and probably a protein rather than RNA or DNA. The factor is only weakly effective in causing differentiation of salivary gland epithelium in culture and is ineffective in inducing the formation of kidney tubules or cartilage from their respective precursors. The experiments of Grobstein indicate the existence of a spectrum of protein factors, each of which is more or less specific for the differentiation of one kind of cell.

CRITICAL PERIODS IN EMBRYOGENESIS

Classic experiments with insects nearly 50 years ago had established the principle that there are sensitive periods in development during which any of a variety of external insults would lead to similar specific changes in development. An external factor like a heat shock, a cold shock, ultra-violet or x-rays, and so forth, would bring about a specific scalloping of the insect's wing if it were applied at a certain time in development, but would be ineffective if applied earlier or later. This same sort of principle was subsequently shown to hold for rats and mice. Pregnant mice subjected to low oxygen tension or to cortisone treatment at a specific time in development — roughly at day 12 — have a high proportion of offspring born with cleft palate and harelip. It is generally found that this critical period coincides with, or just precedes, the time in development when morphologic differentiation of the primordium of the structure takes place. The same principle was dramatically shown to apply in human development with the demonstration of congenital abnormalities produced by rubella and similar viral infections early in development or by the ingestion of a drug such as thalidomide at a specific time in development.

Although the advances in molecular biology provide a reasonable explanation of the means by which a given protein or enzyme may be produced in response to genetic information coded in the DNA in the chromosome, it does not explain as yet how different cells of a multicellular organism end up having quite different spectra of protein constituents. We are also quite far from understanding the biochemical basis for structural differences in cells and for morphogenesis — for the arrangement of cells in specific patterns to yield a finger, an eye, or any other part of the body.

ABNORMALITIES OF CHROMOSOMES

The normal human karyotype for males and females is shown in Figure 6–14. Cells from the bone marrow, blood or skin are incubated until they begin to undergo mitosis and then are treated with colchicine to stop the process in the metaphase. The cells are placed in hypotonic solution, causing them to swell and enabling the chromosomes to spread out so that they can be visualized more readily. The preparation is fixed and stained and the chromosomes are photographed. Each chromosome is cut out of the photographic print and aligned so that the homologous pairs are placed together. Chromosomes are identified by their length, by the position of the centromere, and by the presence of knobs or satellites. The gradation in size between several of the pairs is quite fine, and although the largest chromosome is some five times as long as the smallest chromosome, there are only slight differences between some of the intermediate-sized ones. Sometimes chromosome pairs are simply allotted to one of seven different groups.

Any mechanism as complex as that underlying mitosis or meiosis may occasionally malfunction, resulting in some change in structure or in number of the chromosomes. A piece of a chromosome may be broken off and lost (a *deletion*). Some malfunctioning of the process of crossing over may lead to a section of the chromosome being taken out, turned end-to-end and reinserted

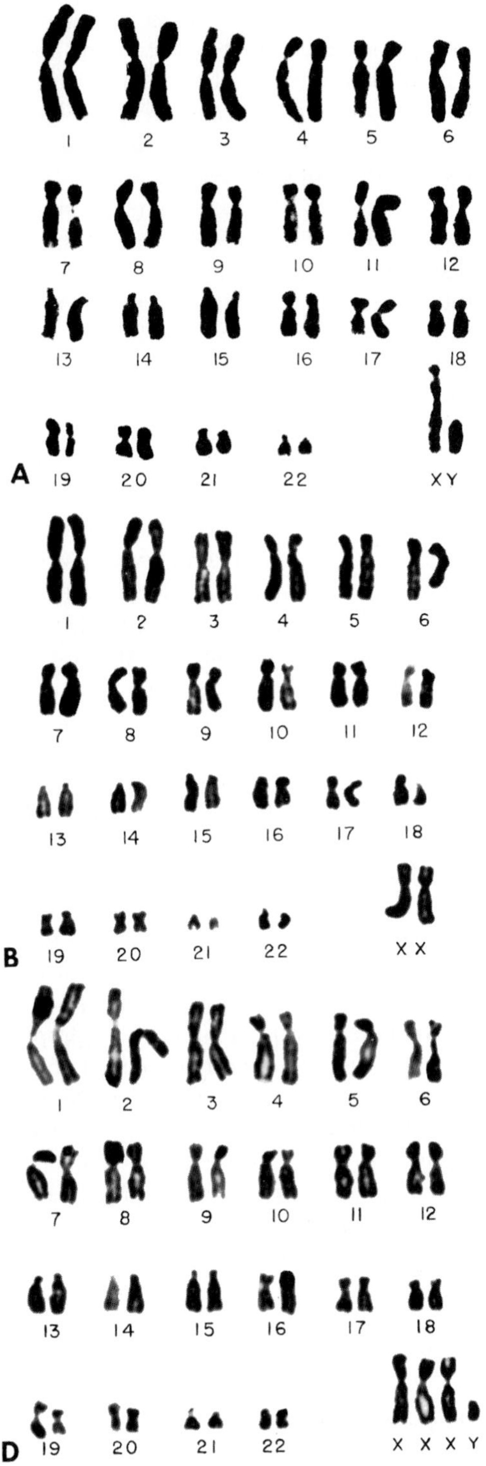

FIGURE 6–14. Human chromosomes. *A*, Normal male. *B*, Normal female. *C*, XO condition—gonadal dysgenesis. *D*, XXXY—an unusual example of Klinefelter's syndrome; the typical individual with Klinefelter's syndrome has an XXY pattern of chromosomes. (Photographs courtesy of Dr. Melvin Grumbach. From Villee, C. Biology, 6th Ed. W. B. Saunders Co., Philadelphia, 1972.)

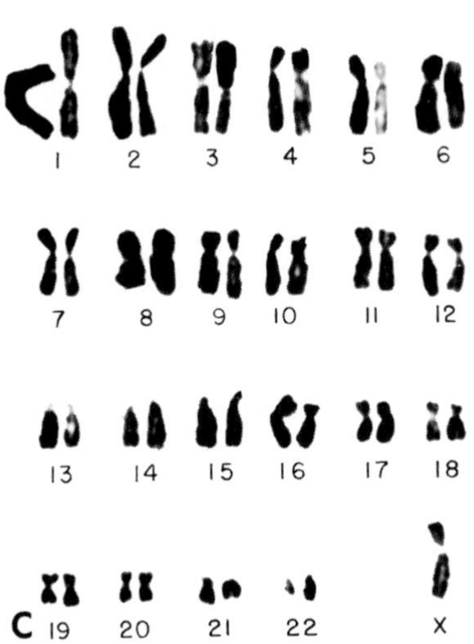

in the wrong order (*inversion*). Thirdly, a piece of one chromosome may be broken off and inserted on a different, nonhomologous chromosome (*translocation*). Frequently, pieces are broken off of two different chromosomes and reinserted in a reciprocal fashion, resulting in two translocations.

A pair of homologous chromosomes may fail to undergo normal synapsis or to separate normally during meiosis. As a result, both members pass to one gamete, and the other gamete is left with no representative of that pair of homologous chromosomes. This phenomenon, termed *non-disjunction*, may occur in either sex chromosomes or autosomes. Other errors in the meiotic process may result in the production of individuals having an entire extra set of chromosomes. For example, a severely defective male child was found to be a triploid individual with a total of 69 chromosomes. He had 66 autosomes, two X chromosomes and one Y chromosome. This zygote was probably formed by the fertilization of a normal haploid egg by an unusual diploid sperm or by the fertilization of an unusual diploid egg by a normal haploid sperm.

When two X chromosomes fail to separate during oogenesis, both might enter the egg nucleus, leaving the polar body with no X chromosome. Alternatively, the two joined X chromosomes might go into the polar body, leaving the female pronucleus with no X chromosome. Non-disjunction of the XY chromosomes in the male could lead to the formation of sperm with both an X and a Y chromosome, or sperm with neither an X nor a Y chromosome. Chromosomal non-disjunction may occur during either the first or the second meiotic division. It may also occur during mitotic division, leading to the establishment of a clone of abnormal cells in an otherwise normal individual.

Non-disjunction of the XY or XX chromosomes will lead to individuals with either Klinefelter's syndrome or Turner's syndrome. Individuals with Klinefelter's syndrome may be nearly normal males in outward appearance, but have small testes and produce few or no sperm. Their seminiferous tubules are usually aberrant in appearance, and the individuals may have gynecomastia. Klinefelter's syndrome usually becomes apparent only after puberty, when the small testes and gynecomastia bring the individual to the attention of his physician. Patients with Klinefelter's syndrome have cells with a chromatin spot, and at one time were thought to be XX individuals, genetic females. However, examination of their karyotype revealed the presence of 47 chromosomes with two X and one Y.chromosome in addition to the 22 pairs of autosomes. The fact that they are nearly normal males in external appearance emphasizes the strong male-determining effect of the Y chromosome.

The external genitalia of individuals with Turner's syndrome are feminine, but are those of an immature female. Uterus, vagina and fallopian tubes are present; the gonads may be absent. The cells of individuals with Turner's syndrome have no chromatin spot in the nucleus, which would suggest that they are genetically male. However, their karyotype reveals only 45 chromosomes. They have 22 pairs of autosomes plus one X chromosome but no Y chromosome. This type of individual again emphasizes the importance of the Y chromosome in determining male characters.

Karyotype screening of large populations has revealed the existence of a variety of other abnormalities of the sex chromosomes, such as "super-females"—women with 22 pairs of autosomes and three X chromosomes. The remarkable feature of these women is that their outward appearance and their reproductive tract are very nearly those of a normal female, and they may be fertile.

The *chimera* was a monster described in Greek mythology as a fire-breathing animal with a lion's head, a goat's body and a serpent's tail. The term has been taken over by biologists to refer to an individual containing populations of cells of two or more genotypes derived from different zygotes. This mixture of cell lines could arise by transplantation, by vascular anastomosis in the chorion or by double fertilization of egg and polar body and the inclusion of both meiotic products into a single developing embryo.

In contrast, a *mosaic* is an individual with cell populations of more than one genotype, all of which are derived from a single zygotic genotype by mutation or events such as somatic mutation, somatic crossing over, the loss of chromosomes during mitosis or mitotic non-disjunction of chromosomes. Thus, chimerism is a condition in which genomes derived from two separate sources coexist in the same individual. In contrast, mosaicism is a condition in which all of the cells of the individual are derived from a single fertilized egg but include two or more lines of cells with different genotypes which result from genetic changes during somatic cell division.

In the past decade, methods have been devised by which mammalian blastocysts may be fused and then reimplanted in a suitably prepared uterus to develop into a chimeric adult.

On occasion, maternal cells in the human may pass to the developing fetus, but only very rarely do such cells become permanently established in the offspring. Melanomas, lymphomas, myelomas and leukemic cells from the mother are very rarely transmitted to the fetus, and it appears that the fetus has the immunologic capacity to reject these foreign tissues. Very rarely, maternal lymphocytes may cross the placenta and become established in the fetus. These maternal cells usually disappear within the first month or two of life. Fetuses that receive intrauterine transfusions for the treatment of erythroblastosis may have chimeric lymphocytes that can be demonstrated postnatally for 16 months or more.

The existence of blood chimeras in cattle twins was recognized by Owen in 1945. This occurs when the placentas of the two twins lie close together in the uterus and the two blood streams undergo fusion. Blood cells from each twin pass into the other and both are chimeras. A similar phenomenon occasionally occurs in human twins. This is also presumed to occur by way of placental anastomoses between the two fetal blood streams but this has not actually been observed. By appropriate techniques, blood chimeras can be detected and the fraction of blood cells of one type or the other can be determined. An individual may have blood cells half of which are type O and half of which are type A, or the ratio may be more than 99 per cent of one kind and less than 1 per cent of the other. In some case reports, it has been found that the chimeric population of red cells gradually disappears, presumably by some immunologic mechanism.

Cytologic examination has revealed individuals that are mixtures of diploid and triploid or diploid and tetraploid cells. In testing the cells from the various body tissues, even with the use of genetic markers such as enzymes, pigments and hemoglobins, it may be impossible to differentiate a chimera from a mosaic. Whole body chimerism may occasionally exist in man, but it is difficult to demonstrate this unequivocally. Cytologic examinations have demonstrated that chimerism may occur spontaneously in many different kinds of mammals. It may involve only small portions of the body, only a few types of cells, or almost all the tissues, as in whole body chimeras. The existence of chimeras and mosaics has provided much useful information regarding sex differentiation and the immunologic aspects of development.

Individuals suffering from mongolism or Down's syndrome have abnormalities of the face, eyelids, tongue and other parts of the body and are greatly retarded in both physical and mental development. The term "mongolism" was originally applied because affected individuals often show a fold of the eyelid similar to that present in members of the Mongolian race. Mongolism is a relatively common congenital malformation (0.15 per cent of all births). It is one hundred-fold more likely in the offspring of women 45 years or older than in the offspring of mothers under 19. The occurrence of mongolism is independent both of the age of the father and of the number of preceding pregnancies in the mother. The karyotype of mongoloids reveals a small extra chromosome number 21, making a total of 47. The presence of this extra chromosome probably arises by non-disjunction in the maternal oocyte.

It is not clear why the DNA transcribing system does not simply ignore the redundant bit of genetic information and produce cells identical to those of the normal individual, but the presence of this extra chromosome leads to the complex of physical and mental differences that are characteristic of individuals with mongolism. Whether the extra genes in the third chromosome 21 lead to the production of an extra amount of certain enzymes and whether this is the basis for the abnormal physical and mental development is not known. In any organism in which a chromosome or a part of a chromosome has been added or deleted (genetic imbalance), comparable developmental effects are observed.

It had been known for decades that in the fruit fly the tendency for non-disjunction to occur increases greatly with maternal age, and this appears to be true in human females. Mongolism and similar traits due to trisomy should be inherited as though they were determined by a dominant gene. A mongoloid would form gametes, half of which have the normal complement of 23 chromosomes and half of which have 24 chromosomes with two chromosome 21's. In the rare instances in which mongoloids have had offspring, they have produced normal and mongoloid children in about equal proportions.

An individual with an extra chromosome, giving him three of one kind of chromosome, is said to be trisomic, and an individual lacking one of a pair is said to be monosomic. Thus, mongoloids are trisomic for chromosome 21, and individuals with Turner's syndrome are monosomic for the X chromosome. When non-disjunction occurs during meiosis, all of the cells of the zygote resulting from the fertilization of the abnormal gamete will have the abnormal chromosome complement. However, non-disjunction may also occur during mitosis after the zygote has been formed, and such an individual would then have two or more cell lines, with different numbers of chromosomes. This is termed *chromosomal mosaicism*. Such individuals containing some cells with 45, 46 or 47 chromosomes might be normal or very nearly normal in phenotype, and would be detected only if they underwent extensive karyotyping.

Individuals who are trisomic for chromosome 18 have a condition somewhat similar to mongolism. They suffer mental retardation and have a peculiarly shaped skull, a small mandible and prominent occiput, malformed ears, a receding chin, and other skeletal and cardiovascular defects. Individuals who are trisomic for one of the medium-sized chromosomes in group 13 to 15 also show mental retardation and multiple abnormalities. Some of the abnormalities are like those found in people who are trisomic for chromosome 18 or 21, but the presence of eye defects (such as anophthalmia) and of neurologic abnormalities (such as fits and hypotonia) is unique.

The oral-facial-digital syndrome appears to result when a portion of chromosome 6 is translocated to chromosome 1, so that an individual has a partial trisomy for chromosome 6. Individuals with the oral-facial-digital syndrome have defects of the upper lip, palate and mouth, a forked tongue, and stubby toes with short nails. They may or may not show mental retardation. These and other chromosomal abnormalities are summarized in Table 6–3.

ABNORMALITIES OF GENES

The 1:1:1 relationship among gene, enzyme and biochemical reaction was first described in man by an English physician, A. E. Garrod. His investigation of alkaptonuria revealed that it is inherited by recessive genes which lead to an interference in the metabolism of tyrosine. The urine of affected patients turns dark on standing because of the presence of homogentisic acid, which undergoes oxidation and is polymerized to a melanin-like compound. His term, "inborn error of metabolism," has been adopted to describe those genetically determined biochemical disorders in which a deficiency of some specific enzyme leads to a specific metabolic block with specific pathologic consequences. Most of the inborn errors of metabolism described thus far are inherited as autosomal recessives, although a few are

Table 6–3.   SOME HUMAN CHROMOSOMAL ABNORMALITIES

Name	Genetic Features	Clinical Aspects
Turner's syndrome (gonadal dysgenesis)	XO	Short stature, streak ovary, juvenile female genitalia, poorly developed breasts
Klinefelter's syndrome	XXY	Gynecomastia, small testes
Triple X females	XXX	Two "Barr bodies" present, fairly normal females but secondary sex characteristics may be poorly developed
Down's syndrome	Trisomy 21	Epicanthal folds, protruding tongue, hypotonia, mental retardation
Trisomy 18	Trisomy 18	Mental retardation, multiple congenital malformations
D trisomy	Trisomy 15	Mental retardation, severe multiple anomalies, cleft palate, polydactyly, central nervous system defects, eye defects
Translocation mongolism	15/21, 21/22, or 21/21 translocation	Mongolism, clinically similar to trisomy 21
Philadelphia chromosome	Deletion of one arm of chromosome 21	Chronic granulocytic leukemia
Oral facial digital syndrome	Translocation of part of chromosome 6 to 1	Defects of upper lip, palate, and mouth, stubby toes with short nails
Cri du chat syndrome	Deletion of short arm of chromosome 15	Mental retardation, facial anomalies

sex-linked. Because of this, it follows that the parents of the affected individual and, on the average, two-thirds of his apparently unaffected brothers and sisters will be heterozygous for that gene. In some, but not all, inborn errors of metabolism, it is possible to detect the heterozygous carriers by some characteristic deviation from normal in the biochemical step affected. Some of the more common inborn errors of metabolism are summarized in Table 6–4.

Whether the inherited biochemical variations which involve the structure of proteins other than enzymes should be termed inborn errors is not universally agreed upon. It seems clear, however, that the same sort of relationship between the mutant gene and the abnormal protein product exists in both alkaptonuria and the abnormal hemoglobin molecules found in sickle cell anemia. In the latter, the specific biochemical abnormality was pinpointed in 1956 when Ingram reported that only one of the 287 amino acids present in the hemoglobin molecule was changed. Normal adult hemoglobin is composed of two identical α chains and two identical β chains, which are folded and fitted together to form a globular molecule with a molecular weight of 68,000. The α chain is composed of 141 amino acids and the β chain of 146 amino acids. The human genotype also contains genes for the synthesis of γ chains and δ chains. Fetal hemoglobin, which is present in the blood of the developing fetus but disappears shortly after birth, is composed of two α and two γ chains. A small fraction of the total hemo-

Table 6–4.   SOME INHERITED ENZYMATIC DEFICIENCIES IN MAN

Disease	Enzyme Involved	Effects of Deficiency
von Gierke's disease	Liver glucose-6-phosphatase	Accumulation of glycogen in liver and kidney. Glycogen has normal molecular structure
Pompe's disease	Amylo-α-1,4-glucosidase	General accumulation of glycogen in tissues. Glycogen has normal structure
Forbe's disease	Amylo-1,6-glucosidase	Glycogen accumulates in liver and heart; glycogen molecules abnormal, with short outer branches
Andersen's disease	Amylo-(1,4⟶1,6)-trans-glucosylase	Glycogen accumulates in liver; glycogen is abnormal, with long inner and outer branches
McArdle-Schmid-Pearson disease	Muscle phosphorylase	Glycogen accumulates in skeletal muscle. It has normal structure
Hers' disease	Liver phosphorylase	Glycogen accumulates in liver. It has normal molecular structure
Lactose intolerance	Lactase	Abdominal cramps, diarrhea
Galactosemia	Galactose-1-phosphate uridyl transferase	Galactose phosphate accumulates in tissues. Vomiting, mental retardation
Fructose phosphatemia	Fructose-1-phosphate aldolase	Accumulation of fructose-1-phosphate by liver and kidneys, vomiting, acidosis
Pentosuria	L-xylulose dehydrogenase	L-xylulose excreted in urine
Adrenocortical hyperplasia	Steroid 11-hydroxylase	Virilization of genitalia in both males and females, hypertension
Adrenocortical hyperplasia	Steroid 21-hydroxylase	Virilization of genitalia in both males and females
Hurler's syndrome	β-galactosidase	Clouding of cornea; accumulation of lipids and mucopolysaccharides
Lipoprotein lipase deficiency	Lipoprotein lipase	Deposition of fat as xanthomas
Tay-Sachs disease	Hexose aminidase A	Generalized gangliosidosis
Albinism	Tyrosinase	Impaired synthesis of melanin; very white skin and hair
Alkaptonuria	Homogentisic acid oxidase	Homogentisic acid excreted in urine; urine darkens on standing
Tyrosinosis	p-Hydroxyphenylpyruvic acid oxidase	p-Hydroxyphenylpyruvate in urine
Phenylketonuria	Phenylalanine hydroxylase	Accumulation of phenylalanine in tissues; mental and physical retardation
Argininosuccinic acidemia	Argininosuccinase	Argininosuccinic acid in urine; mental retardation
Histidinemia	Histidase	Increased histidine in blood
Hartnup's disease	Tryptophan 2,3-dioxygenase	Psychiatric disturbances, mental retardation
Maple syrup urine disease	Branched chain keto acid decarboxylase	Deficiency of enzyme that decarboxylates keto acids corresponding to valine, leucine and isoleucine. Deficient myelination. Mental retardation. Impaired growth
Orotic aciduria	Orotidine-5-phosphate pyrophosphorylase	Orotic acid accumulates and is excreted in urine
Cystinuria	Deficient transport system for tubular reabsorption of diamino compounds	Cystine, lysine, arginine and ornithine excreted in urine
Cystinosis	Not known	Decreased utilization of cystine and other amino acids. Cystine accumulates in tissues
Niemann-Pick disease	Sphingomyelinase	Accumulation of sphingomyelin in tissue
Refsum's disease	Phytanic acid hydroxylase	Cerebellar ataxia
Lesch-Nyhan syndrome	Hypoxanthine guanine phosphoribosyl transferase	Behavioral disturbances, aggressiveness, self-mutilation
Glucose-6-phosphate dehydrogenase deficiency	Glucose-6-phosphate dehydrogenase	Hemolytic anemia
Crigler-Najjar syndrome	Uridine diphosphoglucuronate transferase	Inability to conjugate bilirubin; jaundice, kernicterus
Hyperammonemia	Ornithine transcarbamylase	Decreased urea formation; accumulation of ammonia

globin in the adult—about 2 per cent—is made of two α chains and two δ chains.

In addition to these normal hemoglobins, a variety of abnormal hemoglobins have been described, each characterized by the substitution of one amino acid in one of these chains. Sickle cell anemia is characterized by hemoglobin with two normal α chains, but the amino acid in position 6 from the N-terminal end of both the β chains is valine instead of the normal glutamic acid. The substitution of a valine for a glutamic acid results in an altered electric charge of the molecule and decreases its solubility. Another type of hemoglobin, hemoglobin C, produces an anemia that is somewhat milder than sickle cell disease. By a remarkable coincidence, exactly the same amino acid in the β chain is involved—the one at position 6—but in hemoglobin C, the glutamic acid is replaced by lysine.

Since it is now possible to relate specific triplet codons with specific amino acids, it can be shown that all of the substitutions of one single amino acid for another that are known from the more than 100 kinds of abnormal hemoglobins discovered can be explained by substituting a single nucleotide in the triplet codon for the usual one. For example, the replacement of glutamic acid (GAA) by valine (GUA) or by lysine (AAA) could be explained in each instance by a single nucleotide replacement.

An unusual type of hemoglobin is found in hemoglobin Lepore, which has a longer β chain than normal and appears to be composed of an overlapping portion of a normal β chain and a normal δ chain. This may have arisen through unequal crossing over between the two genes which code for these two polypeptides.

In contrast to other hereditary anemias, thalassemia does not have an abnormality in the primary structure of its hemoglobin, in the sequence of amino acids making up the α and β chains. Instead, there is a genetic block in the rate of synthesis of either the α chain or the β chain of adult hemoglobin. The genetic block in thalassemia results in red cells with a low content of hemoglobin; however, the continued normal synthesis of the other chain results in their accumulation within the red cell and their combination into unstable aggregates, which precipitate within the cell and cause its premature destruction. In thalassemia there are both hypochromic anemia and hemolytic anemia.

DIAGNOSIS OF GENETIC ABNORMALITIES

Genetic abnormalities may become manifest during early intrauterine life or not until late in adult life, as is the case with Huntington's chorea. The possibility of ameliorating or alleviating the effects of a genetic abnormality is obviously greater if the abnormality can be detected as early as possible in life. Thus, efforts have been made over the years to detect genetic abnormalities such as phenylketonuria at birth, so that the infant can be placed on a diet with a minimal amount of phenylalanine, which minimizes the damage to the central nervous system. In the past 15 years, as physicians have become bolder in approaching the fetus in utero, it has become possible to diagnose a number of genetic abnormalities in fetal life.

The first attempts at intrauterine diagnosis were carried out in 1955 and 1956, when four different groups of investigators demonstrated the feasibility of diagnosing the sex of the fetus by determining the presence or absence of sex chromatin in cells recovered from amniotic fluid. Some of the first attempts used amniotic fluid collected at or near term, when the membranes were ruptured for the induction of labor. As it has been shown that amniocentesis can be carried out successfully with minimal risk to mother and fetus, physicians have carried out these analyses at earlier and earlier stages. Early in fetal life, there is relatively little amniotic fluid, and the difficulties of obtaining it with no danger to mother or fetus are consequently greater. A collaborative prospective study of 1040 women undergoing amniocentesis and 992 matched controls was reported in October, 1975, by the National Institute of Child Health and Human Development. These studies showed no adverse effects of the procedure on the women and no significant differences in the rate of fetal loss, prematurity, status of newborn, birth defects or the developmental status of the offspring at one year of age.

A commonly used technique is to insert a needle through the skin of the lower abdominal wall and through the wall of the uterus into the uterine lumen. A syringe is attached and amniotic fluid is withdrawn. The amniotic fluid is centrifuged to sediment the cells, which are transferred to slides, fixed and stained. They can then be analyzed directly for the presence or absence of sex chromatin, permitting a diagnosis of sex of the fetus. This can be of great usefulness in diagnosing the presence of severe sex-linked hereditary diseases, such as hemophilia or muscular dystrophy, and in helping to decide whether the pregnancy should be interrupted.

Alternatively, the cells can be incubated for a suitable period, treated with colchicine and stained so that the complete karyotype can be determined. This, of course, can be very important in the diagnosis of Down's disease and similar chromosomal aberrations.

The number of genetic abnormalities that can be detected by intrauterine diagnosis has increased remarkably in the last few years, and now includes many point mutations as well as chromosomal aberrations detectable by their karyotype. If enough cells can be collected from the amniotic fluid, the activity of a variety of specific enzymes can be measured. Enzyme deficiencies can be detected by incubating cells recovered from the amniotic fluid with the appropriate substrate and measuring the product (Table 6–5). An increased concentration of α-feto protein in amniotic fluid has been found in pregnancies with infants having defects in the development of the neural tube. This may prove to be useful as a diagnostic tool. We can expect further major advances in the diagnosis and treatment of conditions in the fetus.

Table 6–5. DIAGNOSIS OF HEREDITARY METABOLIC DISORDERS BEFORE BIRTH BY AMNIOCENTESIS*

Disorder	Inheritance	Clinical Manifestations	Accumulated Products	Deficient Enzyme Activity	Prenatal Diagnosis Possible
Disorders of lipid metabolism					
Fabry's disease	X*	Purple skin papules; renal failure; cardiac and ocular involvement	Ceramidetrihexoside	Ceramidetrihexoside galactosidase	?*
Gaucher's disease	A*	Infantile and adult forms; hepatosplenomegaly, erosion of long bones, neurologic involvement, anemia and thrombocytopenia	Glucocerebroside	Glucocerebrosidase	+*
G$_{M1}$gangliosidosis (generalized gangliosidosis)	A	Mental retardation from birth, unusual facies, hepatosplenomegaly and skeletal changes	G$_{M1}$ganglioside and ceramide tetrahexoside visceral mucopolysaccharide of keratan sulfate type and a sialomucopolysaccharide	β-galactosidase A, B and C; B and C only	+
G$_{M2}$gangliosidosis (Tay-Sachs disease)	A	Onset at age 5 to 6 months, degenerative neurologic disorder with cherry-red spot within macula, progressing from normal state to apathy, hypotonia, profound psychomotor retardation and death	G$_{M2}$ganglioside and its asialo derivative	Hexosaminidase A

Hexosaminidase A and B | ++*

+ |
Metachromatic leukodystrophy	A	At least two forms: degenerative neurologic disease progressing from normal state to weakness, ataxia, hypotonia, mental retardation and paralysis; excessive sulfatide in urine	Sulfatide	Arylsulfatase A (sulfatidase)	++
Niemann-Pick disease	A	Four types: hepatosplenomegaly; variable skeletal and neurologic involvement	Sphingomyelin	Sphingomyelinase	++
Refsum's disease	A	Cerebellar ataxia, peripheral polyneuropathy, retinitis pigmentosa, and other cardiac, skin, neurologic and skeletal changes	Phytanic acid	Phytanic acid α-hydroxylase	+

*A, autosomal recessive; D, autosomal dominant; X, sex-linked recessive; ++, prenatal diagnosis made; +, prenatal diagnosis possible; ?, prenatal diagnosis questionable.

(Table continued on the following page)

Table 6–5.   DIAGNOSIS OF HEREDITARY METABOLIC DISORDERS BEFORE BIRTH BY AMNIOCENTESIS* (*Continued*)

Disorder	Inheritance	Clinical Manifestations	Accumulated Products	Deficient Enzyme Activity	Prenatal Diagnosis Possible
Mucopolysaccharidoses					
Hurler's syndrome	A	Gargoyle-like facies, early clouding of cornea, early psychomotor retardation, increased linear growth in first year; then decline to become dwarfed, hepatosplenomegaly, kyphosis, joint stiffness, excessive dermatan sulfate and heparitin sulfate in urine	Dermatan sulfate and heparitin sulfate	Specific β-galactosidase	++
Hunter's syndrome	X*	Gargoyle-like facies less obvious and seen later than in Hurler's syndrome, clear cornea, psychomotor retardation, increased linear growth in first year; declines later, hepatosplenomegaly, joint stiffness, excessive dermatan sulfate and heparitin sulfate in urine	Dermatan sulfate and heparitin sulfate	Unknown	?
Maroteaux-Lamy syndrome	A	Severe skeletal changes, cloudy cornea, normal intellect; excessive dermatan sulfate in urine	Unknown	Unknown	?
Morquio's syndrome	A	Severe skeletal changes, with dwarfism, cloudy corneas, aortic regurgitation, intellect usually within normal range, excessive keratan sulfate in urine	Unknown	Unknown	?
Sanfilippo syndrome	A	Severe mental retardation, joint stiffness, excessive heparitin sulfate in urine	Unknown	Unknown	?
Scheie's syndrome	A	Coarse facies, stiff joints; usually normal intellect, aortic regurgitation, excessive dermatan sulfate in urine	Unknown	Unknown	?

Amino acid and related disorders

Argininosuccinic aciduria	A	Mental retardation, trichorrhexis nodosa, ammonia intoxication	Argininosuccinic acid	Argininosuccinase	+
Citrullinemia	A	Ammonia intoxication, mental retardation	Citrulline	Argininosuccinic acid synthetase	?
Cystinosis	A	Failure to thrive, rickets, glycosuria, aminoaciduria, cystine deposition in tissue	Cystine	Unknown	+
Homocystinuria	A	Dislocated lenses, skeletal abnormalities, vascular thrombosis, mental retardation	Methionine and homocystine	Cystathionine synthase	+
Hyperammonemia Type II	A	Ammonia intoxication, mental retardation	Ammonia	Ornithine carbamyl-transferase	+
Hyperlysinemia	A	Mental retardation, muscular asthenia (may be normal)	Lysine	Lysine-ketoglutarate reductase	?
Hypervalinemia	A	Mental retardation, failure to thrive	Valine	Valine transaminase	?
Ketotic hyperglycinemia	A	Ketoacidosis, protein intolerance, developmental retardation	Glycine	Propionyl CoA carboxylase	?
Maple syrup urine disease					
Severe infantile	A	Ketoacidosis, neurologic abnormality, mental retardation, early death	Valine, leucine, isoleucine, and their keto acids	Branched-chain keto acid decarboxylase	++
Intermittent	A	Ketoacidosis, neurologic abnormality, mental retardation, early death	As above	As above	+
Methylmalonic aciduria	A	Acidosis, lethargy, failure to thrive, early death	Methylmalonic acid, glycine, homocystine, and cystathionine	Methylmalonyl CoA isomerase or vitamin B_{12} coenzyme (decreased carbon dioxide production from propionate)	++
Ornithine-α-ketoacid transaminase deficiency	A	Liver disease, renal tubular defect, mental retardation	Ornithine	Ornithine-α-keto acid transaminase	

Disorders of carbohydrate metabolism

*A, autosomal recessive; D, autosomal dominant; X, sex-linked recessive; ++, prenatal diagnosis made; +, prenatal diagnosis possible; ?, prenatal diagnosis questionable.

(Table continued on the following page)

Table 6–5.   DIAGNOSIS OF HEREDITARY METABOLIC DISORDERS BEFORE BIRTH BY AMNIOCENTESIS* (*Continued*)

Disorder	Inheritance	Clinical Manifestations	Accumulated Products	Deficient Enzyme Activity	Prenatal Diagnosis Possible
Fucosidosis	A	Severe progressive cerebral degeneration, intense spasticity, thick skin and excessive sweating, increased salinity of sweat	Fucose containing heteropolysaccharide	α-fucosidase	+
Glycogen-storage disease (Type II)	A	Failure to thrive, hypotonia, hepatomegaly, cardiomegaly	Glycogen storage	α-1,4-glucosidase	++
Glycogen-storage disease (Type III)	A	Hepatomegaly, cardiomegaly, hypoglycemia	Abnormally structured glycogen	Amylo-1,6-glucosidase	+
Glycogen-storage disease (Type IV)	A	Familial cirrhosis with splenomegaly	Abnormally structured glycogen	Branching enzyme	+
Galactosemia	A	Cirrhosis, cataracts, mental retardation and failure to thrive	Galactose	Galactose-1-P-uridyl transferase	+
Mannosidosis	A	Gargoyle-like facies, psychomotor retardation, accelerated growth in infancy, hypotonia, mild hepatosplenomegaly	Mannose and glucosamine containing heteropolysaccharide	α-mannosidase	+
Pyruvate decarboxylase deficiency	A	Intermittent cerebellar ataxia and choreoathetosis, with elevated urinary alanine	Pyruvic acid, alanine and lactate	Pyruvate decarboxylase	?
Glucose-6-PO_4 dehydrogenase deficiency	A	Hemolytic anemia		G-6-PO_4 dehydrogenase	+
Miscellaneous disorders					
Acatalasemia	A	Recurrent anaerobic infections of gums and oral tissue	Unknown	Catalase	?
Adrenogenital syndrome	A	Virilization or pseudohermaphroditism, adrenal insufficiency with salt loss, hypertensive cardiovascular disease, pregnanetriol and 17-ketosteroids in urine		C21, C11 or C17 steroid hydroxylase	++

Disorder	Inheritance	Clinical features	Cellular/biochemical finding	Enzyme defect	Prenatal diagnosis
Chediak-Higashi syndrome	A	Photophobia, decreased pigmentation of skin, hair and eyes, increased susceptibility to infections	Cellular inclusions	Unknown	?
Congenital erythropoietic porphyria	A	Photosensitive dermatitis, anemia, splenomegaly, hypertrichosis, massive porphyrinuria	Uroporphyrin I and coproporphyrin I in tissues	Cosynthetase	?
Cystic fibrosis	A	Recurrent pulmonary infection, malabsorption, failure to thrive, increased sodium chloride in sweat	Mucopolysaccharide storage or increased production in cultured fibroblasts	β-glucuronidase deficiency in skin components	?
I-cell disease	A	Gargoyle-like facies, dwarfism from birth, gingival hyperplasia, psychomotor retardation	Acid mucopolysaccharide and glycolipids	Reduced β-glucuronidase and excessive acid phosphatase	?
Lesch-Nyhan syndrome	X	Self-mutilation, choreoathetosis, spasticity, mental retardation	Uric acid	Hypoxanthine-guanine phosphoribosyltransferase	++
Lysosomal acid phosphatase deficiency	A	Failure to thrive, progressive neuromuscular involvement, hypoglycemia, seizures and hepatomegaly	Unknown	Lysosomal acid phosphatase	++
Marfan's syndrome	D	Connective-tissue disorder with skeletal, cardiovascular and ocular signs	Hyaluronic acid in cultured fibroblasts	Unknown	?
Orotic aciduria	A	Infantile megaloblastic anemia, orotic acid in urine	Unknown	Orotidylic pyrophosphorylase and decarboxylase	?
Xeroderma pigmentosum	A	Photosensitive dermatitis, skin cancers	Unknown	DNA "repair enzyme"	+

*A, autosomal recessive; D, autosomal dominant; X, sex-linked recessive; ++, prenatal diagnosis made; +, prenatal diagnosis possible; ?, prenatal diagnosis questionable.

GENETIC COUNSELING

The physician who is not a medical geneticist may be called upon to give genetic counseling. The most frequent occasion in which genetic prognosis is requested is that in which parents have had one abnormal child and are concerned about the risk of abnormality in a subsequent child. Other individuals who have a parent or some other member of the family affected with a hereditary disease may seek advice about the probability of the appearance of that trait in their offspring.

Advice to prospective parents is best given in terms of the probability that any given offspring will have a particular condition. This requires that the family histories be carefully made and verified. If appropriate tests are available for detecting heterozygous carriers of the condition, these should be made. Only a small fraction of diseases are inherited by a single pair of genes, but in these diseases the probabilities are easily calculated. For example, if one parent is affected with a trait that is inherited as an autosomal dominant, such as perineal muscular atrophy or Huntington's chorea, the probability that any given offspring will have the disease is 0.5. The birth of one child with a trait such as albinism or phenylketonuria inherited as a recessive to two normal parents establishes the probability of 0.25 that any subsequent offspring will be affected. For a disease inherited by recessive genes located on the X chromosome, such as hemophilia, the probability depends on whether the father or the mother has the disease or is a carrier for it. A normal woman and an affected male will have daughters that are carriers (heterozygotes) and sons that are normal. The probability that a carrier woman and a normal male will have an affected son is 0.5, and the probability that they will have a carrier daughter is also 0.5.

More often, one is dealing with conditions in which the method of inheritance is unknown or is doubtful, and then an estimate of the probability of appearance of a given trait in a future child can be obtained from a table of empiric risk. It is difficult to give a precise probability, because the genes involved may show incomplete penetrance or varying degrees of expressivity. Several traits appear to be inherited in different ways in different pedigrees. For example, retinitis pigmentosa appears to be inherited as a sex-linked trait in some pedigrees and as an autosomal trait in others.

Mental deficiency, epilepsy, deafness, congenital heart disease, anencephaly, harelip, spina bifida and hydrocephalus are examples of the kinds of deficiencies about which inquiries are frequently made. It is important to consider whether an environmental factor may have played a role in the appearance of the previous abnormal child. Did the mother have some infectious disease during pregnancy? Was she on any sort of drug therapy? Was she subjected to any radiation? By carefully dissecting out possible environmental contributions, the counselor can make a better estimate of the probability that genetic factors may be involved and a better estimate of the probability of recurrence of the particular trait in some subsequent offspring.

In summary, it is clear that human inheritance is governed by the same set of laws that govern inheritance in all other organisms. Rapid progress is being made in detecting inherited traits by the recovery of cells using amnio-centesis, by analyzing specific enzyme deficiencies in those recovered cells maintained in cell culture and in identifying specific human chromosomes by means of fluorescent dyes, such as quinacrine mustard. The prospect of genetic engineering, that is, of replacing a specific defective human gene by a normal gene introduced in some fashion into the nucleus, still seems to lie far in the future.

SUGGESTED SUPPLEMENTARY READING

Bergsma, D. (Ed.). Conference on the Clinical Delineation of Birth Defects. The National Foundation, New York, 1969.

   Contains the papers presented at this interesting conference and the transcript of the discussion which followed.

Carlson, E. A. The Gene: A Critical History. W. B. Saunders Co., Philadelphia, 1966.

   A stimulating presentation of the development of the major concepts of genetics, well worth reading for its broad overview of the subject.

Herskowitz, I. H. Genetics, 2nd Ed. Little, Brown and Company, Boston, 1965.

   A standard text, with a section in the back in which many of the classic papers in genetics are reprinted in part or in their entirety.

Milunsky, A. Prenatal genetic diagnosis. New Eng. J. Med., *283*:1370, 1970.

   A well-written summary of the newer discoveries in human genetics made with such techniques as amniocentesis, cell culture, karyotyping and enzyme assays.

Strickberger, M. W. Genetics. Macmillan Company, New York, 1968.

   One of the standard texts in genetics, with a broad coverage of the field.

Thompson, J. S., and Thompson, M. W. Genetics in Medicine. W. B. Saunders Company, Philadelphia, 1966.

   An excellent, concise presentation of genetic principles and their applications to medical problems.

Turpin, R., and Lejeune, J. Human Afflictions and Chromosomal Aberrations. Pergamon Press, London, 1969.

   An excellent source book for information regarding human genetics.

Chapter 7

SPERMATOZOA, FERTILIZATION, FERTILITY AND INFERTILITY

SPERMATOGENESIS

Testicular Maturation

At birth the male testes contain recognizable spermatogonia which may have been stimulated to develop by maternal or placental hormones. As in the female (p. 19) primordial germ cells migrated from the yolk sac to the genital ridge very early in development, and subsequently divided to form the precursors of the cells of the seminiferous tubules.

The pituitary of the newborn, freed of feedback inhibition by placental estrogens, produces a postnatal surge of gonadotropins in both male and female infants. In the male there is a sharp rise in serum FSH and LH in the first few weeks, followed by a rise in circulating testosterone to 200–300 ng per 100 ml (Figure 7–1). By four months of age gonadotropin levels return to the normal childhood range and testosterone levels soon follow, decreasing to a mean value of 7 ng per 100 ml. by about age seven months, a value identical to that in prepuberal children of both sexes. In females a striking and prolonged rise in FSH and a lesser rise in LH occur, but there appears to be no steroidogenic response by the ovary. The physiologic significance of these endocrine changes in very early infancy is not yet clear. The testis remains quiescent from about age six months until puberty.

The cause of puberty in the male is presumably similar to that described earlier for the female, except that the secretions of FSH and LH by the male pituitary are tonic, continuous and noncyclic. FSH stimulates spermatogenesis in the seminiferous tubules; LH is primarily responsible for stimulating the interstitial Leydig cells which secrete androgens. There is indirect evidence to suggest that testosterone produced in the Leydig cells regulates the development of the adjacent seminiferous tubules. The normal histologic appearance of the adult testis is shown in Figure 7–2.

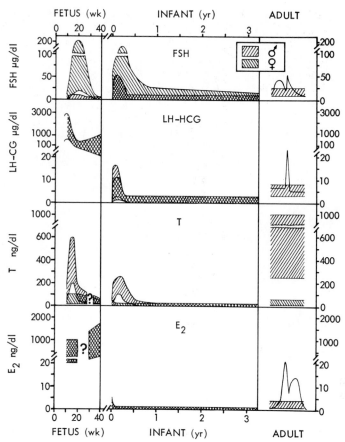

FIGURE 7–1. Concentrations of gonadotropins, testosterone (T) and estradiol (E_2) in the bloods of male and female fetuses and infants. (Reproduced, with permission, from C. Faiman, *in* Sexual Endocrinology of the Perinatal Period, J. Bertrand and M. Forest (Eds.). Inserm, Paris, 1974.

The seminiferous tubules of several mammals have been shown to contain an androgen binding protein that has a high affinity for testosterone and dihydrotestosterone. Androgen binding protein is produced in the Sertoli cells, and its production is stimulated by FSH. Testosterone is able to maintain both the production of androgen binding protein and the process of spermatogenesis in the hypophysectomized animal if the treatment is begun immediately after hypophysectomy. Spermatogonia have been shown to have cytoplasmic and nuclear receptors for androgens which have high affinity for both testosterone and dihydrotestosterone. The properties of these receptors are similar to those in the epididymis and ventral prostate but different from the androgen binding protein. The androgen binding protein is secreted into the lumen of the seminiferous tubules, and appears to take up testosterone and deliver it, via the excurrent ducts, directly to the cells in the epididymis.

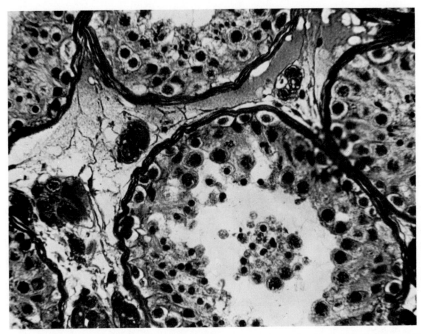

FIGURE 7-2. Photomicrograph of normal adult human testis. (From Lloyd, Human Reproduction and Sexual Behavior, Lea & Febiger, Philadelphia, 1966.)

Formation of Spermatozoa

After maturation at puberty, the testis continues its dual function of secreting hormones and producing sperm throughout the remainder of its life. The development of spermatogonia into spermatocytes is followed by the two meiotic divisions which result in spermatids. The spermatids then undergo a prolonged process of differentiation into the mature sperm (Figure 7-3). A spermatid is a mature gamete with a haploid number of chromosomes, but to become a functional sperm it must undergo a complex process of differentiation. The nucleus shrinks in size, becoming the head of the sperm, and most of the cytoplasm, containing the endoplasmic reticulum and ribosomes, is shed. Some of the Golgi bodies aggregate at the anterior end of the sperm and form the acrosome. The two centrioles of the spermatid move to a position posterior to the nucleus. The proximal centriole assumes a position in a small depression on the surface of the nucleus at right angles to the axis of the sperm; the distal centriole gives rise to the axial filament of the sperm tail. This filament consists of two central longitudinal fibrils and a ring of nine pairs, or doublets, of longitudinal fibrils surrounding the two, all composed of the protein tubulin. This pattern of the nine plus two axoneme is common to cilia and flagella as well as to the sperm tail. The two fibers in the doublets are not identical; the A fiber is a complete circle, whereas the B fiber is shaped like the letter C and is joined to the adjacent A fiber. The fact

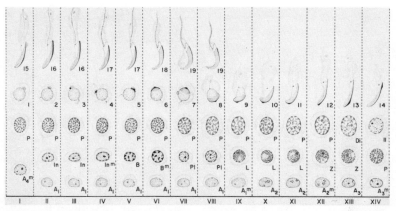

FIGURE 7–3. Diagram of the cellular composition of the 14 stages of the cycle of sperm formation in the seminiferous epithelium of the rat. Each column with a Roman numeral represents the cell types present in one of the cellular associations found in a cross-section of a seminiferous tubule. These associations succeed one another in time in any given area of the seminiferous epithelium so that, following cellular association XIV, cellular association I reappears. *A*, type A spermatogonia; *In*, intermediate spermatogonia; *B*, type B spermatogonia; *Pl*, preleptotene spermatocyte; *L*, leptotene spermatocyte; *Z*, zygotene spermatocyte; *P*, pachytene spermatocyte; *Di*, diakinesis; *II*, secondary spermatocytes; 1 to 19, steps in spermiogenesis. (From Dym and Clermont, Am. J. Anat., *128*:265, 1970.)

that the arms of the doublets were missing in the nonmotile sperm produced by a human patient suggests that these arms are the site of the generation of the force that provides for the beating of the tail. The mitochondria move to the point where the head and tail meet and form a spiral middle piece surrounding the axoneme. Oxidative phosphorylation in the mitochondria of the middle piece provides the ATP required for the beating of the tail. A thin sheath of cytoplasm remains surrounding the mitochondria in the middle piece and the axial filament of the tail. The haploid spermatozoa contain either an X or a Y chromosome, and function in determining the sex of the fertilized egg.

Anatomy of the Spermatozoon

The entire spermatozoon, like other cells, is contained within a plasma membrane. The outer surface of the plasma membrane includes a lipoglycoprotein which bears a negative charge, owing to its content of neuraminic acid. This surface coat contains antigens which may be responsible for the specificity of the egg-sperm interaction. The top of the nucleus is the head-cap, which covers the acrosome (Fig. 7–4). The acrosome contains hydrolytic enzymes such as acrosin (a proteolytic enzyme) and hyaluronidase. It was shown by Blandau that sperm heads contain enzymes that will dissolve gelatin.

Human sperm have an unusual degree of heterogeneity in their shape and structure. Variable amounts of cytoplasm may remain at the posterior

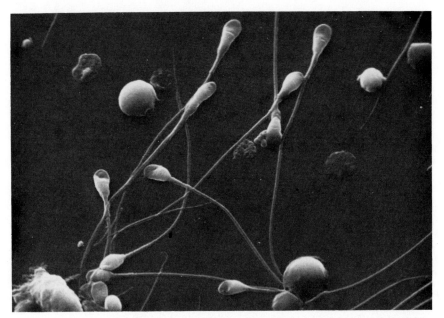

FIGURE 7–4. Scanning electron micrograph of human sperm showing head, midpiece and tail. (Micrograph courtesy of Drs. J. E. Flechon and E. S. E. Hafez.)

end of the head. The chromatin in the nucleus of the sperm has a coarse granular appearance that becomes more compacted in the epididymis. Human sperm have variable nuclear cavities or vesicles. The proximal centriole remains, but the distal centriole disappears, permitting the inference that the centriole is not necessary for sperm motility; however, it is necessary for the development of the tail. The nine radially arranged doublets are connected by nexin bridges (Figure 7–5). The doublets are found to contain an ATPase termed dynein, which mediates the beating of the tail. It appears that these filaments are not contractile; instead, they slide one on another, and thus are analogous to the sliding of the filaments of actin and myosin in contracting skeletal muscle. In the more anterior part of the tail the nine plus two axoneme is surrounded by outer dense fibers which are probably elastic and resilient rather than contractile. They are composed of proteins and triglycerides.

The mitochondria are organized into a tightly coiled helix of 10 to 15 turns, containing elongated mitochondria in which the process of oxidative phosphorylation can generate $\sim P$ for the beating of the tail. The beating of the tail is two-dimensional in the proximal portion, but the distal end undergoes three-dimensional helical undulations that cause the spermatozoon to rotate as it swims forward.

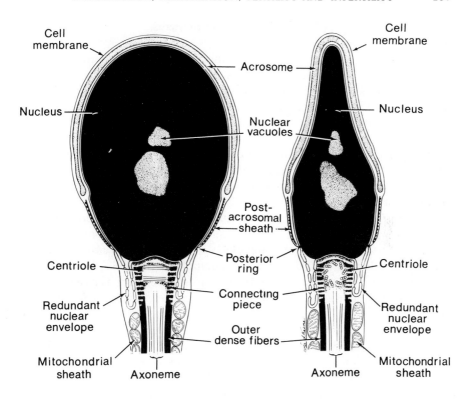

Cell membrane

Acrosome

Cell membrane

Nucleus

Nuclear vacuoles

Nucleus

Post-acrosomal sheath

Centriole

Posterior ring

Centriole

Redundant nuclear envelope

Connecting piece

Redundant nuclear envelope

Outer dense fibers

Mitochondrial sheath

Axoneme

Axoneme

Mitochondrial sheath

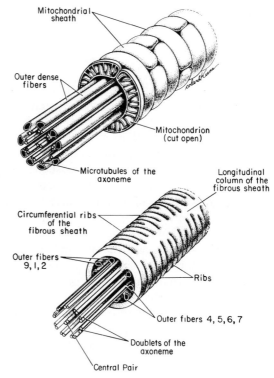

Mitochondrial sheath

Outer dense fibers

Mitochondrion (cut open)

Microtubules of the axoneme

Longitudinal column of the fibrous sheath

Circumferential ribs of the fibrous sheath

Outer fibers 9, 1, 2

Ribs

Outer fibers 4, 5, 6, 7

Doublets of the axoneme

Central Pair

FIGURE 7–5. The structure of human spermatozoa, based on analyses of electron micrographs. *Top:* sections of the head of the sperm in two planes. *Middle:* diagram of the structure of the middle piece. *Bottom:* diagram of the tail showing the 9, 1, 2 doublets of the axoneme. (From Pedersen, Henning and Fawcett: Functional anatomy of the human spermatozoon. *In* Hafez, E. S. E., ed.: The Human Semen and Fertility Regulation in the Male. The C. V. Mosby Company, St. Louis, 1976.)

Metabolism of Spermatozoa

The mature spermatozoon can neither divide nor differentiate further, having lost its RNA apparatus for the synthesis of proteins. It cannot repair any worn-out components. Most of its biochemical capacities are directed toward maintaining an ATP concentration adequate to supply its requirements for motility for a limited period of time. Only motile sperm are capable of fertilizing mammalian ova. Once sperm are deposited in the female reproductive tract they may remain motile for as much as a week, but they probably retain their fertilizing capabilities for only three days. In the resting nonmotile state within the male reproductive tract they may survive for several months.

Spermatozoa do not contain any nutritional reserves except for some lipids, and they depend upon fructose and other extracellular substrates as a source of nutrients. The major metabolic processes are glycolysis and oxidative phosphorylation; the rate of metabolism parallels the degree of motility. Within the acrosome are hydrolytic enzymes such as hyaluronidase and acrosin, which appear to be important in the process of fertilization. Hyaluronidase causes dispersion of the cumulus and corona radiata, and acrosin may be of importance in penetrating the cervical mucus as well as the zona pellucida of the ovum. The mitochondria surrounding the midpiece contain a complete cytochrome system, various dehydrogenases, and the complete complement of enzymes needed for glycolysis and respiration. The ATPase dynein, which provides the energy for motility, is located on the arms of the doublets of the axoneme.

Seminal Plasma

The plasma in which the sperm are suspended is secreted by a number of accessory glands, including the seminal vesicle, prostate, and bulbourethral glands and, to a lesser extent, by the epididymis and vas deferens. It differs in many respects from blood plasma and other body fluids. Certain amino acids, especially glutamic acid, are present in high concentration, and there are a number of unique amines such as spermine and spermidine which impart a characteristic odor to the semen. The major carbohydrate present is fructose, and its rate of disappearance is a function of the density and motility of the sperm. The prostate gland secretes citrate, acid phosphatase, spermine, spermidine, zinc and magnesium, whereas the primary products of the seminal vesicles are fructose and some 13 varieties of prostaglandins. The first portion of the ejaculate mainly contains sperm, prostatic secretions and the very viscous mucoproteins secreted by the bulbourethral glands. The second portion contains the secretions of the seminal vesicle.

The major function of the seminal plasma is to counter the acidity of the vagina. Sperm are very quickly destroyed by a low pH; this fact is used in preparing spermicidal foams and jellies, which are acidic. Human semen coagulates immediately after ejaculation and normally liquifies again within 10 minutes. The coagulating proteins originate in the seminal vesicles. The coagulum appears as a network of fibrous strands within which spermatozoa are trapped. Failure to liquify promptly may lead to infertility. When the coagulum liquifies, the fibers become disorganized and turn into spherical material. Factors which bring about lysis are enzymatic and are secreted by the prostate and bulbourethral glands. Lysozyme, α-amylase, plasminogen activator and a chymotrypsin-like proteinase secreted by the prostate are present in the first portion of the ejaculate. The coagulation and liquefaction of the semen are very different processes from the coagulation of blood and the liquefaction of a blood clot. The immunoglobulins, IgG and IgA, have also been detected in seminal fluid and appear to be secreted by the prostate.

The Clinical Analysis of Semen

The only method of evaluating the fertility of a male, other than his ability to produce pregnancy, is by the examination of a sample of his semen. This should be one of the first steps in the study of an infertile couple. To be adequate for evaluation, the specimen should be obtained by masturbation directly into a clean bottle, preferably after three days of continence, and should be examined within two hours. The volume, color, degree of lique-faction and viscosity are noted, and sperm density is determined with a hemocytometer. The percentage of sperm which are motile is determined, and of those which are moving, the percentage which make forward progression is recorded. A stained preparation is then examined to determine the number of normal shapes found, but it must be admitted that standards of normality vary a great deal from one laboratory to another.

In general, the minimal standards for a "probably fertile" specimen are these: (a) over 20 million spermatozoa per ml., (2) over 50 million in the total specimen, (3) over 50 per cent motile, (4) more than 50 per cent of the motile sperm make forward progression, and (5) more than 70 per cent conform to normal morphology. If any one of these standards is not met, particularly with two or more samples, the man is suspected of being infertile, but only when motile spermatozoa are totally absent can it be truthfully stated that he is sterile. The five attributes are somewhat like the five cards of a poker hand; no one can describe a hand which has a zero chance of winning.

FERTILIZATION

Sperm Maturation and Transport

The mitotic divisions of the spermatogonia serve to maintain the population of spermatogonia while providing cells which develop into sper-

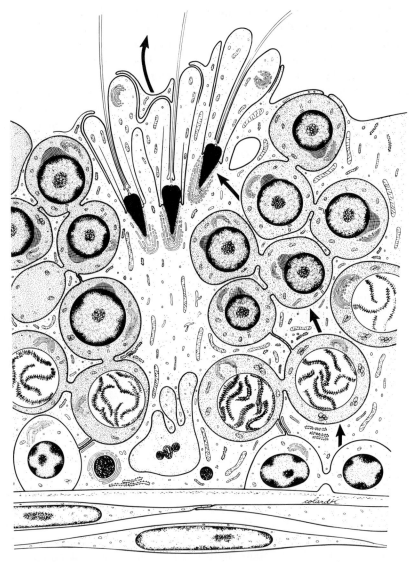

FIGURE 7-6. Diagram showing the development of spermatogonia into spermatocytes and spermatids. Three late spermatids, nearly mature sperm, are shown imbedded in the Sertoli cell with their tails extending into the lumen of the seminiferous tubules. The arrows indicate the path of movement of the cells as they develop from spermatogonia, present in the basal layer, into spermatocytes and spermatids before being shed as mature sperm. (Courtesy of Dr. Don Fawcett.)

matocytes. The spermatogonia divide repeatedly but remain attached by cytoplasmic bridges. The cells remain attached while developing into spermatocytes and undergoing meiotic divisions to form spermatids. During the early stages of sperm formation there is a high rate of spontaneous degeneration or attrition, just as there is in the formation of ova. While in the early spermatocyte stage the sperm cells begin to move from the basal compartment of the seminiferous tubule to the portion of the tubule next to the lumen. The end result of this process is the formation of a large number of spermatids, all of which are connected by intercellular bridges. During the differentiation of spermatids to form spermatozoa, much of the cytoplasm flows posteriorly, forming bulges around the base of the tail. The tubulin begins to polymerize at the end of the centriole to form the central axoneme. The developing spermatids are attached to the Sertoli cell, and when sperm maturation is complete the Sertoli cell expels the sperm but retains the excess cytoplasm. The stalk connecting the spermatozoon to the Sertoli cell eventually breaks and the residual cytoplasm attached to the Sertoli cell undergoes autolysis by the lysosomes in the Sertoli cell (Fig. 7–6).

The sperm are moved down the seminiferous tubules to the rete testis, in part by the pressure of additional sperm forming behind them and aided, at least in some species, by contraction of myoid elements in the walls of the seminiferous tubules. The sperm leaving the testis are not fertile, but undergo a gradual maturation process as they pass through the epididymis. The sperm in the epididymis are not motile; they are moved by contractions of the muscular coat in the wall of the epididymis. The sperm are moved down the vas deferens by peristaltic contractions of the walls, and finally are ejaculated through the urethra during orgasm by the contractions of the ampulla and the ejaculatory duct. The human ejaculate has a volume of 2 to 5 ml. and contains from 60 to 100 million sperm per ml. The ejaculate contains an appreciable number of immature, senescent and abnormal spermatozoa, but it appears that only the healthiest and most vigorous sperm are able to complete the journey to the upper end of the fallopian tube.

When the sperm have been deposited against the external os of the cervix, a small percentage of them gain access to the cervical mucus and, if this is favorable for sperm migration, the sperm will ascend rapidly into the uterine cavity. Some sperm are found in the crypts in the wall of the cervix and in the endometrial glands in the uterus. It has been demonstrated that sperm may reach the upper end of the fallopian tube within five minutes of the time they have been deposited in the vagina, but the mechanism of this rapid transport is still far from clear. Progesterone renders the cervical mucus unfavorable for sperm migration; therefore, during the luteal phase of the cycle, or if the woman is taking an oral contraceptive containing progestin, it is unlikely that many spermatozoa will reach the uterine cavity.

Capacitation

In 1951 Chang and Austin independently observed that rabbit spermatozoa must be in the female reproductive tract for four to six hours before they gain the ability to fertilize. This final step in the maturation of the sperm was called "capacitation." These investigators also found that sperm underwent "decapacitation" when returned to seminal plasma; that is, the sperm lost their capacity to fertilize but could regain this ability by another sojourn in the uterus or fallopian tube. The process of capacitation is as yet poorly understood, but it appears to be enzymatic in nature. It has been suggested that the epididymis produces an inhibitor of a trypsin-like enzyme found in the sperm acrosome, and that this is the basis of the decapacitation phenomenon.

Syngamy

The union of egg and sperm is referred to as syngamy. This process, consisting of a number of steps, is initiated by the fusion of the egg and sperm and involves the subsequent interaction and combination of maternal and paternal chromosomes within a single nucleus. Two types of syngamy can be distinguished: in the first, found in the sea urchin, the male and female pronuclei fuse to give a uninucleate zygote nucleus before the first cleavage division occurs. In the second type, found in *Ascaris,* the male and female pronuclei do not fuse but, instead, line up side by side on the metaphase plate. The first cleavage division occurs, and combination of male and female chromosomes does not occur until the nuclei of the two-cell stage appear. In mammals the male and female pronuclei typically interdigitate very closely, but they do not actually fuse. The pronuclear membranes degenerate and mitosis proceeds. The temporal relationship of fertilization and meiosis varies considerably in different species of animals. The sperm may enter the ovum when the latter is in meiotic prophase, in the first meiotic metaphase, the second meiotic metaphase or even after meiosis is complete. At the time of fertilization many sperm usually become attached to the surface of the egg. The cortical granules in the egg cytoplasm release proteases that appear to take part in the penetration of the sperm. It is not yet clear what feature of the egg prevents all of the sperm except the very first from entering.

In the sea urchin the initial adherence of the sperm to the egg involves fertilizin, a component of the jelly coat on the egg that can agglutinate sperm and that combines with an antibody-like material on the sperm surface, termed "antifertilizin." In mammals and certain other animals the union of the egg and sperm surfaces initiates the "acrosome reaction" in which the contents of the acrosome, including its hydrolytic enzymes, are released. The plasma membrane of the acrosome disappears and the lytic enzymes released appear to facilitate the penetration of the egg membrane. The plasma membranes of the egg and sperm fuse and disappear so that the nucleus of the sperm head comes to lie within the cytoplasm of the egg.

As observed by Dickmann, the zona pellucida of the rabbit and pig egg contain an outer granular layer and an inner clear layer. Although a number of sperm may penetrate the zona pellucida only one penetrates into the cytoplasm, leaving a space through the zona. When this has happened the zona reaction occurs, and this prevents the further penetration of any other sperm, but the physical-chemical nature of the zona reaction remains obscure.

FERTILITY AND INFERTILITY

Requisites for Normal Fertility

Pregnancy can only occur when healthy spermatozoa are deposited at the cervical os and can ascend through a receptive mucus into a normal, patent fallopian tube at the proper time to fertilize a healthy ovum, which must then implant itself upon a suitably prepared endometrium. This statement also forms the basis for a rational and sequential investigation of infertility. The achievement of normal pregnancy is a most stringent test of the anatomic and functional integrity of the reproductive and endocrine systems of both sexes.

Fecundity (the ability of a couple to produce a live child) reaches its peak at about age 25 in both males and females. After 30, the fecundity of women declines steadily, and it declines most rapidly after the age of 40. Fecundity is also related to the frequency of coitus, being highest when intercourse occurs four or more times per week. In the absence of contraception, about 25 per cent of couples will conceive within the first month, 60 per cent within six months and 80 per cent within a year of regular exposure. It is reasonable, therefore, to recommend an infertility study for those who desire children but do not conceive within 12 months of exposure.

The Study of Infertility

Between 15 and 18 per cent of couples in the United States present a problem of infertility. Following a complete investigation, about half of those presenting themselves for study will not be found to have any identifiable factors to account for their infertility, and in these the prognosis is relatively poor; less than 10 per cent ever achieve pregnancy. Even when all findings are normal, one is not justified in telling the husband and wife that they are both normal because this is obviously not true and only raises false hopes. The situation simply results from our inability to detect certain causes of infertility, which are therefore not amenable to correction.

When factors probably responsible for the problem are found, they are frequently multiple. About one-third are attributable to the husband and two-thirds to the wife. In private practice, etiologic causes found in women are about equally divided between cervical, tubal and hormonal factors; when identified, over half may be corrected. The therapy of male infertility is much less rewarding. In the lower socioeconomic classes, the endocrine abnormalities are less frequent and the incidence of tubal occlusion due to gonorrheal disease or postabortion infections is increased.

Procedure for an Infertility Study

The basic study for infertility can be completed in three or four properly organized office or clinic visits, and may perhaps be described most simply by an illustrative case.

A young woman, age 24, has been married for three years and has not conceived. She practiced contraception sporadically for only the first six months of her marriage. A complete history is taken, with special emphasis on the menstrual history; sexual history, including frequency of premarital and postmarital coitus and use of vaginal lubricants or douches; illnesses and surgical operations; the history of pelvic infections; the use of drugs; and all other items which might be pertinent to her fertility. A general physical and pelvic examination is performed, including cervical and vaginal cytology and microscopic examination of the cervical mucus and vaginal secretions. She is then supplied with a basal body temperature chart and instructed in its use. She is referred to the laboratory for a blood count, urinalysis and a serum thyroxine determination. Written instructions are transmitted to her husband regarding the proper collection of a semen sample for laboratory investigation, which must be completed before further investigations are carried out.

The patient is asked to return near the conclusion of her next menstrual period. Assuming that the semen analysis indicates probable fertility of the husband, the patency of the fallopian tubes is determined by the controlled insufflation of carbon dioxide, popularly known as the Rubin test. Numerous devices are available which control the rate of flow, record the pressures obtained, and limit the maximum pressures to 200 mm. Hg. After an initial rise to somewhere between 60 and 150 mm. Hg, the pressure should fall with a free flow of the gas, easily heard by auscultation of the abdomen. Upon assuming the erect posture, the free carbon dioxide should rise within the abdominal cavity and, by irritation of the subdiaphragmatic area, result in a transient, referred pain in the shoulder. Many gynecologists prefer hysterosalpingography as the primary test for tubal patency.

She is asked to return at the estimated time of ovulation, and to have coitus some four to 18 hours before the scheduled examination. At this time, a postcoital examination of the cervical mucus is done (the Sims-Huhner test). Mucus is obtained from the upper portion of the cervical canal with the aid of a dressing forceps, and its physical characteristics are noted. If the patient is truly near the time of her ovulation, the mucus can be pulled from the slide into a long thread and will then retract into its original globule. It should have the viscosity and clear appearance of fresh egg white, and a drop that is allowed to dry on a slide at room temperature should form a delicate fern pattern. Microscopically, there should be an abundance of motile spermatozoa, with from 5 to more than 20 in each microscopic field (high dry objective), demonstrating active forward progression.

Should no definite sterility factors be found thus far, she is asked to return six to eight days after the estimated time of ovulation, and this estimate is checked at the time of her visit by reference to her basal body temperature chart (Figure 3–10). Endometrial biopsies are obtained from the uterine fundus by the use of a small suction curet, and these are submitted to a gynecologic pathologist for evaluation of the degree of pro-gestational development (see Figure 3–8).

This comprises the basic or minimal work-up for infertility; additional tests or procedures will depend upon the findings. Further studies might include a history and physical examination of the husband, with laboratory work to include an evaluation of his thyroid function as well. A visualization of the uterine cavity and tubes with a radiopaque medium, or even a direct visualization of the internal pelvic organs by culdoscopy or laparoscopy, may be indicated. Ovulatory failure would lead the physician to conduct a number of endocrine studies, and a lack of tubal patency would lead him to a consideration of corrective surgery. The student must be referred to current textbooks on infertility for a full discussion of therapy. The problem of the infertile couple is worthy of every physician's attention, and each case is deserving of a thoughtful and complete study.

ARTIFICIAL INSEMINATION

In several species of domestic animals, such as cattle and sheep, the artificial insemination of the female with sperm samples is widely practiced for economic and genetic reasons. It is less expensive to use a sample of sperm from a bull donor than to maintain a bull to service a herd of cows. A bull that has demonstrated its ability to sire cows with high milk fat produc-tion is used as the donor of semen to inseminate entire herds. It is said that no cow in New England has seen a bull in many years! Artificial insemina-tion has also been used in the human for a number of years. In some cases the sperm have been placed within the uterine cavity; this may be helpful in dealing with infertility due to the nature of the cervical mucus. If a man suffers from oligospermia it has sometimes proved successful to collect and concentrate several samples of semen and then use artificial insemination to introduce the concentrated sperm sample. Where the husband has been found to be infertile, the artificial insemination of donor sperm has been used, taking care to match the genetic characteristics of the donor with those of the husband as closely as possible. It is now possible to store and freeze human sperm, and subsequently thaw and use them in artificial insemination, just as frozen bull sperm have been thawed and used in artificial insemina-tion in cattle. Semen banking has received some public attention as a possible "fertility insurance" for a man who is contemplating vasectomy. The number of conceptions obtained with frozen, thawed human semen has not been very large, but the progeny have been perfectly normal. The motility of the sperm is somewhat depressed by the freeze-thaw process and is a function of the freezing technique and the particular cytoprotective agents used. The number of congenital anomalies in babies conceived by freeze-preserved semen is less than in the general population, so there is no suggestion that freezing the sperm and then thawing and using it may incur the risk of an increased number of congenital abnormalities.

SUGGESTED SUPPLEMENTARY READING

Fawcett, D. W. The Structure of the Mammalian Spermatozoon. Internat. Rev. Cytology, 7:195, 1958.

Hafez, E. S. E., and Evans, T. N., eds.: Human Reproduction. Harper and Row, Hagerstown, Md., 1973.
Section I on parameters of human fertility extends the details of topics discussed in this chapter.

Mastroianni, L., *in* Romney et al., Gynecology and Obstetrics. McGraw-Hill, New York, 1975.
Chapter 24 on fertility disorders presents a complete, though concise, summary of etiologic factors in infertility.

Metz, C. B., and Monroy, A. (Eds.). Fertilization, Vol. I. Academic Press, New York, 1967.
The introduction by Albert Tyler, Nelson's discussion of sperm motility and Mann's discussion of sperm metabolism are particularly pertinent. The process of fertilization is also well summarized by Metz, the Colwins, and Dan.

Wolstenholme, G. E. W., and O'Connor, M. Preimplantation Stages of Pregnancy. Little, Brown & Company, Boston, 1965.
This is a Ciba Foundation Symposium, with the usual inter-disciplinary discussion. Austin's discussion of ultrastructural changes in the egg during fertilization and Dickmann's presentation on sperm penetration of the mammalian egg are especially pertinent to the topics of this chapter.

Chapter 8

PLACENTA AND GRAVID UTERUS: STRUCTURE AND FUNCTION

PREIMPLANTATION STAGES OF DEVELOPMENT

Fertilization of the ovum ordinarily occurs in the distal part of the uterine tube within 24 hours after ovulation. The first cleavage occurs some time during the next 36 hours, during which time the conceptus is slowly transported through the tube into the uterine cavity. The ovum retains its clear zona pellucida until it enters the uterine cavity, but has, of course, lost its corona radiata of granulosa cells before passing through the narrow interstitial portion of the tube. Transport of the fertilized egg is believed to be due to the muscular peristaltic movements of the tube rather than to activity of the cilia.

The egg cleaves into two, then four (Figure 8-1), then eight, and then 16 cells, at which point it forms a solid ball (morula), certain cells of which are destined to become the embryo. A fluid-filled cavity forms, transforming the morula into a blastocyst. The wall of this tiny hollow ball is composed of primitive cytotrophoblastic cells, which are usually smaller than the basophilic embryonic cells. There may be a coating of sticky mucus derived from uterine fluid, although this has not been directly observed in the human. Six days after ovulation, the embryonic pole of the blastocyst has attached itself to the endometrium, usually near the mid-posterior or mid-anterior portion of the uterine fundus. Up to this stage, nourishment of the conceptus has occurred by the diffusion of gases and nutrients from the surrounding uterine fluid. Fluid within the rabbit blastocoele has a high content of both bicarbonate and carbon dioxide, but a normal pH.

IMPLANTATION

It is a full week after ovulation before the blastocyst penetrates the endometrium deeply and becomes implanted. The precise cause of the

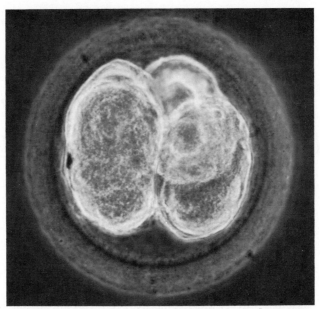

FIGURE 8–1. A four-cell human egg recovered from a removed fallopian tube 36 hours after an isolated coitus. (Courtesy of Dr. A. J. Margolis.)

implantation is not known, but it is under the control of a delicate balance between estrogen and progesterone. In the rat, in which delayed implantation may be produced experimentally by castration followed by daily injections of progesterone, a minute single dose of estrogen is required to cause implantation. Implantation almost always occurs close to a maternal capillary, as though some type of a tropism toward maternal blood is involved.

Within a few hours, the cytotrophoblast begins its invasion, forming an outer coat of syncytium as it does so. Only now do the endometrial stromal cells become large and pale (the decidual reaction). Endometrial cells are destroyed and some of the cytoplasm is probably incorporated within the trophoblast by a process of fusion or phagocytosis. Strange as it may seem, it is more difficult for the trophoblast to penetrate the uterus than other organs to which blastocysts have been transplanted in rats and mice. There are many who feel that the purpose of decidualization is to limit trophoblastic invasion, whereas others feel that the decidua serves the purpose of histotrophic nutrition. Wynn believes that both functions are served. In any event, the ultimate limitation of trophoblastic invasiveness remains a mystery. One of the early stages of implantation in the human species is shown in Figure 8–2.

FORMATION OF THE EARLY PLACENTA

Cytotrophoblastic cell columns progress to the deep layer of the decidua basalis. During this process, the cytotrophoblast (Langhans cells)

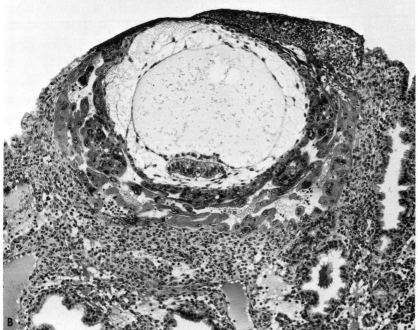

FIGURE 8–2. *A*, Surface view of implantation site of a 12.5 day ovum. *B*, A mid-cross-section of the 12.5 day ovum and surrounding endometrium. (From Hertig and Rock, Amer. J. Obstet. Gynec., *47*:149, 1944. C. V. Mosby Co., St. Louis. Published with permission of the Carnegie Institution of Washington.)

gives rise to the outer coat of syncytiotrophoblast, and the solid tubes which reach almost to the myometrium become anchoring villi. The advancing cell columns tap the endometrial venules and capillaries, forming lacunae filled with sluggishly circulating maternal blood, the forerunner of the intervillous space.

These primary villi branch to form secondary villi, and later, through the process of sprouting, free-floating tertiary villi form. These will become the major surfaces for feto-maternal exchange. The mesoblast, formed from the primitive trophoblast, invades the cell columns to form a central stromal core. Within this core, capillaries form *de novo*. In the body stalk, which is to become the umbilical cord, the umbilical vessels advance into the mesoblastic cores and are tapped by the villous capillaries to establish the fetal circulation. In the area of placental attachment, the branching villi resemble trees; hence, the term *chorion frondosum* ("leafy") is applied to this area, to distinguish it from the *chorion laeve* ("smooth") which covers the remainder of the expanding conceptus. When the amniotic sac forms and expands, it pushes the chorion laeve against the opposite wall of the uterus, and the villi atrophy, leaving "ghost villi" within the chorion, which is the outer layer of the fetal membranes.

The study of the development of the placenta between 40 and 80 days after conception falls between the areas of the embryologist and the placentologist. The process is poorly understood, yet it is during this period when the major fetal cotyledons form and the maternal circulation is established. Reynolds has pieced together the available evidence and believes that the following events occur.

FORMATION OF THE MATERNAL VASCULAR UNITS

About six weeks after conception, the trophoblast invades some 40 to 60 spiral arterioles (about 20 in the rhesus monkey), and each vessel may be tapped in several places. The resulting pulsatile outflow, entering under a pressure of 40 to 50 mm. Hg, separates the chorionic plate from the basal plate and pushes the free-floating villi into sort of a tent shape, with a hollow center sparsely populated by villi. The thick walls of the tent, held down to the basal plate by the anchoring villi, are composed of closely packed secondary and tertiary villi, and since the fetal capillary pressure within the villous stroma is about 20 mm. Hg, they are erectile and kept distended. There are a dozen or so large tents, or major maternal vascular units, and 40 to 50 smaller ones. Inasmuch as the chorion frondosum contained some 200 primary and anchoring stem villi, about 150 are left as relatively functionless fetal cotyledons squeezed between the major maternal vascular units. As the tents crown and extend upward to the fetal surface of the placenta, these remaining cotyledons pull the maternal decidua upward into septa which more or less surround the vascular units.

The timetable for these various events is summarized in Table 8–1.

Table 8–1.   DEVELOPMENT OF THE HUMAN PLACENTA*

Days After Ovulation	Important Morphological-Functional Correlations
6 to 7	Implantation of blastocyst
7 to 8	Trophoblast proliferation and invasion. Cytotrophoblast gives rise to syncytium
9 to 11	Lacunar period. Endometrial venules and capillaries tapped. Sluggish circulation of maternal blood
13 to 18	Primary and secondary villi form; body stalk and amnion form
18 to 21	Tertiary villi, 2 to 3 mm. long, 0.4 mm. thick. Mesoblast invades villi, forming a core. Capillaries form *in situ* and tap umbilical vessels which spread through blastoderm. Feto-placental circulation established. Sluggish lacunar circulation
21 to 40	Chorion frondosum; multiple anchored villi which form free villi like "chandeliers." Chorionic plate forms
40 to 50	Cotyledon formation 1. Cavitation. Trophoblast invasion opens 40 to 60 spiral arterioles. Further invasion stops. Spurts of arterial blood form localized hollows in chorion frondosum. Maternal circulation established 2. Crowning and extension. Cavitation causes concentric orientation of anchoring villi around each arterial spurt, separating chorionic plate from basal plate 3. Completion. Main supplying fetal vessels for groups of second order vessels are pulled from the chorioallantoic mesenchyme to form first order vessels of fetal cotyledons 4. About 150 rudimentary cotyledons with anchoring villi remain, but without cavitation and crowning ("tent formation"). Sluggish, low pressure (5 to 8 mm. Hg) flow of maternal blood around them
80 to 225	Continued growth of definitive placenta. Ten to 12 large cotyledons form, with high maternal blood pressures (40 to 60 mm. Hg) in the central intervillous spaces; 40 to 50 small to medium cotyledons and about 150 rudimentary ones are delineated. Basal plate pulled up between major cotyledons by anchoring villi to form septa
225 to 267	Cellular proliferation ceases, but cellular hypertrophy continues until term

*(Adapted from Reynolds, S. R. M. Amer. J. Obstet. Gynec., *94*:432, 1966.)

GROSS ANATOMY OF THE MATURE PLACENTA

The "roof" of the placenta is called the chorionic plate. Through it course the major branches of the umbilical arteries and veins. The "floor" of the placenta, which is periodically indented by the maternal septa, is called the basal plate. These septa may or may not reach the area of the chorionic plate, so that separation of the maternal vascular units is not complete, and small quantities of maternal blood may flow from one unit into another, or may flow toward the periphery of the mature placenta, where there are villous-free spaces referred to as the "marginal sinus." A

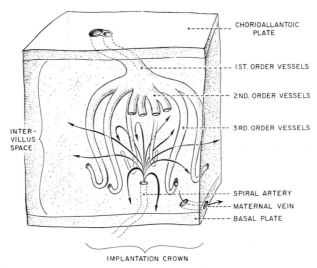

CHORIOALLANTOIC PLATE

1ST. ORDER VESSELS

2ND. ORDER VESSELS

3RD. ORDER VESSELS

INTER-VILLUS SPACE

SPIRAL ARTERY

MATERNAL VEIN

BASAL PLATE

IMPLANTATION CROWN

FIGURE 8-3. Schematic representation of the fetal and maternal vascular units. (From Reynolds, Amer. J. Obstet. Gynec., *94*:425, 1966. C. V. Mosby Co., St. Louis.)

schematic representation of the fetal and maternal vascular units is shown in Figures 8-3 and 8-4.

The DNA content of the total placenta ceases to increase when the fetus has attained a weight of about 2500 gm., or near the thirty-fourth week, indicating a cessation of cellular growth. The RNA content and placental weight continue to increase, however, and at term the placenta weighs about one-sixth the weight of the fetus, or approximately 500 gm.

The general appearance of the delivered placenta is shown in Figure 8-5. There are innumerable variations in shape and size, most of them of little physiological significance. In about 1 per cent of cases, the umbilical vessels may divide before reaching the chorionic plate, a condition called *velamentous insertion,* and this is hazardous for the fetus because the vessels are more easily torn, especially when they course over the lower uterine segment (vasa previa). A corrosion specimen of the fetal vessels of the term placenta is shown in Figure 8-6.

The Umbilical Cord

Under normal circumstances, the umbilical cord contains two arteries and a single vein. Rarely, a single umbilical artery is present, resulting either from primary aplasia or from atrophy of one artery during early development. This anomaly, occurring about once in 200 deliveries, is accompanied by a 15 to 20 per cent incidence of cardiovascular abnormalities in the fetus. The vessels are surrounded by a mucopolysaccharide material called Wharton's jelly, which is probably identical with inter-

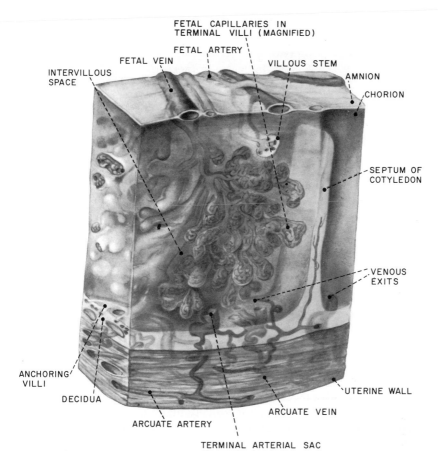

FETAL CAPILLARIES IN
TERMINAL VILLI (MAGNIFIED)

FETAL ARTERY

FETAL VEIN

VILLOUS STEM

INTERVILLOUS
SPACE

AMNION

CHORION

SEPTUM OF
COTYLEDON

VENOUS
EXITS

ANCHORING
VILLI

DECIDUA

ARCUATE ARTERY

ARCUATE VEIN

UTERINE WALL

TERMINAL ARTERIAL SAC

FIGURE 8–4. Schematic drawing of the utero-placental circulation. The terminal villi are greatly magnified for purposes of illustration. (Drawing by R. Sweet, based on sketch by E. W. Page.)

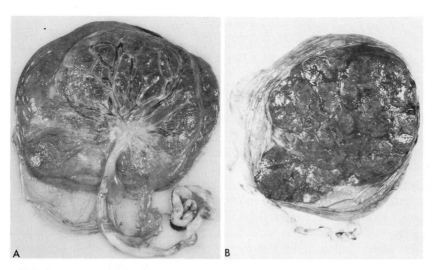

FIGURE 8–5. Photograph of a term placenta. *A,* Fetal surface. *B,* Maternal surface.

175

FIGURE 8–6. Corrosion specimen showing the elaborate fetal vasculature of a term placenta.

cellular ground substance elsewhere in the body. The cord is covered by amnion, but not by chorion, and it is likely that a considerable exchange of water and solutes occurs between the cord and the amniotic fluid. The umbilical cord varies from 30 to 90 cm. in length, with an average of 55 cm., but extreme cases in which it was less than 1 cm. or as long as 200 cm. have been reported. In one-fifth of all deliveries, the cord is looped once around the fetal neck (without increased fetal risk), and in 1 per cent of deliveries there are true knots in the cord (with a doubling of the perinatal loss).

The Villus

The functioning unit of the placenta, insofar as exchange is concerned, is the chorionic villus. Its gross appearance is shown in Figure 8–7. The total villous surface area may be estimated by planimetry, and according to Aherne and Dunnill, it averages 11 square meters at term. This figure, of course, is based upon a smooth surface, but ultramicroscopic studies reveal countless microvilli streaming from the surface of the syncytium, so that the effective surface may be enormously increased.

Light microscopy of villi in early pregnancy (Figure 8–8) reveals two distinct layers, an outer syncytial coat and an inner layer of cytotrophoblast. It is now certain that the latter gives rise to the former by a process of proliferation, followed by the formation of transitional or daughter cells, which then coalesce, with a loss of the cellular membrane. As the placenta matures, there is a steady reduction in the number of cytotrophoblastic cells,

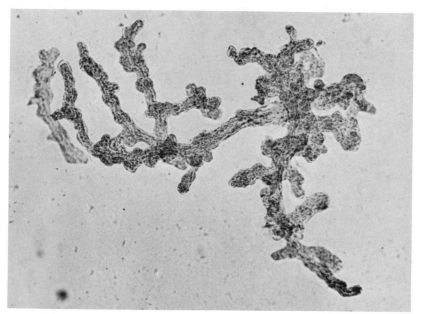

FIGURE 8–7. Photograph of a teased-out human chorionic villus. (Courtesy of Dr. M. C. Carr.)

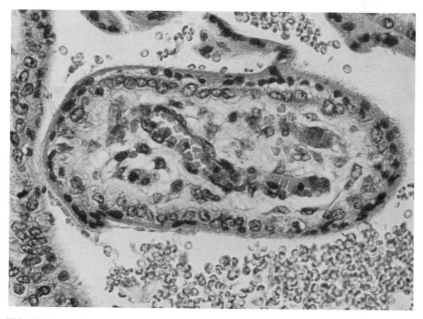

FIGURE 8–8. Cross-section of a villus in early pregnancy. (Courtesy of Dr. M. C. Carr.)

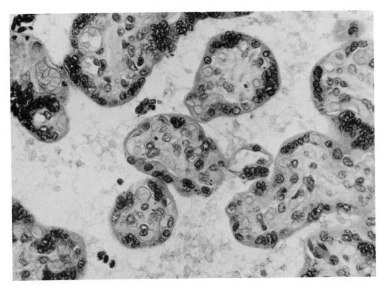

FIGURE 8-9. Cross-sections of villi from a term placenta. Note the apparent disappearance of the cytotrophoblastic layer. (Courtesy of Dr. M. C. Carr.)

until at term they are difficult to find unless one employs special stains or the electron microscope. The appearance of the villi in cross-section at term is shown in Figure 8-9.

At times the syncytial cytoplasm is so thinned out over portions of the villus that the stromal core appears naked. There are varying degrees of clustering of the syncytial nuclei, which in the early placenta may represent true sprouting. In the full-term placenta, however, formation of knots has been attributed to local hypoxia, and, according to Tominaga and Page, is a reversible phenomenon. Fox has shown that placental explants subjected to hypoxia reveal a proliferation of the cytotrophoblast, together with a degeneration of the syncytium, suggesting that the Langhans cells may thrive under relatively anaerobic conditions, and that this proliferation secondary to hypoxia may represent a means of repair following hypoxic damage.

These knots are constantly breaking off, and they may be found in the maternal blood leaving the uterus. Presumably the trophoblastic fragments are filtered out by the maternal lung and lysed there, for they cannot be recovered in the peripheral circulation. The effect of this trophoblastic deportation upon the maternal organism is not understood, but when the process is excessive—a possibility in preeclampsia and eclampsia—it may well initiate a process of intravascular coagulation in the greater circulation.

The ultramicroscopic appearance of a villus margin at term is shown in Figure 8-10.

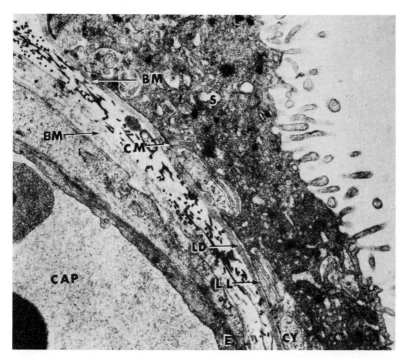

FIGURE 8–10. Electron microscopic appearance of a chorionic villus at term. The basement membrane (BM) consists of a lamina lucida (LL) and a lamina densa (LD) and lies immediately adjacent to the cytoplasmic membranes of both the syncytiotrophoblasts (S) and the cytotrophoblasts (CY). Portions of two red cells may be seen within the capillary (CAP), which is lined by endothelium (E). Note the microvilli which extend into the intervillous space. (From Verbeek, Robertson and Haust, Amer. J. Obstet. Gynec., 99:1136, 1967. C. V. Mosby Co., St. Louis.)

Ultrastructure of Trophoblast

The syncytiotrophoblast is a plastic ameboid mass containing nuclei which are not separated by plasma membranes, so that in one sense, the entire mass, covering some 11 square meters of villous surface, is a single "cell." The cytoplasm is rich in rough and smooth endoplasmic reticulum, Golgi bodies, and other organelles associated with protein synthesis and export. The surface in contact with maternal plasma is characterized by microvilli which contain pinocytotic vesicles, indicating that pinocytosis, the "drinking in" of whole droplets, is one method of placental transport. Mitotic figures in the nuclei have never been observed.

Within the syncytium are vesicles which communicate with conduits in the endoplasmic reticulum, presumably forming a continuous channel between the intervillous space and the fetal capillary, through which materials might be processed in either direction. The large mitochondria are probably related to the sources of energy required for the active transfer of amino acids and other nutrients. The smooth endoplasmic reticulum is allegedly concerned with the segregation and transport of non-protein materials, whereas the rough endoplasmic reticulum synthesizes poly-peptides, which are then assembled in the Golgi apparatus and packaged into secretory granules.

The Langhans cell, or cytotrophoblast, is basically a simpler structure with a large nucleus and prominent nucleoli and mitochondria, but the cytoplasm is almost free of endoplasmic reticulum. The ultrastructure is like that of an embryonic, poorly differentiated cell specializing in growth and reproduction, but not in the elaboration of proteins or endocrine products for export.

ORIGIN OF PLACENTAL HORMONES

From the standpoint of the electron microscopist, the syncytiotropho-blast is the mature form and is therefore the site of production of steroid hormones, chorionic somatomammotropin (hPL) and human chorionic gonadotropin (hCG). Furthermore, fluorescent antibody studies have located hCG within the syncytium, and not in the cytotrophoblast, al-though this might be interpreted as storage rather than production. Never-theless, the rate of production of hCG correlates with the volume of cyto-trophoblast rather than with the volume of syncytium, a fact which is yet to be explained satisfactorily. During the process of transformation from cyto- to syncytiotrophoblast, there are transitional cells which may be the primary source of hCG.

THE FETO-MATERNAL JUNCTION

In placentas with a highly invasive trophoblast, such as that of man, there is considerable necrosis of adjacent endometrium and a deposition of fibrinoid material. Thus, in the mature placenta, trophoblast is regularly separated from maternal cells by this amorphous layer. Kirby believed that this may form an antigenic barrier which might protect the villi from im-munologic rejection. Wynn points out, however, that this amorphous "barrier" is not present in a number of mammalian species. A more or less continuous layer of fibrinoid in the decidua of the basal plate is known as Nitabuch's layer, and it is at this level that placental separation normally occurs in the third stage of labor. As stated by Wynn, the areas of junction between fetal and maternal cells constitute "an active placental battleground on which there are heavy casualties on both sides."

THE VILLOUS STROMA

In the villous core of the early placenta, branching fibrocytes are separated by a loose matrix of ground substance. The mature villus has a denser stroma with closely packed spindle cells and more voluminous capillaries. Within the stroma are round cells with vesicular nuclei and vacuolated cytoplasm, called Hofbauer cells, which are thought by some to be phagocytic in nature.

Traditionally, the placenta is supposed to be free of nerves. Jacobson and Chapler, however, have described nerve trunks and fine fibrils surrounding the fetal capillaries, and ganglia identical in staining qualities and appearance with autonomic nerve structures elsewhere in the body. In a variety of species, these have been found in association with the umbilical artery of the cord, and presumably connect with the fetus. If they do, indeed, exist, their function is unknown.

THE INTRALOBULAR CIRCULATION

The trophoblast invades maternal arterioles to establish the utero-placental circulation. In so doing, it protrudes into the arteries, even those within the myometrium, obstructing the circulation in some. This effectively reduces the number of arterial openings into the intervillous space. It is noteworthy that, despite the abundance of maternal veins draining the basal plate, these are not invaded in a like manner by trophoblast, suggesting the possibility that the primitive trophoblast might possess the inherent property of seeking sources of oxygen. The venous exits draining the maternal vascular units are numerous and appear to be randomly distributed over the basal plate. A subdecidual venous plexus is, like the spiral arterioles, surrounded by myometrium; hence uterine contraction compresses both the arteries and veins and profoundly affects the utero-placental circulation.

As shown by direct punctures of the intracotyledonary spaces in the pregnant monkey (Reynolds et al.), the pressure of the maternal blood is highest (about 17 mm. Hg) in the center of the vascular unit and the blood is arterial in nature. As one probes further toward the periphery of the cotyledon, the pressure is lower (about 10 mm. Hg) and the blood is venous in character. With uterine contraction, all pressures rise and tend to become nearly equal. At the same time, the pressure within the fetal capillaries increases to the same degree as the amniotic fluid pressure rises, thus preventing their collapse.

EFFECT OF UTERINE CONTRACTIONS

Arterial blood enters the intervillous space in a continuous, pulsatile flow and is disbursed upward and laterally around the villi, which act like baffles to produce a mixed, slow stream, favoring the exchange of gases and solutes across the villous surface. The angiographic studies of Ramsey in the monkey, and those of Borell et al. in a few human subjects, reveal that when a bolus of radiopaque material is injected into the aorta, some 15 to 25 "bursts" or "spurts" are noted within the human placenta (fewer in the

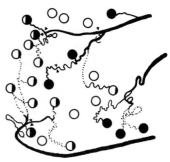

FIGURE 8–11. The complete filling of maternal vascular units, indicated by solid circles, is shown by angiography during uterine relaxation (left). During a uterine contraction (right), only a few are completely filled, and the remainder either are partially filled or cannot be visualized. (Modified from Borell et al., Amer. J. Obstet. Gynec., *93*:51, 1965. C. V. Mosby Company, St. Louis.)

monkey), which quickly form rings as each intervillous lake is diluted by blood free of contrast medium. The dye is then seen streaking out through numerous veins. During the height of a uterine contraction, there is a sharp reduction in the number of spurts visualized (Figure 8–11) and no observable venous escape. Thus, both inflow and outflow of maternal blood are sharply reduced during vigorous contractions of labor.

Ramsey and her co-workers injected both the umbilical and uterine arterial circulations simultaneously in the pregnant monkey, recording the results by cineangioradiography. They found that the fetal cotyledons looked like fine cottony balls. Into the center of these balls came the spurts of maternal blood, which then spread like expanding doughnuts to the periphery of the fetal cotyledons.

MAGNITUDE OF FETO-PLACENTAL BLOOD FLOW

Assali and his group made direct measurements of umbilical vein flow of human fetuses during the middle trimester, and if one extrapolates the results to term, the feto-placental blood flow for a 3200 gm. baby would be about 350 ml. per minute. In all likelihood, the true value in the undisturbed subject is somewhat higher.

MAGNITUDE OF UTERINE BLOOD FLOW

The uterine blood flow in women has been estimated by the nitrous oxide method or by an electromagnetic flowmeter around one uterine artery. Expressed per unit weight of the uterus and its contents, the reported flows are variable (8 to 20 ml. per 100 gm. of uterus), but relatively constant throughout the course of gestation. In a uterus at term, with an average weight fetus, the total uterine flow is from 500 to 750 ml. per minute. The distribution of this flow in the human subject is not known. In pregnant sheep near term, about 85 per cent flows through the placental cotyledons, and the balance supplies the myometrium and endometrium. In women near term, the percentage of uterine blood flow that courses through the intervillous space must be between 60 and 90 per cent, and no doubt this fluctuates from time to time, depending upon the state of myometrial tone. Thus, the feto-placental flow and the utero-placental flow are of similar magnitude. It should be noted that maternal blood is unevenly distributed to the diffusing surfaces of the placenta, which makes the calculation of feto-maternal gradients most difficult.

Factors Influencing Uterine Blood Flow

More than any other single factor, the welfare of the fetus is dependent upon an adequate perfusion by maternal blood of the effective exchange areas in the placenta. Inasmuch as the supply of most essential nutrients, as well as oxygen, is more flow limited than diffusion limited (Meschia, Battaglia et al.) factors which reduce uterine blood flow become extremely important. Chronic reductions result in intrauterine growth retardation, and acute reductions of the utero-placental circulation lead to fetal hypoxia or death.

Beker has demonstrated in several species, including man, that the multiparous uterus has a more adequate circulation than the primigravidous uterus. This may account for the fact that in postmature pregnancies, when the nutritional supply line may become less than adequate for a normal fetal growth rate, primigravidous patients have a higher perinatal mortality than multiparous patients.

Normal uterine blood flow is dependent upon normal perfusion pressures, so that with severe maternal hypotension, such as might occur with spinal anesthesia or secondary to hemorrhage, the uterine flow is seriously compromised. In pregnant ewes, hemorrhagic shock results in not only a decrease of perfusion pressure but also an increased uterine vascular resistance. Restoration of the blood pressure with phenylephrine does not restore the flow to normal levels (Greiss). On the other hand, when Assali and Morris induced acute hypotension in pregnant ewes with high spinal anesthesia, uterine flow fell to low levels but was restored to normal by the administration of metaraminol.

Acute reductions of uterine blood flow occur during uterine contractions. The reduction is proportional to the intensity of the contraction, and is due to an increased uterine vascular resistance resulting from myometrial compression of the uterine arterioles. Fortunately, normal labor consists of intermittent contractions of 70 to 100 seconds' duration so that recurring hypoxic episodes are both mild and of short duration. Tetanic contractions, such as those which might be induced by an overdosage of oxytocin, reduce the flow for longer periods of time and may lead to serious hypoxia or fetal death.

In the sheep, both epinephrine and norepinephrine increase uterine vascular resistance and reduce flow. Unlike the catecholamines, angiotensin leads to an increase of both perfusion pressure and uterine flow.

In women, the anatomic relationships of the viscera in late pregnancy are such that lying in the supine position may lead to compression of the vena cava, interfering with venous return to the heart and reducing cardiac output. This may be accompanied by hypotension, bradycardia, and a sharp reduction in uterine flow, with fetal distress (see Chapter 12). Poseiro has shown that in some patients, uterine contractions may compress the internal iliac artery, reducing the flow to the uterus. It is likely that maximal uteroplacental flows are achieved when a patient is lying on her side and the uterus is relaxed. Although no measurements have been made, it is quite possible that uterine flow is decreased in the erect position and with exercise.

UTERINE GROWTH AND ACCOMMODATION

The non-gravid uterus of a nulliparous woman weighs about 40 gm., that of a multipara weighs from 60 to 70 gm., and the gravid uterus at term weighs about 680 gm. The early growth following implantation is due to the combined effects of estrogen and progesterone, which arise from both the corpus luteum of pregnancy and the developing syncytiotrophoblast, and is due to both cellular proliferation and hypertrophy. The uterus may double in size even though the conceptus is developing in an extrauterine site, such as a tube or the abdominal cavity. Further growth of the uterus is due solely to hypertrophy, and requires an expanding conceptus. At three and a half to four months, the uterine wall has thickened from about 10 to 25 mm., but thereafter the wall thins, until at term it is 5 to 10 mm. in thickness.

The individual myometrial cells increase from 0.05 mm. to between 0.2 and 0.6 mm. in length. There is a progressive increase in the content of actomyosin. The myofilaments, according to Rosenbluth, are oriented obliquely to the long axis of the muscle fiber, an arrangement which allows smooth muscle to exert a relatively large force through a short distance at low velocity. This may also account for the ability of the myometrium to sustain a forceful contraction for a prolonged period of time with a minimal expenditure of energy.

Between the second and fourth month, the shape of the uterine fundus is essentially spherical. Thereafter, conversion to an elliptical or cylindrical shape occurs, and distention occurs primarily in a cephalad direction, reaching the level of the umbilicus at about five months and the level of the xiphoid process of the sternum at eight months. Thereafter, the height of the fundus may decrease slightly as the fetal head sinks into the true pelvis. As emphasized by Reynolds, this conversion from a spherical to a cylindrical shape has important implications regarding intrauterine pressures. In a sphere, tension is a geometric function of the radius of the curvature, whereas in a cylinder it is a linear function. Thus intrauterine pressure (in the absence of contractions) reaches its peak just before mid-pregnancy, decreases to about 10 mm. Hg of amniotic fluid pressure, and remains there until term, despite the continued increase of volume. So it is that when the internal os of the cervix is lacerated or otherwise incompetent, there may be a "blow-out" of the fetal membranes in the middle trimester, with late abortion. This may occur repeatedly until the condition is treated at 12 to 15 weeks of pregnancy by placing a snug purse-string suture of heavy, nonabsorbable suture material around the internal os, an operation which may salvage the pregnancy.

Plasticity of the Uterus

The cervix is composed of 90 per cent connective tissue and collagen, with 10 per cent or less of myometrium. The reverse is true of the uterine fundus, and the area of transition from connective tissue to myometrium may occupy a centimeter or so of the lower uterine segment. The connective tissue framework of the uterus, particularly of the cervix, undergoes a change in its viscoelastic plasticity during pregnancy, so that at term it combines the properties of a rubber band with those of salt-water taffy. With a steady slow pull, the cervix will stretch with but little rebound, and will thus progressively dilate.

To summarize briefly, the occurrence of a successful implantation is in itself the most stringent measure of the integrity of a woman's endocrine system. Following implantation, a most remarkable series of events is set into motion. The development of the placenta, its gradual modifications designed to facilitate transport, its progressive assumption of endocrine dominance, its aging and its eventual death all within the space of months constitute a chronicle hardly less amazing than embryogenesis itself. In the meantime, the uterus must undergo transformations of form and function which are unparalleled by any other adult organ. It must, on the one hand, be able to retain the conceptus for nine months or so, and on the other hand, be able to adapt itself for a prompt evacuation of its contents without injury to the fetus, and then return by involution to its prepregnant status.

SUGGESTED SUPPLEMENTARY READING

Boyd, J. D., and Hamilton, W. J. The Human Placenta. W. Heffer & Sons, Ltd. Cambridge, England, 1970.
  This is a beautiful volume summarizing the life work of two outstanding British anatomists. There are 99 color plates and 348 black and white illustrations.
Harris, J. W. S., and Ramsey, E. M. The morphology of human utero-placental vasculature. Contr. Embryol. Carneg. Inst., *38*:43, 1966.
Hartman, C. G. (Ed.). Mechanisms Concerned with Conception. Pergamon Press, New York, 1963.

Particularly pertinent are the chapters by C. R. Austin, "Fertilization and Transport of the Ovum," and by B. G. Boving, "Implantation Mechanisms."

Morris, J. A. Vascular physiology of the uterus. In Wynn, R. M. (Ed.). Cellular Biology of the Uterus. Appleton-Century-Crofts, New York, 1967.

Wynn, R. M. Morphology of the placenta. In Assali, N. S. (Ed.). Biology of Gestation. Academic Press, New York, 1968.

Particular emphasis is given to electron microscopy of the placenta, but there is a full discussion of implantation as well.

Chapter 9

INTRAUTERINE FETAL NUTRITION AND GROWTH

FETAL METABOLISM

The human fetus requires substrates for the production of energy and for growth. The most important energy-producing metabolic pathways are aerobic (citric acid cycle), requiring oxygen. Traditionally, glucose has been regarded as the primary energy substrate supplied to the fetus by the mother. However, recent studies by Battaglia and his colleagues have shown that glucose uptake measured in the sheep by arteriovenous difference accounted for less than 50 per cent of fetal oxygen consumption in the final month of gestation. This does not permit the inference that all tissues utilize non-glucose substrates for oxidative metabolism. Indeed, present evidence suggests that the fetal brain obtains all of its calories from glucose, except when under unusual conditions of stress or severe starvation. This would imply that other tissues must obtain more than 50 per cent of their aerobic calories from sources other than glucose, or that non-glucose substrates are converted into glucose (gluconeogenesis) for subsequent catabolism. For example, in fetal liver (a very important organ for gluconeogenesis) the conversion of lactate, pyruvate or amino acids (such as alanine) into glucose is a very active process, which could supply glucose for other tissues, especially the brain.

Amino acids may be one major source of calories in the fetus. Battaglia estimates that about 25 per cent of total oxygen consumption in the sheep fetus utilizes the carbon chains of amino acids as substrate. Stated another way, as much as 40 per cent of the amino acids crossing the placenta to the fetus are destined not to be incorporated into fetal proteins but to be metabolized. The carbon chain enters the citric acid cycle and is metabolized to carbon dioxide. The nitrogen is ultimately returned to the maternal circulation as urea. Amino acids are actively transported from mother to fetus, and

187

the large protein reservoir of the mother would ensure a constant supply of this substrate even in starvation. Maternal starvation in sheep produces a prompt and significant fall in glucose concentrations in both maternal and fetal blood. In contrast, the urea concentrations rise and remain above baseline values, suggesting that amino acids are being utilized for energy.

Another important source of calories in the human fetus may be free fatty acids. Placental transfer of free fatty acids has been observed *in vivo* in sheep, rabbits, guinea pigs and primates, and *in vitro* in perfused human placenta. This transfer is primarily gradient-dependent; thus conditions which predispose to high concentrations of free fatty acids in maternal blood (e.g., diabetes mellitus) would result in enhanced transport of free fatty acids to the fetus. In species having relatively high transfer of free fatty acids (guinea pig, human), the newborn has a large adipose tissue mass (10 per cent and 16 per cent of body weight, respectively). In animals with poor free fatty acid transfer (rat), the adipose tissue mass composes only 1 per cent of the body weight of the newborn. Undoubtedly most of the free fatty acid transferred to the fetus in the fed state is converted to triglyceride and stored in adipose tissue. Thus free fatty acids would not serve as substrates for oxidative metabolism to any significant degree under normal conditions.

Other experiments by Battaglia and his colleagues, using the catheterized fetal sheep preparation, indicate that lactate is produced in the placenta from glucose and secreted to the fetus, where it is used as a substrate. Their calculations suggest that it is nearly as important as glucose as an energy source for the fetus (1.2 gm. lactate per kg. fetus per day versus 1.5 gm. glucose per kg. fetus per day). Whether this is peculiar to the sheep or true for mammals in general remains to be seen.

Insulin, of course, is required for metabolism of glucose by certain tissues, and its source is the fetal pancreas. There is general agreement that no significant quantities of maternal insulin reach the fetus. Under normal circumstances, the fetal supply of glucose is relatively constant, night and day, and the responses of the fetal pancreatic islets to fluctuations of glucose concentrations in the blood are absent or at least blunted when compared with those in the adult. This is not so, however, with the newborn of a diabetic mother, because the blood insulin levels are higher and the insulin response to hypoglycemia is greater than it is in the normal newborn. Infants of diabetic mothers are sometimes accelerated in growth, and it is presumed that this is due to (1) recurring episodes of hyperglycemia, (2) increased transfer of free fatty acids from mother to fetus and (3) hypersecretion of insulin in the fetus. The insulin would serve to promote the transfer of glucose and amino acids into muscle cells, favoring deposition of glycogen and protein. Insulin would also promote transfer of free fatty acids and glucose into adipocytes. The glucose in the adipocyte can serve as a precursor of glycerol phosphate (Figure 15–3), which combines with free fatty acids to form triglyceride. Thus insulin can serve as a growth-promoting agent in the fetus, favoring the accumulation of protein, glycogen and triglyceride.

So long as all essential amino acids are available to the fetus in appropriate proportions and all energy needs are met, protein synthesis by the fetus should proceed at normal rates for any given period of growth. In experimental animals, single deficiencies of one amino acid in the maternal diet will slow fetal growth as well as produce anomalies, but this situation is unlikely to exist in human populations. Severe limitations of utero-placental blood flow will, on the other hand, limit the supply of all amino acids, as well as of glucose and other foodstuffs essential for growth, and will result in fetal undernutrition.

As compared with most mammals, the human fetus after midpregnancy is generously endowed with stores of glycogen and of body fat. Unlike those of the adult or even the newborn, the epithelial tissues of the fetus (skin, lung, renal epithelia, etc.) have significant stores of glycogen which may play a role in carbohydrate homeostasis. At birth, the liver contains about 100 mg. of glycogen per gram, and the total lipids in the fetus average 161 gm./kg. These reserve fuels are utilized by the newborn during the fasting state, but the fetus *in utero* is in a constantly fed state. Although both the glycogen and lipid stores are in a dynamic state, undergoing constant turnover, it is unlikely that these hoards of potential energy are used as energy sources under normal circumstances.

MECHANISMS OF PLACENTAL TRANSFER

With the exception of a few metals, such as iron, all gases and nutrients move across the placental membrane in both directions in quantities which are, for the most part, considerably greater than the fetal requirements for metabolism and growth. The trophoblastic layer is the parenchyma of the membrane and is probably the limiting factor for rates of diffusion and active transport. Other components of the placental membrane, such as the underlying connective tissue layers and the endothelium of the fetal capillaries, may be limiting membranes only for large molecules and cells.

There are few compounds, endogenous or exogenous, that are unable to traverse the placenta in detectable amounts. The placenta may act as a true barrier when the molecule has a certain size, configuration and charge (e.g., heparin), or when enzymes within the trophoblast are able to alter the molecular structure during transport (e.g., simple amines), or when the compound, though small, is firmly bound to some maternal cell (e.g., carbon monoxide). With these exceptions, it may be predicted that drugs or other chemicals present in maternal plasma will also be found in fetal plasma. Most problems involving placental transfer, therefore, become quantitative rather than qualitative in nature. Is a given substance transported in a quantity which is of physiological or pharmacological significance? Is its transport diffusion-limited or flow-limited, or both? Can a limitation of its transport be a factor in causing fetal malnutrition? These are the types of questions which modern placentologists are trying to answer.

There are four common mechanisms of transport across the placenta:

diffusion, facilitated diffusion, active transport and pinocytosis. These are not peculiar to the placenta, of course, but are mechanisms common to other epithelial membranes, such as the gastrointestinal or renal tubular epithelium. Many of our deductions about human placental transfer are derived by analogy from experiments with various rodents, which may or may not be applicable to the human species. Fortunately, the problems of intrauterine fetal nutrition and growth in the rhesus monkey are so similar to those of the human subject that much of the information derived from the higher primates is probably applicable to man.

Simple Diffusion

Most small molecules move across a membrane because of chemical or electrochemical gradients. When these gradients cease to exist, then the rates of exchange across the membrane become equal in both directions. The quantity of solute transferred by simple diffusion is described by the Fick diffusion equation;

$$\frac{Q}{t} = \frac{K\,A\,(C_1 - C_2)}{L}$$

where $\frac{Q}{t}$ is the quantity transferred per unit of time, K is the diffusion constant for the substance in question, A is the total surface area available for exchange, $C_1 - C_2$ is the concentration gradient across the membrane which exists at the time of the transfer, and L is the thickness of the membrane.

One of the difficulties in utilizing such an equation is that the values for C_1 and for C_2 cannot be readily measured. Each value, of course, is affected by the rates of blood flow on both sides of the membrane. In sheep, it has been shown that close approximation of C_1 for oxygen is one-half the sum of the concentrations in the uterine artery and in the uterine vein.

In the experimental animal, the rate of umbilical blood flow and the rate of uterine blood flow can be determined simultaneously by the infusion of antipyrine into a fetal vein. Under these circumstances, the quantity of a solute utilized by the fetus can be determined by multiplying the rate of flow by the concentration difference between the umbilical vein and the umbilical artery. By the same token, the amount of solute extracted by the entire uterus may be determined by multiplying the rate of uterine blood flow by the concentration difference between the uterine artery and the uterine or ovarian vein. The difference between the values obtained by these two methods represents the amount of solute utilized by the myometrium and the placenta.

It is generally agreed that oxygen and carbon dioxide traverse the placenta by simple diffusion alone. The average chemical gradient $(C_1 - C_2)$ is about 20 mm. Hg for oxygen and about 5 mm. Hg for carbon dioxide. The oxygen gradient is high not only because of the lower diffusion constant (K), but also because almost 40 per cent of the oxygen extracted by the uterus is utilized by the placenta and myometrium. Both gradients are higher

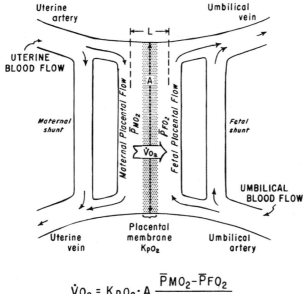

$$\dot{V}O_2 = KpO_2 \cdot A \frac{\overline{P}MO_2 - \overline{P}FO_2}{L}$$

$\dot{V}O_2$ = rate of oxygen transfer
KpO_2 = placental diffusion constant for oxygen
A = placental surface area
$\overline{P}MO_2$ = average oxygen tension in maternal placental blood
$\overline{P}FO_2$ = average oxygen tension in fetal placental blood
L = average transplacental diffusion distance for oxygen

FIGURE 9–1. Schematic diagram of the placental circulations using oxygen diffusion as an example. (From Metcalfe et al., Physiol. Rev. *47*:782, 1967.)

than one would calculate from the total surface area because the utero-placental blood is unevenly distributed to the areas of potential exchange. A schematic representation of oxygen transfer is shown in Figure 9–1. The placental membrane actually approaches the efficiency of the lung for gas transfer. The quantity of oxygen reaching the fetus is primarily flow-limited rather than diffusion-limited, so that when we consider fetal hypoxia we must think of factors which diminish either the uterine blood flow or the fetal blood flow, unless the area for exchange is suddenly diminished by such an event as placental separation. Under normal conditions, the human uterus near term extracts between 20 and 35 ml. of oxygen per minute. Something over half of this is utilized by the fetus, and the balance is used by the uterus and placenta.

Facilitated Diffusion

Some compounds, notably glucose, are not transported across the placenta against a chemical gradient, and yet the rate of transfer is significantly more rapid than would be the case with simple diffusion. By analogy with other membranes, such as the red cell, it is assumed that the trophoblast contains a glucose transport system, a "down-hill rapid transit," which may become saturated at very high concentrations. Under normal conditions, the concentration of glucose in the fetal plasma is about 60 to 70 per cent of the maternal concentration. This down-hill gradient results chiefly from the rapid utilization of glucose, the primary source of energy for fetal metabolism.

Active Transport

In the steady state, when the net transfer of a compound to the fetus occurs against a chemical gradient, one must assume either that it is more firmly bound by fetal blood than by maternal blood (which is rarely true) or that an active, energy-requiring transport system exists in the trophoblastic membrane. This appears to be the case with the essential amino acids and with the water-soluble vitamins. Common features of any active transport system are these: (a) the system may be saturated at very high concentrations; (b) structurally similar molecules may compete for transport; (c) transfer may be inhibited by metabolic poisons which reduce the available energy; and (d) the transport process is stereo-specific. In the case of the human placenta, stereo-specificity for amino acids has been demonstrated by showing that the net transfer of natural L-histidine is more rapid than the transfer of its optical isomer D-histidine. This demonstration, plus the facts that the average amino acid concentration in fetal plasma is twice that of the maternal plasma and that there is no greater binding by fetal blood, provides strong evidence for the existence of an active transport system.

Pinocytosis

This process, observed chiefly by electron microscopy, is a "drinking in" or engulfing by the cellular membrane of intact microdroplets of plasma which appear as vacuoles within the cytoplasm. The material is somehow sorted out, sometimes channeled through ultramicroscopic canals, frequently transported unchanged into the fetal capillaries, and at other times (as with foreign antigens) destroyed. The process is considered to be of importance in the placental transfer of globulins, lipoproteins, phospholipids, and other molecules too large for diffusion and for which no other active transport mechanism exists. In all likelihood, the amounts transported are sufficiently small that they are not of much significance for fetal nutrition, but they may be of considerable importance from the immunologic standpoint.

Other Modes of Transport

There are times, particularly during labor, when intact fetal red blood cells escape into the maternal circulation, presumably through microscopic breaks in the placental membrane. This is the basis for isoimmunization (see Chapter 11). Labelled maternal red blood cells have also been found in the fetal circulation, which is difficult to understand because the pressure within the fetal capillaries is greater than that within the intervillous space, at least with the patient recumbent and the uterus at rest. When the woman stands or when the uterus contracts, however, the hydrostatic pressure gradients may be transiently reversed.

Some cells, such as maternal leukocytes, or organisms, such as *Treponema pallidum* (causing syphilis), may traverse the placenta under their own power. A large number of viruses may infect the fetus, and it is not known precisely how they cross the placenta, although the pinocytotic process might suffice to carry a few viral particles, which then multiply rapidly within fetal cells.

PLACENTAL TRANSPORT OF SPECIFIC MATERIALS

The compound exchanged most rapidly and freely between mother and fetus is water, but the factors which govern the net transfer of water in either direction are multiple and complex. A "bulk" transfer of water may occur in response to minor changes either of hydrostatic pressure or of osmotic gradient, and when this occurs there will be a "solute drag," so that any given solute may be transferred in amounts which are in addition to its calculated diffusion rate. Furthermore, each solute transported to and utilized by the fetus liberates its molecules of hydration, which further complicates any calculations relative to water exchange. Finally, the oxidation of glucose and other nutrients results in the production of water molecules.

Electrolytes are also freely exchanged across the placenta, each at its own specific rate. Although a "sodium pump" may be operating in the placental membrane, the concentrations of electrolytes in maternal and fetal plasma are essentially identical.

There is little or no transfer of maternal cholesterol, triglycerides or phospholipids. There is, on the other hand, a definite — though probably relatively small — transport of free fatty acids, possibly enough to account for much of the fat which is ultimately deposited in the fetus.

Albumin and the alpha and beta globulins of maternal origin reach the fetus in very small quantities, but many of the gamma globulins, notably those of the IgG (7S) class, are readily transported to the human fetus despite their large molecular size. They serve to protect the fetus passively until, after birth, the newborn gradually replaces the maternal immunoglobulins with its own. Protein hormones of maternal origin do not reach the fetus in significant amounts, except for a slow transport of thyroxine and triiodothyronine. Unconjugated steroid hormones traverse the placenta rather freely, unless they are specifically and firmly bound to proteins, in which case they have slow transport. Steroids conjugated with glucuronide or sulfate cross the placenta with difficulty unless the ester bond is cleaved by placental enzymes.

The water soluble vitamins are concentrated in the fetal plasma and are transported by unique mechanisms. For example, ascorbic acid as such cannot cross the placental membrane, but dehydroascorbic acid crosses readily and is then reduced to ascorbic acid by the fetus. Riboflavin is accumulated in fetal blood by a mechanism which involves its conversion from flavin adenine dinucleotide to free riboflavin in fetal plasma by placental enzymes.

RATE OF FETAL GROWTH

After the initial lag period, which lasts until the appearance of the primitive streak (which also corresponds to the time when placental function becomes established), the rate of fetal growth in both mammals and birds conforms to a cubic law as related to time. The equation is:

$$W = a \, (t - t')^3$$

where W is the weight of the fetus on any given day of gestation, a is a constant expressing the rate of supply of nutrients per unit of fetal surface area, t is the day of gestation and t' is the lag period. For man, the value of a is 0.24×10^{-6} and the value of t' (when calculated from the first day of the last menstrual period) is 36 (Payne and Wheeler). For large mammals, the value for a is considerably larger — 16 times as great in the cow, for example. A decrease in the value of a toward the end of gestation will result in intrauterine growth retardation.

The rate of growth is enormous at first, and progressively slows as pregnancy progresses. For example, in the first month after fertilization, the human zygote increases a million times in weight, whereas in the last month of gestation the rate of increase is 0.3.

The rate of growth for the human fetus during the last trimester is shown in the curves constructed by Gruenwald (Figure 9–2). It will be noted that the rate deviates from the straight line after 38 weeks (except in Sweden) and particularly deviates after 40 weeks, when the fetus is postmature. This is presumed to reflect inadequate fetal nutrition toward the end of pregnancy, and may be due to changes in the placenta or to a failure of the uteroplacental circulation to keep pace with the total nutritional requirements. Note also that the deviation is greater in smoking mothers and among mothers of low socioeconomic status, and that the effects of the two are additive.

It is particularly noteworthy that in multiple pregnancy, a deviation from the straight line of growth occurs when the combined weight of the fetuses is about 3500 to 4000 grams. Thus, the birth weight of an individual triplet at term is considerably below that of a twin, and the birth weight of a twin is considerably below that of a single fetus. Again, it is evident that the total requirements of twins or triplets in late pregnancy exceed the capability of the nutritional supply line.

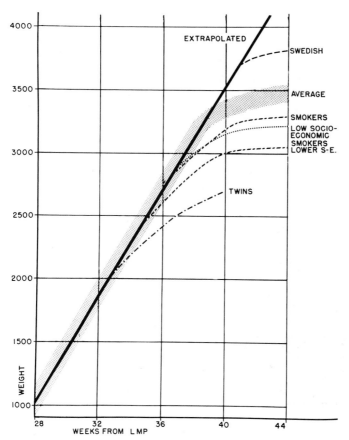

FIGURE 9-2. Deviations from the straight line growth curve for the human fetus. "Average" is for the United States. Note that the effects of low socioeconomic class and of smoking are additive. (From Gruenwald, Amer. J. Obstet. Gynec., *94*:1112, 1966. Courtesy of C. V. Mosby Co., St. Louis.)

FACTORS WHICH MAY REDUCE THE RATE OF FETAL GROWTH

Maternal Nutrition

In most polytocous animals in which the total weight of the litter is large relative to the weight of the mother, undernutrition has a marked effect on fetal birth weight. This is noted only to a minor degree in the higher primate species, including man. In all species, the effects of maternal undernutrition on fetal growth are only operative in about the last one-sixth of gestation. On many occasions in the past during periods of war or famine, reduced birth weights have been reported, but few observers have differentiated between the effects of an increased prematurity rate and the influence of diet. Such a differentiation was made by Gruenwald and his Japanese coworkers, who

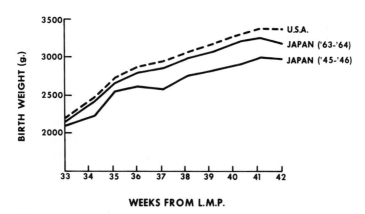

FIGURE 9–3. Fetal weight curves in Japan during a year of severe food deprivation compared with those during a year of prosperity. (Modified from Gruenwald et al., Lancet, *1*:1026, 1967.)

studied fetal weights for 1945 to 1946 (a period of severe food deprivation) and for 1963 to 1964 (a year of good nutrition for the population as a whole). Figure 9–3 compares the results with the average fetal weights for the United States. At 33 weeks, no differences are noted, and at term (40 weeks) the mean difference between the years of war and peace is only 300 gm., or about 10 per cent.

Overnutrition of the mother (total weight gains exceeding 26 pounds) produces birth weights which are about 10 per cent greater. It may be concluded that marked under- or over-nutrition of the mother affects fetal growth rates to a slight degree, and then only during the last six or eight weeks of pregnancy.

Severe *malnutrition,* referring now to a poor quality of diet rather than to the total caloric content, may well increase the frequency of such obstetrical complications as prematurity, premature separation of the placenta, megaloblastic anemia or iron deficiency anemia and eclampsia. Poverty may be associated with undernutrition, malnutrition, or both, and it is difficult to separate one from the other. In a study of 445 consecutive autopsies on stillborn or newborn infants, Naeye et al. have shown that poverty of the parents, as measured by U.S. poverty-index tables (which are actually generous compared with those of other countries), resulted in infants who were 15 per cent smaller than those of non-poor parents and showed multiple anatomic evidences of prenatal undernutrition.

The subject of optimal maternal nutrition will be discussed in Chapter 13.

"Placental Insufficiency"

Intrauterine fetal growth retardation may occur as the result of pathologic changes in the placenta, such as gross infarction, avascularity of the chorionic villi, excessive deposition of fibrin in the intervillous space, and so forth. The net effect of these changes is to reduce the total effective surface area for exchange. It is very difficult, however, to separate the effect of these changes from the effect of a reduced maternal blood flow to the placenta. In some instances of chronic maternal disease, such as glomerulonephritis, it is likely that maternal vascular changes in the uterus are primary and the placental lesions are secondary. In view of this difficulty in distinguishing between faults of fetal and maternal tissues, it would be more fair to use the term "uteroplacental insufficiency" than to ascribe the difficulty to the placenta alone.

Fetal size and placental size are related, of course, but one is not necessarily due to the other. Each organ reaches its maximal cell number, as measured by total DNA content, at a different period of pregnancy. In the human placenta, for example, the DNA content no longer increases after the fetus weighs about 2500 gm., although the total size and the RNA content continue to increase until term. In the fetal brain, on the other hand, the number of cells continues to increase until some six months after birth. Therefore, severe undernutrition both before and after birth may limit the numbers of brain cells and produce neurologic defects. The effects of fetal starvation upon various organs and tissues depend upon the period of gestation in which the starvation occurs. In most instances of "small for date" infants, the placenta is small because of a reduced number of cells, indicating that the deprivation occurred prior to the thirty-fifth week of pregnancy.

The villous surface area for exchange may be estimated by planimetric techniques. It has been found that there is a very good correlation between villous surface area and fetal weight, as shown in Figure 9–4. At term, the estimated area of exchange is between 11 and 15 square meters. No infant ever survived when the villous surface area was less than 5 square meters.

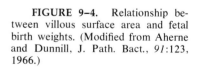

FIGURE 9–4. Relationship between villous surface area and fetal birth weights. (Modified from Aherne and Dunnill, J. Path. Bact., *91*:123, 1966.)

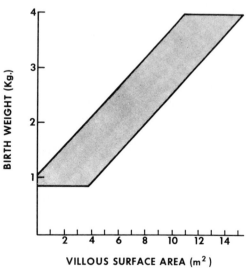

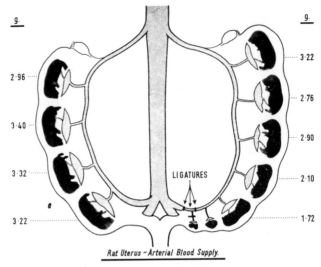

Rat Uterus – Arterial Blood Supply.

FIGURE 9–5. Experimental production of intrauterine growth retardation in the rat by ligation of one uterine artery. Figures are the mean fetal weights in grams. (From Wigglesworth, J. Path. Bact., *88*:1, 1964.)

Impaired Utero-placental Blood Flow

The various factors which may impair the utero-placental circulation have been discussed in the previous chapter. Irrespective of whether the transport of a particular nutrient may be diffusion-limited to a large extent, it is clear that a chronic reduction of uterine blood flow will result in intra-uterine fetal starvation. Experiments by Wigglesworth in the rat clearly indicate this (Figure 9–5). When the uterine artery is ligated on one side, the blood supply must then come only from the anastomotic ovarian artery. Under these circumstances, there is a progressive reduction in fetal weight as one proceeds from the cephalad to the caudal (or cervical) end of that uterine horn. It is as though the first fetus extracted all of the nutrients it required from the maternal blood, leaving less for the second fetus, and so on.

Chronic uterine ischemia is a prominent cause of intrauterine fetal growth retardation in the human species, but since it also leads to placental changes, it is part of the syndrome we label utero-placental insufficiency.

Endocrine Factors

It might be surmised that growth hormone would affect the rate of fetal growth, but this does not appear to be the case. Growth hormones of maternal and placental origin do not reach the fetus in appreciable amounts. That fetal growth hormone has little or no influence on fetal weight is suggested by the fact that idiopathic hypopituitary dwarfs have normal birth weights, and so do anencephalic, apituitary monsters.

It would appear that fetal weight is not significantly influenced by the fetal thyroid, inasmuch as the mean birth weight in a series of 49 cretins was reported to be 3600 gm. (8 pounds).

Abolition of the fetal adrenal glands in sheep does not affect birth weight at term. Anencephalic human fetuses have atrophy of the fetal adrenal cortex, yet attain normal birth weights.

Insulin of fetal pancreatic origin is necessary for normal glucose metabolism; hence, if we were to single out one fetal hormone which regulates the rate of growth, we would be left with insulin.

Chronic Maternal Hypoxia

Data from Colorado indicate that pregnant women residing at elevations above 9000 feet had three times as many low birth weight babies as the general population of the state. Women with heart disease associated with cyanosis also give birth to infants which are small for date. Whether the decreased fetal growth rate under these circumstances is due to a deficiency in oxygen supply *per se* is not known, because there might be associated circulatory factors which affect the supply of nutrients.

The neonatal growth rate of mice is retarded despite adequate nutrition when conditions of high altitudes are simulated. Unlike starvation, in which the cytoplasmic cell mass is reduced in size, chronic hypoxia results in large cells which are fewer in number. One of the best indices of true undernutrition is an increase in the ratio of brain weight to liver weight.

Obscure Maternal Factors

When a mother produces an infant whose weight is two standard deviations or more below the average for gestational age, then most of the siblings also have low birth weight, and no sibling is large for gestational age. The obverse is also true; that is, the siblings of an infant who was large for gestational age are also above average in weight and are almost never more than one standard deviation below the mean. It is also worthy of note that the birth weight of the mother is correlated with the birth weight of her infants, suggesting that there are genetic factors which in some way influence fetal growth rate. This is also suggested by the differences of birth weights in various races. Women in the lower socioeconomic order have smaller infants for the same gestational age, and so do mothers who smoke as compared with non-smokers (Figure 9–2). How these effects are mediated is not known.

Women of small stature, or with small heart size or reduced cardiac output also have infants of lower birth weight. The interpretation of such data is difficult, but it is tempting to generalize and to guess that all of the obscure maternal factors noted above are associated with a reduction in the value for a in the fetal growth equation $W = a (t - t')^3$, that is, a reduction in the rate of intrauterine nutrition per unit of fetal surface area. The most likely cause of such a reduction is some degree of utero-placental insufficiency, which in turn may be brought about in many ways.

A mathematical analysis of the three equations which determine the rate of fetal nutrition (two direct Fick principles plus the Fick diffusion equation) reveals that there are eight primary possible factors influencing fetal growth rate. These are listed, together with a clinical example for each.

1. Decreased rate of utilization per unit of fetal weight. Examples: A primordial dwarf or chronic fetal infection.

2. A reduced concentration of some essential nutrient in the uterine artery. Example: Maternal malnutrition.

3. A chronic reduction of umbilical blood flow. This would be rare except in the case of monozygotic twins with vascular anastomoses.

4. An abnormally high rate of placental utilization of oxygen or nutrients. No clinical examples known.

5. A reduced velocity of active placental transport. Example: A loss of the ability to concentrate amino acids in the fetal plasma does occur with preeclampsia. An impairment of active transport might occur as the result of drug inhibition (nicotine?), but no examples have been demonstrated.

6. An increase in the total surface area of the fetus (or litter). Examples: Twins, triplets.

7. A decrease in the effective placental area for transport. Examples: Placental infarction, intervillous coagulation.

8. A reduction in the rate of uterine blood flow. Examples: Maternal renal disease, preeclampsia, increased uterine tonus, reduced maternal cardiac output.

Factors number 2, 6, 7 and 8 are undoubtedly of more importance in the causation of intrauterine growth retardation than the other four. When confronted with a puny infant at term, there is a great deal of imprecision in the assignment of a cause, and much research is needed in the field of antenatal fetal growth and welfare.

SUGGESTED SUPPLEMENTARY READING

Dawes, G. S. Foetal and Neonatal Physiology. Year Book Publishers, Inc., Chicago, 1968.
Although most of the data discussed in this monograph are derived from animals, particularly sheep, it is an excellent reference for such topics as oxygen transfer, the placenta and fetal growth, energy metabolism before and after birth, and the fetal circulatory and respiratory systems.

Hoet, J. J. Normal and abnormal foetal weight gain. In Wolstenholme and O'Connor (Eds.): Foetal Autonomy. J. & A. Churchill, Ltd., London, 1969.
This is a Ciba Foundation symposium, and the entire volume is pertinent to fetal growth.

Page, E. W. Human fetal nutrition and growth. Amer. J. Obstet. Gynec., *104*:378, 1969.

Seeds, A. E., Jr. Placental transfer. In Barnes, A. C. (Ed.). Intra-uterine Development. Lea & Febiger, Philadelphia, 1968, Chap. 5.
This discussion amplifies the summary of placental transport contained in this chapter. The entire volume is a good reference source in the area of fetal development.

Waisman, H. A., and Kerr, G. Fetal Growth and Development. McGraw-Hill Book Co., New York, 1970.
This report of a conference contains many reviews pertinent to the problems of fetal growth.

Chapter 10

DISORDERS
OF EMBRYONIC
DEVELOPMENT

The major genetic anomalies which affect embryonic development were described in Chapter 6. The remaining disorders which mar the normal progression of early pregnancy are concerned with unfavorable environmental influences, abortions, ectopic implantation or trophoblastic disease. A discussion of diseases which may affect the fetus or newborn is found in Chapter 15.

ENVIRONMENTAL INFLUENCES

History

In the nineteenth century, most of the laity and even some physicians believed that the baby could be marked by maternal impressions. As late as 1881, the Obstetrical Gazette of Cincinnati published a case report about a Mrs. Wilkins, who developed a craving for oysters while she was pregnant. Fearing that this craving might mark her child, she clapped her hand to her buttock and prayed that should the mark occur, it would be there. The child, of course, was graced with the mark of a well-formed oyster upon his buttock!

From about 1910 until 1940, environmental influences on prenatal development were essentially denied, and fetal anomalies were simply classified as genetic in origin. In 1940, however, Gregg demonstrated that even mild cases of rubella in the mother could result in deformities of the fetus. In the same year, Warkany showed that anomalies could be produced in animal fetuses by specific dietary deficiencies. Thus began the present period, in which scientists have been assiduously seeking out drugs, chemicals, radiation effects, viruses and other hazards which might be teratogenic. Qualitatively speaking, the placenta does not act as a barrier to noxious materials except under rare circumstances; when we speak of the environment, we refer to the chemical milieu surrounding embryonic cells during the period of organogenesis.

The only two viruses which are definitely known to produce fetal anomalies are those of rubella and cytomegalic inclusion disease. Other viruses, such as those of smallpox, herpes simplex, mumps, rubeola and hepatitis, may infect the embryo or fetus and cause serious disease or death. Toxoplasmosis, acquired during pregnancy, may also cause serious disease in the fetus.

Rubella

Much of our statistical information about the effects of rubella upon the embryo was gathered during the epidemic of 1964–1965, which resulted in the birth of an estimated 10,000 defective children in the United States.

The earlier in pregnancy the woman contracts rubella, the greater the chance of defective organ development in the fetus. During the first two or three weeks after conception, the frequency of malformation is about 50 per cent; in the second month it is about 25 per cent and in the third month 7 per cent. The eyes, ears and heart are the organs most frequently affected, but about 10 per cent of the defects involve mental retardation.

During the epidemic, serologic methods for establishing the clinical diagnosis were not available. Since then, several antibody tests have become available. It is now known that 85 per cent of the adult population is immune. Serologic testing for rubella antibodies is of importance for three reasons: to identify susceptible women and immunize them before pregnancy, to establish the clinical diagnosis by serial testing after a presumed exposure has occurred in pregnancy and to identify susceptible hospital personnel in order to exclude them from contacts in the nursery.

After the fourth month, the fetus may become infected with the rubella virus and develop the *rubella syndrome,* which includes thrombocytopenic purpura, pneumonia, changes in the long bones, and occasionally hepatitis, myocarditis or encephalitis. The infected newborn may harbor the virus for periods up to six months or longer and becomes a serious source of cross-infections. Fortunately, the modern development of an effective vaccine promises the eventual end of this scourge to embryos.

The history or lack of history of rubella infection is a notoriously unreliable index of immunity. Young women of childbearing age as well as those already pregnant should be screened with the hemagglutination inhibition (HI) test, which is now widely available.

Live rubella virus vaccine was licensed and made available in June, 1969. It should not be administered during pregnancy because of the theoretical risk that the vaccine itself might infect the fetus, but it may be given in the postpartum period. Because the live virus may persist for two or three months following vaccination, it should not be given to women who may conceive during the ensuing three months, so oral contraception, when permissible, is advised for that period of time. The vaccine is highly effective, although not devoid of transient side effects. Its use in adults should be limited to those women who are not already immune, as demonstrated by the HI test. The total duration of immunity is not yet known.

Cytomegaloviruses

There are at least three serotypes of viruses that cause characteristic enlargement of cells which contain the inclusions, giving rise to the term *cytomegalic inclusion disease.* Congenital infection may result in microcephaly, chorioretinitis, cerebral calcifications with seizures, progressive anemia and hepatomegaly. Until recently, the disease was recognized only at autopsy, but increasing numbers of surviving children with milder congenital infections are being recognized.

The viruses are ubiquitous, and large segments of the population have apparently had subclinical infections. Cytomegalovirus antibodies were found in 30 per cent of a group in Boston and in 60 per cent of a sampling in Philadelphia. Active disease can most easily be determined by recovery of the virus from the urine. In one study of 200 pregnant women, the virus was recovered in the urine of seven, and of these, one newborn had viruria but was not deformed. The frequency of embryonic or fetal infection is not known, but congenital anomalies due to cytomegalovirus are not common.

Irradiation

Cellular mutations may result from ionizing radiation, such as alpha particles, beta and gamma rays, x-rays, protons and neutrons, which interact with and alter DNA molecules. Mutations in somatic cells may lead to malformations, if they occur during the critical periods of organogenesis, or to the formation of tumors or leukemias. Mutations in gametes may lead to hereditary alterations which are not apparent until subsequent generations are born.

All of us are subject to ionizing radiation from cosmic rays and other natural sources. During the first 30 years of life (by which time half of all parents have completed their families), we receive 4.3 r (Roentgen units) of natural radiation at sea level (greater amounts at higher altitudes) and an additional average of 3.0 r from diagnostic x-ray studies. Animal studies have shown that it is the total accumulated radiation of hereditable material as well as the rate of radiation which determines the rate of genetic mutations.

The "threshold dose" of ionizing radiation needed to produce somatic defects in an embryo depends upon the stage of development and the organ concerned. In general, a total body dose (which is always the case) of 5 to 10 r greatly increases the probability of a somatic defect if delivered during the first six weeks after implantation. We are aware of one well-documented case in which 20 r were delivered to a woman each week for three weeks in an attempt to induce ovulation (an old method now abandoned), and the third treatment was administered six days after ovulation, as indicated by a continuous basal body temperature chart. A viable child was born with 18 known defects involving every system.

The genetic effects of ionizing radiation are well known in fruit flies and mice, but are most difficult to assess in the human race. In 1958, Kaplan reported on the children and grandchildren of 644 women who had received 60 r to their ovaries for the treatment of prolonged amenorrhea. Of 545 living children born to 318 of these women, only three were abnormal, and of 45 grandchildren of the treated women, all were normal. A study of pregnancies subsequent to exposure to the atom bombs in Japan revealed a significant change in the sex ratio, suggesting the induction of sex-linked lethal mutations, but no conspicuous genetic effects in the children. Nevertheless, we are warned by animal experimentalists that increased mutation rates may be anticipated in the third or fourth generations. We are urged, therefore, to eliminate all non-essential x-ray exposures to the gonads until the reproductive period of life has ended.

According to Hammer-Jacobsen, x-ray pelvimetry at term (AP and lateral views) delivers 0.18 r to the maternal ovaries and 0.7 r to the fetal ovaries. Pelvimetry is required in about 5 per cent of obstetrical patients, but is no longer done as a purely elective procedure. Inasmuch as the only time that we can be sure that a young woman is not pregnant is the pre-ovulatory phase, all elective diagnostic radiology involving the abdomen should be confined to that part of the menstrual cycle.

Chemicals and Drugs

The majority of fetal abnormalities are the result of an interaction between genetic and environmental factors. Thus, a chemical agent which might not be teratogenic for the majority of embryos may increase the penetrance of a genetic defect in some. Even monozygotic twins do not have an identical chemical environment. When one twin, for example, is born with a harelip and cleft palate, only a third of the genetically identical siblings demonstrate the anomaly.

Some drugs may induce anomalies identical to those which are known to be hereditary. Conversely, true genetic defects may simulate somatic defects which can be induced by teratogenic drugs or viruses. To further complicate the assessment of drug effects, there are wide species differences, so that man himself, in the final analysis, is the only suitable test object. The thalidomide disaster is a case in point. The human embryo is exquisitely sensitive to the drug, and the ingestion by the mother of a single 100 mg. tablet of this sedative a month after conception may deprive her infant of his limbs; yet pregnant rats tolerate the drug without the production of fetal anomalies.

It is only when a drug tends to produce a specific defect, as thalidomide does phocomelia, that its teratogenic nature is readily recognized. The only other drug so highly teratogenic is the folic acid antagonist, aminopterin or amethopterin. In therapeutic doses given in the first trimester, it usually kills the embryo, but if it should not, it has been our experience that every fetus had been grossly deformed. Testosterone and certain synthetic progestins related to testosterone may cause external masculinization of a female embryo, or may affect the LH cycling center in the female fetal hypothalamus if given later in the pregnancy. Various chemotherapeutic agents used in cancer therapy appear to have caused malformations, but since they are so seldom used in pregnancy, we are dependent upon isolated case reports.

Diethylstilbestrol administered to gravidae during the first half of pregnancy has been associated with the subsequent development of vaginal cancer in their daughters (see Chapter 18). Dexamphetamine, tolbutamide and diphenylhydantoin have been shown to produce major anomalies in animals, and thus are potentially harmful in humans. A recent prospective study from the University of California at Berkeley indicates that meprobamate and chlordiazepoxide (Librium) given during the first six weeks of pregnancy resulted in a severe anomaly rate of 12.1 and 11.4 per cent, respectively, a significant increase over the 4.6 per cent rate for all other drugs, and the 2.6 per cent rate when no drugs were used.

There are many drugs which may affect the fetus adversely without producing malformations. Antithyroid drugs may produce congenital goiter, quinine may lead to deafness, hexamethonium may lead to fetal ileus, synthetic vitamin K or sulfonamides given during labor may produce hyperbilirubinemia, tetracycline stains fetal bones and deciduous teeth yellow, Coumadin has caused hemorrhagic disorders in the fetus, and chronic cigarette smoking interferes with fetal growth. There are many others, and the list grows yearly.

About 6 per cent of 1-year-old infants are found to have a congenital defect. If a commonly used drug such as aspirin were to cause congenital malformations in one-quarter of these, the association would never be recognized. Inasmuch as the pregnant woman consumes an average of four different drugs and is exposed to numerous chemicals, such as hair sprays, pesticides, cleaning fluids, trace metals, and carbon monoxide, we may never learn how many of the quarter million malformed children born in the United States each year are the result of a toxic chemical environment during fetal life.

ABORTION

The term abortion is applied to all pregnancies which terminate before the period of fetal viability, which is arbitrarily defined as a fetal weight of less than 500 gm. Fetuses weighing between 500 and 1000 gm. are called *immature,* and those weighing 1000 to 2500 gm. are referred to as *premature.* In discussing spontaneous abortions with patients, we frequently use the lay term miscarriage, because the word abortion may be associated with illegal procedures in the layman's mind.

Classification

There are several descriptive adjectives applied to abortions, and each has its usefulness. First of all, one may classify abortions as spontaneous or induced, and the latter are subdivided into legal (or therapeutic) and illegal. Depending upon the presence or absence of infection, abortions are also described as septic or non-septic. Habitual abortion is a term used to describe a third consecutive spontaneous loss. A missed abortion is one in which the embryo has been known to be dead for a month or longer (some authorities prefer two months as a criterion). The phrase middle trimester abortion is frequently used to distinguish those which occur after the twelfth week, because the etiologic factors differ from those of early abortion.

SPONTANEOUS ABORTION

About 15 per cent of all pregnancies between four and 22 weeks after the last menstrual period terminate spontaneously. Prior to the fourth week, i.e., within the first two weeks after conception, the pregnancy wastage is high, but the rate is unknown because most women do not know that they are pregnant.

Well over 50 per cent of spontaneous abortions are due to embryonic defects. Gross chromosomal defects have been found in as high as 60 per cent in one series and as low as 22 per cent in another. The XO abnormality of the sex chromosomes is commonly found, and inasmuch as these cells are chromatin-negative, this led earlier workers to assume that male fetuses were more commonly aborted than females. Almost any serious maternal disease or trauma may lead to spontaneous abortion, but falls, automobile accidents and psychic trauma rarely are responsible. It must be remembered that the abortion is preceded by death of the embryo by several weeks in most instances, and those events which are temporally associated with the onset of cramps and bleeding are rarely related causally. Endocrine deficiencies have frequently been blamed but are rarely at fault.

Just as there are stages of labor and delivery, spontaneous abortion passes through stages which are known as threatened, inevitable, incomplete and complete. *Threatened abortion* occurs when there is cramping or bleeding or both, but examination reveals no dilatation or effacement (thinning) of the cervix. The best treatment is simply rest and reassurance. The patient should be told that completely normal pregnancies are rarely lost, and that if the embryo is defective its loss will be a most desirable event.

Inevitable abortion signifies that the amniotic sac has ruptured or that the cervix has thinned out and has already dilated. Efforts are then made to speed the event with oxytocic drugs. *Incomplete abortion* is common, and may be associated with profound blood loss. The continued bleeding is due to the retention of placental tissue in the lower fundus or cervix, preventing adequate uterine contractions, and this tissue must be removed as an emergency measure. A *complete abortion* is difficult to diagnose with certainty unless the uterine cavity is explored, but may be surmised if, after the passage of tissue, cramps cease and bleeding is reduced to a scanty flow. Complete abortions are common prior to the ninth week, but between nine and 13 weeks, the anchoring villi become well established and portions of placenta commonly remain in the uterus, requiring a dilatation and curettage for completion.

From a pathologic standpoint, the first event which leads to the bleeding of a threatened abortion is hemorrhage into the decidua basalis, with local tissue necrosis following. This acts as a foreign body to stimulate myometrial contractions, which may cause further placental separation. The earliest phase of a threatened abortion is shown in Figure 10-1, the photograph of a uterus containing twins. The uterus was removed for the purposes of therapeutic abortion and sterilization, and some cramping and

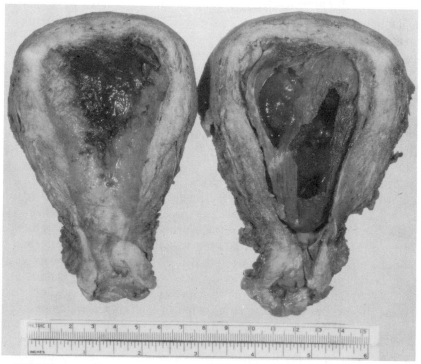

FIGURE 10-1. Uterus containing twins at 2 months of gestation. On the left, an area of decidual hemorrhage may be seen, indicating threatened abortion.

bleeding had begun 30 minutes prior to the hysterectomy. A small area of hemorrhage may be seen in the decidual area at the top of the uterine cavity.

Many theories have been advanced to explain *habitual abortion*. Most of them involve recurring maternal factors, but it has been most difficult to establish the etiology in any given patient. Hypothyroidism has been blamed in some cases, with beneficial results claimed from replacement therapy. Some type of progesterone deficiency has been implicated by the finding of low urinary excretions of pregnanediol prior to the repeated abortion, and this has led to the widespread use of progesterone therapy in an attempt to salvage the pregnancy, but the reported results are difficult to evaluate. One difficulty arises from attempting to estimate the chances of abortion without therapy. If a patient has a spontaneous abortion in her initial pregnancy, her chances for aborting the second time are about doubled. Thereafter, however, the chance of repetition does not appear to increase progressively, as was originally believed by Malpas or by Eastman, but appears to remain at about one in four. Many psychiatrists believe that some women repeatedly abort because of deep-seated psychologic factors leading to a rejection of motherhood, but this does not satisfy the physiologist who wants to know how the pregnancy is anatomically rejected. Nevertheless, some of the best clinical results in the management of women who repeatedly abort have been produced by tender loving care and psychotherapy.

Late or *middle trimester abortions* in which the fetus is normal and weighs up to 500 gm. are most often the result of some uterine abnormality. A prominent cause is the "incompetent internal os of the cervix," which is most often due to a prior laceration resulting from a surgical dilatation or to a traumatic delivery. As noted in Chapter 8, before the uterus converts from a spherical to an ellipsoid shape in the middle three months, intrauterine pressure is at its maximum, and if the internal os is dilated to a centimeter or more at this time, the membranes may balloon out and rupture, like the inner tube of a defective tire casing. At times, the entire fetus, with placenta and membranes intact, may be aborted, as shown by the four months specimen in Figure 10–2.

Fortunately, if the condition is recognized, a purse-string suture of stout, non-absorbing material may be placed around the cervix before the membranes rupture, or even before the cervix dilates. The safest time for such a prophylactic procedure is between the thirteenth and fifteenth week of pregnancy. The method was devised by McDonald in Australia and by Shirodkar in India and is often referred to by their names. In past years, prophylactic surgical repair or circular scarring or suturing was done prior to pregnancy in an attempt to prevent further pregnancy losses, but some of these procedures may have contributed to infertility, and in any event, should pregnancy not occur, the procedure was obviously useless. It is for this reason that the majority of obstetricians prefer to carry out the "circlage" just after the first trimester is completed, when the dangers of spontaneous abortion from other causes or from the procedure itself are

FIGURE 10–2. Photograph of a 4-month fetus with intact membranes and attached placenta.

minimal, but before the usual time of loss from an incompetent cervix, which is between 16 and 28 weeks.

Even though rare, the second most common cause of late abortions is a congenital anomaly of the uterine corpus, such as a double or bicornuate, septate or arcuate uterus which does not readily accommodate to the expanding conceptus. The diagnosis can be made during the non-gravid state by hysterography, and with some types of uteri, a surgical reconstruction may permit successful term pregnancies.

INDUCED, ILLEGAL ABORTION

The frequency of illegal abortions in the United States is not known, but it is at least as common as spontaneous abortion, and may range from 15 to 25 per cent of live births, depending upon the population studied. The infection rate with induced, illegal abortion is high; indeed, septic abortions are very rare in the spontaneous variety. Self-induced or "criminal" abortions account for about one-third of all maternal deaths in this country, mostly from infection or air embolism. Deaths from hemorrhage in the first trimester of pregnancy are extremely uncommon because, unlike post-partum hemorrhage, which may proceed rapidly to a fatal outcome, the bleeding from an incomplete abortion slows down or ceases when severe hypotension ensues. A fatal internal hemorrhage may occur, of course, from a gross perforation of the uterus.

Fashions for the self-induction of an abortion change from year to year and vary from one locale to another. The fact remains that a high percentage of illegal abortions become infected, especially with coliform or Welch bacilli, which may cause endotoxic shock and which have a high fatality rate. Because the bacterial flora in septic abortion so often are mixed, the recommended therapy is intravenous antibiotic administration in massive doses, such as 60 million units of penicillin per day. Evacuation of the necrotic and infected placental material is also an important part of the management.

A reduction of the death toll from septic abortion depends upon education of the public regarding its hazards, expansion of family planning services so that every pregnancy can be a wanted pregnancy, and a continued use of legal abortion to replace illegal abortion.

LEGAL ABORTION

Since the momentous Roe and Doe decisions of the United States Supreme Court in January, 1973, which declared restrictive state abortion laws unconstitutional, the number of legal abortions in this country has progressively increased. The estimated number of legal abortions performed in the United States during 1975 exceeds one million. Expressed as rates, this is about 22 abortions for every 1000 women of reproductive age, or about 250 abortions per 1000 live births. The decision to abort is made solely by the patient with approval by the physician, so that therapeutic abortion committees are no longer utilized. The Supreme Court decisions stated that after the first trimester states may regulate the abortion procedure to safeguard maternal health, and may prohibit abortions after the period of viability, defined as 24 to 26 weeks of gestation. The fact that a Boston physician was convicted of manslaughter in 1975 for an abortion death of a fetus at this late period of pregnancy suggests the hazard of trying to define "viability" with any precision.

TECHNIQUES OF LEGAL ABORTION.   Inasmuch as legal abortion is now the most common surgical procedure performed in this country, a brief description of the methods employed is pertinent. During the first 12 weeks after the last menstrual period the methods employed are (1) "menstrual extraction," (2) uterine aspiration and (3) dilatation and curettage (D & C). During the second trimester the methods utilized are (1) intra-amniotic injections of various substances, (2) hysterotomy and (3) hysterectomy.

"Menstrual regulation," or *"menstrual extraction,"* is a euphemistic term for probable early abortion. It is performed during the first two weeks after a missed menstrual period by the use of a flexible, plastic, 5 mm. cannula and some suction device, with or without paracervical anesthesia. Studies indicate that at 5 days overdue on menses 35 per cent, and at 14 days 65 per cent, of women prove to be pregnant. This suggests that from

65 to 35 per cent of the procedures, which carry some slight risk of morbidity, are unnecessary.

Uterine aspiration is the preferred technique for pregnancy terminations in the first trimester. Under paracervical block or general anesthesia, the cervix is dilated with a Hegar or Hanks dilator to a size 1 mm. larger than the cannula to be used. After sounding the uterine cavity for depth, a flexible or rigid plastic cannula, 5 to 10 mm. in diameter, is introduced and attached to an electrical or manual pump which delivers about 25 inches of mercury-negative pressure. The procedure takes from 1 to 5 minutes, the usual blood loss is between 25 and 200 ml. and the mortality rate has been 0.4 per 100,000 abortions.

Dilatation and curettage, the conventional method prior to 1970, has a case mortality record of 2.6 per 100,000 abortions, and is therefore not as safe as uterine aspiration.

Amnioinfusion is widely used for the termination of pregnancy between 16 and 24 weeks after the last menstrual period. Hypertonic glucose or urea have been used, but the widest experience in the United States has been with the instillation of 20 per cent sodium chloride. The mean abortion interval from injection to termination is about 30 hours. The procedure is occasionally complicated by disseminated intravascular coagulation or death from hypernatremia due to intravenous instillation. The overall mortality is about 18 per 100,000 abortions.

In recent years, the preferred method has been the intramniotic injection of 40 mg. of prostaglandin $F_{2\alpha}$, or dinoprost tromethamine. This technique has the following advantages over the hypertonic saline method: (1) there are no changes in the coagulation system; (2) the risks of hypernatremia and increased body sodium are avoided; (3) the injection-abortion interval is shortened to less than 20 hours; and (4) the fetal tissues are suitable for genetic studies, such as tissue culture. Disadvantages include higher cost and gastrointestinal side effects in about 50 per cent of women.

Hysterotomy is sometimes resorted to when an amnioinfusion has failed, but its mortality rate of 270 per 100,000 operations makes it a far more hazardous procedure. The same is true of hysterectomy, which should be done only when there is also a definitive indication for removal of the uterus.

ECTOPIC PREGNANCY

The term ectopic refers to implantation in any site other than the uterine fundal cavity. It includes interstitial and cervical sites, which are still uterine, as well as tubal, ovarian and abdominal locations, which constitute extrauterine pregnancies. Over 90 per cent of ectopic pregnancies are within the fallopian tube. About 60 per cent of these are in the ampullar portion or fimbriated end, about 25 per cent in the isthmic portion and 5 per cent in the interstitial part; the remainder are rare forms of cervical, abdominal or ovarian pregnancies.

The incidence varies from one in 80 to one in 250 pregnancies, depending upon the socioeconomic level of the population studied. If a woman has had one tubal pregnancy, her chances of having another on the opposite side are about 15 to 20 per cent. Despite the low incidence, ectopic pregnancy accounts for about 8 per cent of all maternal mortalities in the United States.

The causes of ectopic pregnancy include all factors which may impede or delay the normal transport of the fertilized ovum into the uterine cavity. Chief among these factors is prior inflammatory disease of the tubes, which can be found microscopically in at least a third of the tubes removed for ectopic pregnancy. Since the introduction of antibiotic drugs for the treatment of gonorrheal salpingitis, the incidence of tubal pregnancy has risen sharply, because in the pre-antibiotic era, gonococcal infections ordinarily resulted in sealed fallopian tubes and sterility. There are some who believe that areas of ectopic endometrium (endometriosis) predispose to extrauterine implantation, but this is difficult to prove. Some tubal pregnancies may, theoretically, be caused by the transabdominal migration of a fertilized ovum, permitting it to develop more fully before entering the tube. Transmigration is revealed when the corpus luteum of pregnancy is found in the ovary opposite to the involved tube.

Endometrial Reaction

According to one series, in the presence of an extrauterine pregnancy, the endometrium forms a decidual reaction in 44 per cent and a peculiar enlargement of the epithelial cells with bizarre nuclei in 25 per cent. The latter is referred to as the "Arias-Stella phenomenon," named after the man who described the change in 1954. Uterine bleeding occurs at some time in three out of four extrauterine pregnancies, and is sometimes misinterpreted by the patient as being a normal period. Thus, the symptom of amenorrhea does not appear in about a third of the histories of women with ectopic gestations. A curettage is sometimes done for diagnostic problems, especially when the pregnancy test is negative, and the finding of decidual tissue without villi or with the Arias-Stella change is highly suggestive of extrauterine pregnancy.

Clinical Course

The natural history of an ectopic pregnancy depends largely upon its location. In tubal isthmic implantations, rupture through the wall of the tube is the rule, with extensive intraperitoneal hemorrhage. Patients may then present the classic picture of a ruptured ectopic. A stabbing unilateral lower abdominal pain is followed by vertigo or fainting. If the patient then lies down, the free intraperitoneal blood may rise to the diaphragm, irritate the phrenic nerve and produce a referred pain to the shoulder, usually on the right side. She then appears pale and shocky. On pelvic examination, upward motion of the cervix produces exquisite pain, and a boggy, tender mass may be felt in one adnexal region. Under such circumstances, there is little doubt about the diagnosis, and immediate laparotomy must be performed.

The more common ampullar locations ordinarily follow a different clinical course, which may mimic such conditions as salpingitis, appendicitis, a twisted ovarian cyst or a ruptured corpus luteum. When the gestational sac has distended the tube, partial separation of the placental tissue may occur, and spurts of blood escape from the fimbriated end intermittently, sometimes over a period of many days. When the blood coagulates, an irritating substance (possibly serotonin) is released, which causes peritoneal pain lasting for several hours. (Blood containing an anticoagulant, when introduced into the peritoneal cavity, is rarely irritating.) At this stage of threatened tubal abortion, the rapid immunochemical test for pregnancy is usually positive and is of considerable aid in establishing the diagnosis. Another diagnostic aid is culdocentesis, that is, the insertion of a needle through the posterior vaginal fornix into the cul-de-sac for aspiration of non-clottable blood. The gross appearance of a tubal pregnancy is shown in Figure 10–3.

The hematoma which forms in the tubal ampulla once bleeding begins may literally push the conceptus out of the tube. More commonly, the extrusion is incomplete, because the trophoblastic anchoring villi invade the

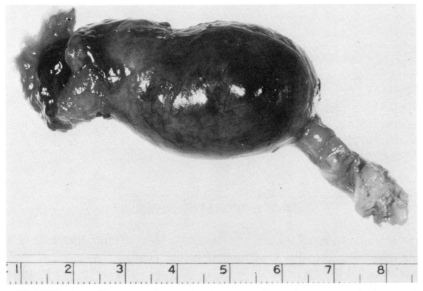

FIGURE 10–3. Gross appearance of an unruptured ectopic pregnancy in the ampullar portion of the tube.

muscularis of the tube. Inasmuch as there is no well-developed decidual layer, an incomplete tubal abortion results, with continued intraperitoneal hemorrhage. On rare occasions when the embryo is still viable, the placenta may continue to grow, spreading out to other pelvic structures, such as the broad ligament, ovary and bowel. This is called a secondary abdominal pregnancy, and some of these have been known to proceed to term with delivery by laparotomy of a living child. The placenta becomes so tenaciously attached to vital structures that its removal may be hazardous, and it is best to leave it *in situ* for subsequent reabsorption.

Diagnosis

Ectopic pregnancies which do not present the classic picture of rupture followed by shock may pose difficult diagnostic problems. Many emergency rooms have a sign on the wall, "Is this an ectopic pregnancy?", to remind all personnel of the possibility. A careful history is the most reliable single aid. The pelvic examination may reveal only tenderness, which is common to many pelvic lesions. The immunochemical hCG test on urine is of value only if it is positive, and even then, of course, it will not distinguish between an intrauterine and an extrauterine pregnancy. Culdocentesis or laparoscopy are often employed in doubtful situations. The patient's temperature and white blood count may be normal or elevated and the hemoglobin or hematocrit may not reflect a recent internal hemorrhage. About one in every five or six operations for "probable ectopic pregnancy" finds some other lesion.

Treatment

The treatment is prompt surgery, with removal of the affected tube and blood replacement. There is no expectant treatment. It is possible, when the patient desires children and the contralateral tube is either missing or diseased, to preserve the affected tube by careful ligation of all bleeding points, leaving the incised portion open. Hysterectomy is recommended in cervical pregnancy, which is rare.

TROPHOBLASTIC DISEASES

Hydatidiform moles and choriocarcinoma constitute the most bizarre and enigmatic diseases encountered in the field of human reproduction. Hertig called the malignant forms "God's first cancer and man's first cure," referring of course to one of the first — and one of the few — cures of cancer by chemotherapy.

Table 10–1. CLASSIFICATION OF TROPHOBLASTIC NEOPLASIA

Gestational or non-gestational (e.g., arising in an ovarian or testicular teratoma)
A. *Clinical diagnosis*
    1. Non-metastatic
    2. Metastatic
      (a) Local (pelvic)
      (b) Extra-pelvic
Other required information
    (a) Evidence
      (i) Morphologic
      (ii) Non-morphologic
    (b) Antecedent pregnancy
      (i) Normal
      (ii) Abortal
      (iii) Molar
    (c) Previous treatment
      (i) Untreated
      (ii) Treated
B. *Morphologic diagnosis*
    1. Hydatidiform mole
      (a) Non-invasive
      (b) Invasive
    2. Choriocarcinoma
    3. Uncertain
Other required information:
    (a) Diagnostic basis
      (curettage, excised uterus, necropsy or other means specified)
    (b) Date of diagnosis
      (with respect to date of onset of treatment)

Classification

In 1965, an International Study Group for Trophoblastic Neoplasia recommended the classification of trophoblastic diseases shown in Table 10–1.

HYDATIDIFORM MOLE

About two-thirds of spontaneous abortions with blighted or absent embryos reveal a grape-like swelling of the villi due to the accumulation of fluid within the stromas. A true mole is a blighted ovum which is retained (like a missed abortion) with a marked cystic swelling of the villi and variable degrees of trophoblastic proliferation. The cysts may vary from 1 to 20 mm. in diameter and dangle from threads of villi attached to main stems or solid masses like bunches of grapes. When Countess Margaret of Henneberg passed a mole in 1276, it was believed that each vesicle was a conceptus, and every cyst was baptized alternately as a boy or girl. The gross appearance of a hydatidiform mole is shown in Figure 10–4.

The genesis of a mole begins about a month after conception, when the feto-placental circulation is normally established. Since the embryo dies or is defective, the chorionic villi do not become vascularized. Because the *incidence* is so high in the Orient (1:173 pregnancies in the Philippines but

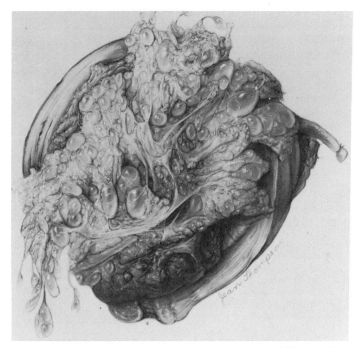

FIGURE 10–4. Drawing of a uterus containing a hydatidiform mole.

only about 1:2000 in the United States), and because the incidence is higher in poor people, there are many who believe that some form of malnutrition is etiologically related.

One puzzling feature of this strange disease is that essentially all benign moles are "female"; that is, the cells are chromatin-positive, and those which have been karyotyped are predominantly of the XX variety. Rarely, half or more of a placenta will have undergone molar degeneration in association with a living fetus, and with few exceptions the fetus has been female. In the case of choriocarcinoma, about one in four has an apparent XY constitution, perhaps because these followed a spontaneous abortion rather than a molar pregnancy.

The *pathology* varies greatly with respect to the degree and type of trophoblastic proliferations. Hertig has classified moles on a histologic basis from the "apparently benign" to the "apparently malignant" grades, but it is generally agreed now that the histologic appearance is not a reliable guide to prognosis. The pathologist's opinion may determine the degree of a physician's worry, but the subsequent management after evacuation of the uterus depends upon the clinical course and the behavior of the chorionic gonadotropin titer (Figure 10–5).

The *clinical course* of a hydatidiform mole depends upon its rate of growth and duration. Spontaneous abortion, almost always incomplete, may occur at any time from one to seven months after conception, the

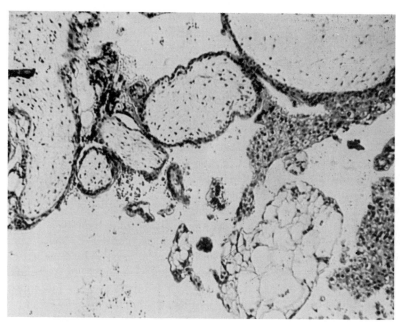

FIGURE 10-5. Microscopic picture of a benign hydatid mole.

average duration being 18 to 20 weeks. In three out of four cases, the size of the uterus is larger than expected, but in 20 per cent of cases, the uterus is smaller than anticipated for the duration of amenorrhea. There will always be uterine bleeding sooner or later, and this, coupled with an unusually large uterus, may first arouse suspicion. There is frequently a progressive anemia which is out of proportion to the external blood loss. The passage of a typical vesicle, of course, establishes the diagnosis. When moles proceed beyond 20 weeks of gestation and when the uterus is larger than it should be (well above the umbilicus), about half of the patients develop preeclampsia. Indeed, the sudden onset of hypertension and proteinuria before the twenty-fifth week of pregnancy is almost of itself diagnostic of a mole. With large moles there may be an excessive deportation of villi to the lungs, and we have seen several patients with radiologic evidence of lungs filled with villous emboli—all patients, incidentally, with severe preeclampsia.

The *diagnosis* during the first three months is most readily made by sonography. The sonogram reveals a characteristic mottled appearance with an absence of the gestational sac. After this time, there is a continued rise of the chorionic gonadotropin titer at a time when the levels should normally be falling, and serial determinations in serum or in 24-hour urine specimens may be a diagnostic aid. After the fourth month, the absence of fetal bones by x-ray examination and the absence of a fetal heartbeat by search with the Doptone instrument are helpful. In addition to sonography, an accurate diagnosis may be made by arteriography or amniography. Enlargement of both ovaries due to theca lutein cysts (from overstimulation by excessive concentrations of hCG) is present in about a third of the patients, but frequently they are not palpated until the uterus has been evacuated.

The *treatment* of a mole, once the diagnosis is established, is either evacuation of the contents or total hysterectomy, depending upon the age and parity of the patient. Evacuation is most easily accomplished by the suction technique commonly used for therapeutic abortion, with intravenous oxytocin administration used concurrently. Dr. Tow of Singapore has shown abdominal hysterotomy, often used in this country, to be followed by the highest morbidity and malignancy incidence; it should probably be avoided if possible. Large theca lutein cysts should be left alone, because the ovaries return to normal. Hysterectomy is the treatment of choice in women over 35 or when further pregnancies are not desired.

Meticulous follow-up of the patient who has had a mole is the most important part of the therapy, since about 3 per cent give rise to chorio-carcinoma and about 15 per cent to an invasive mole. The hCG titer should be measured weekly, and if the hormone rises in quantity, or if it plateaus for three consecutive weeks, especially if the uterus fails to involute and bleeding continues, the suspicion of malignancy is high.

INVASIVE MOLE

This lesion, also called *chorioadenoma destruens,* is histologically the same as the more proliferative hydatidiform moles, except that it invades the myometrium and may metastasize to distant organs. The treatment of invasive mole is the same as for choriocarcinoma.

CHORIOCARCINOMA

This malignant trophoblastic lesion formerly carried a mortality which approached 100 per cent, but since the introduction of chemotherapy with amethopterin (Methotrexate, a folic acid antagonist) and dactinomycin (Actinomycin D), which acts directly on DNA, the lives of three out of four women with choriocarcinoma may be saved.

The genesis of gestational choriocarcinoma may be any type of preg-nancy. In this country, about half follow moles, a quarter follow spontaneous abortion and the rest occur after normal term deliveries or, rarely, after an ectopic pregnancy. Shintani et al. have produced experimental chorio-carcinoma in the rat by the direct application of a carcinogenic agent to retained placentas.

The microscopic appearance of a typical choriocarcinoma is shown in Figure 10–6. It resembles very closely the plexiform pattern of the pure trophoblast seen in a normal 13-day human implantation. Its rapid erosion of blood vessels produces hemorrhagic lesions, which may appear almost purple in color. Choriocarcinoma invariably metastasizes by way of the

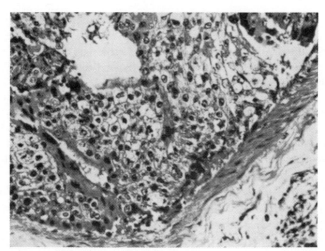

FIGURE 10–6. Choriocarcinoma within a pulmonary artery. (From Bagshawe, Chorio-carcinoma, Edward Arnold, Ltd., London, 1969.)

blood stream, and the most common metastatic sites are the lungs, the vagina and the brain, although any organ may become involved.

The *diagnosis* is suggested by the history of a recent pregnancy, especially a mole, uterine bleeding and enlargement, and a positive hCG test. The radiologic evidence of pulmonary lesions is confirmatory. A diagnostic curettage may yield tumor tissue, but all too often the lesion is buried in the myometrium and the procedure may actually disseminate the lesion. It is not necessary to depend upon curettage for the diagnosis of malignant trophoblastic disease.

The *treatment* consists of primary chemotherapy, as noted earlier. The surgical extirpation of organs has an increasingly limited role. When choriocarcinoma is discovered within four months of the offending pregnancy, and when the hCG titer in a 24-hour urine specimen is less than 100,000 I.U., the "total remission rate" with appropriate chemotherapy may exceed 90 per cent.

A much worse prognosis prevails under four circumstances: (1) when the duration of the lesion is over four months; (2) when the urinary excretion of hCG exceeds 100,000 I.U. per 24 hours; (3) when metastases are present in the liver. brain or bowel; and (4) when the choriocarcinoma is of non-gestational origin, as in males or in primary ovarian tumors. Under these conditions, triple therapy is utilized; that is, the sequential use of methotrexate, actinomycin D and either chlorambucil or Cytoxan.

The usual immunologic pregnancy test performed on urine (e.g., Pregnosticon) is not sensitive below 700 I.U. per liter. Therefore, when this test is negative, the physician should utilize a serum LH radioimmune assay or, if available, a radioimmune assay for the beta subunit of hCG. Women who are being followed after treatment of any type of trophoblastic neoplasm should be placed on oral contraception, not only to prevent pregnancy (which would truly confuse the issue) but to prevent the preovulatory surge of LH hormone which cross-reacts with hCG in the usual LH assay. The concentration of hCG in serum (expressed as LH units) should not exceed 15 milli-I.U. to be considered negative. The program of chemotherapy must be continued well past the time when the hCG concentration falls below that level.

To summarize briefly, the total wastage of human pregnancies is high, but this is one of Nature's methods of reducing the production of defective offspring. The last three decades have revealed many environmental influences which affect human prenatal development, among them the viruses, ionizing radiation, chemicals and drugs. Spontaneous abortion during the first three months is more often than not due to a defective embryo, although habitual abortion and losses during the middle trimester are due primarily to maternal conditions, some of which are remedial. Illegal abortion is a national menace which contributes greatly to the maternal death rate. Legal abortion poses a complex problem in this country and is in an unsatisfactory phase of development for many reasons. The remaining disorders of early gestation, such as ectopic pregnancy and trophoblastic diseases, are less common, but nevertheless constitute serious threats to the individual involved.

SUGGESTED SUPPLEMENTARY READING

Bagshawe, K. D. Choriocarcinoma. Williams & Wilkins Co., Baltimore, 1969.
    A 300-page monograph dealing with the clinical biology of the trophoblast and its tumors.
Fraser, F. C., and McKusick, V. A. (Eds.). Congenital Malformations. Excerpta Medica Amsterdam, 1970.
Hertig, A. T. Human Trophoblast. Charles C Thomas, Springfield, Ill., 1968.
    A renowned obstetrical pathologist discusses, among other things, his experiences with spontaneous abortion and trophoblastic disease.
McElin, T. W. Ectopic pregnancy. In Danforth, D. N. (Ed.). Textbook of Obstetrics and Gynecology. Harper & Row, New York, 1966, Chap. 13.
    This and the preceding chapter on abortion (by M. E. Davis) are concise 15-page summaries.
Schenkel, B., and Vorherr, H. Non-prescription drugs during pregnancy: potential teratogenic and toxic effects upon embryo and fetus. J. Reprod. Med., *12*:27, 1974.
Tuchmann-Duplessis, H. The effects of teratogenic drugs. In Philipp, E. E. (Ed.). Scientific Foundation of Obstetrics and Gynecology. F. A. Davis Co., Philadelphia, 1970.
Woollam, D. H. M. (Ed.). Advances in Teratology. Academic Press, New York, 1967.
    A series of authoritative reviews, beginning with Volume I in 1967.

Chapter 11

ENDOCRINOLOGY AND METABOLIC CHANGES OF PREGNANCY

From the moment that the human blastocyst sinks into the endometrium, both mother and fetus commence a temporal partnership in which each member must share the available energy for growth and sustenance and adapt the controls of metabolism to this short-term parasitism.

FETAL ENDOCRINOLOGY

Early Development of the Fetus

In order to understand the development of the fetal endocrine system, we should first review the development of the embryo as a whole (Table 11–1). Once implanted in the endometrium of the uterus, the developing human embryo begins a series of cellular migrations and alterations that are necessary for the orderly differentiation of the organism. At 2 weeks of age, the human organism consists of a two-layered trophoblast surrounding a mass of primary mesoderm. Within this mesoderm are two cavities: the primitive amniotic cavity and the cavity of the yolk sac. The future embryo lies as a two-layered plate or disc between these two cavities.

A number of splits appear in the primary mesoderm, and as these splits coalesce, the first coelomic cavity is formed. At the same time, some of the primary mesoderm at the posterior end of the embryo condenses to form the body stalk, the connecting link between the developing embryo and the placenta. For the first time, the anterior and posterior ends of the embryo can be differentiated. During the third week of life, ectodermal cells near the midline of the posterior half of the embryonic disc divide rapidly, forming a linear opacity, the primitive streak (Figure 11–1). The cells migrate out

Table 11–1.   DEVELOPMENT OF FETAL ENDOCRINE GLANDS

Age (wk.)	Extent of Development	Adrenal	Gonads	Thyroid
1	Cleavage (days zero to three). Differentiation of trophoblast			
2	Implantation. Bilaminar embryo. Differentiation of syncytiotroph- and cytotrophoblast. hCG detectable		Sex chromatin first seen	
3	Primitive streak stage. Three-layered embryo. hPL isolated from placenta		Germ cells seen	Diverticulum present
4	Cylindrical body, flexed and C-shaped. Rapid organogenesis. Pharyngeal pouches forming. Pancreatic buds present	Anlage first seen	Gonadal ridges first seen	Gland is a stalked endodermal sac. Can synthesize thyroglobulin
5	Umbilical cord organizes	Gradual condensa-tion of cells	Gonadal ridges prominent	Gland bilobed. Duct begins to atrophy
6	Parathyroids solid. Thymic and ultimobranchial sacs prominent	Fetal zone visible as a mass of cells	Gonadal ridges in-vaded by germ cells	
7	Ultimobranchials fuse with thy-roid. Dorsal and ventral pri-mordia of pancreas fuse	Chromaffin cells begin to migrate into cortex	Gonads begin to differentiate	Thyroglossal duct atrophied
8		Definitive cortex forms. Desmolase, 16α- and 17-hydro-xylases present	Testis differentiated. 3β-ol-dehydrogenase in Leydig cells	Follicles form
10	Pancreatic alveoli present. Vaso-pressor and oxytocic activities in fetal pituitary	21- and 11β-hydro-xylases present	Testis can form testosterone	Iodination of pro-tein. TSH in serum
12	First islets seen in pancreas. In-sulin extracted. hGH formed in pituitary		External genitalia dis-tinctive	

from this streak between the ectodermal and endodermal layers of the embryonic disc, forming a trilaminar embryo. The embryonic disc, which is circular initially, elongates. A portion of the ectoderm thickens to form the first rudiment of the nervous system, the neural plate. The edges of the neural plate bend dorsally to form two ridges, the neural folds. As the neural folds rise, the paraxial mesoderm thickens. These thickened masses of mesoderm undergo segmentation to form the somites of the body (Figure 11–2).

The embryo must now fold, and as the head and tail ends of the embryo bend toward each other, the yolk sac is gradually pinched off from the gut, remaining connected by the yolk stalk. Swellings called blood islands appear in the primary mesoderm around the yolk sac. The peripheral cells of these blood islands flatten and join edge to edge to form an endothelium. The inner cells become the primitive erythroblasts. The further develop-

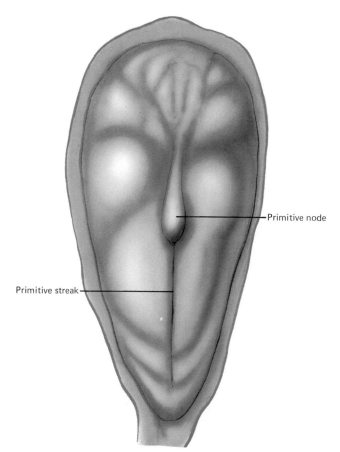

FIGURE 11–1. A dorsal view of a 16-day-old embryonic disc, showing the primitive streak and the primitive node.

ment of the erythrocyte series and its relation to hemoglobin synthesis and oxygenation will be discussed in Chapter 15.

While the neural folds are joining to form the neural tube, cells from their lateral margins condense to form two bands of tissue, the neural crests. Cells from the neural crest migrate ventrally to form the sympathetic ganglia. Other cells migrate from the neural crest around the sides of the aorta, eventually to lie beside (and later invade) the anlage of the fetal adrenal cortex. These cells will become the adrenal medulla. Still other cells of the neural crest become incorporated into the thyroid gland, forming the parafollicular or calcitonin-secreting cells.

By $3\frac{1}{2}$ weeks of age, the human embryo has a cylindrical body attached to the yolk sac at the midgut. The neural groove has closed (except at the ends), and the neural crests are evident. Pharyngeal pouches are forming, and the thyroid diverticulum is present. The lung and liver buds are present.

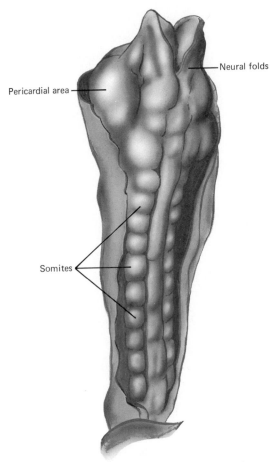

FIGURE 11–2. A dorsal view of a 21-day-old human embryo, showing the neural folds, which have closed anteriorly, the pericardial area and the somites, the precursors of the muscle masses.

The heart tubes have fused and begun to beat. The embryonic coelom is lined with mesoderm.

The development of all the organ systems of the embryo proceeds at a rapid pace during the next few weeks. The body of the 4-week-old embryo is flexed and C-shaped. Rathke's pouch can be observed; the thyroid gland is a stalked endodermal sac, and the liver cords and ducts are forming (Figure 11–3). The mesonephric tubules are rapidly differentiating. Indeed, the mesonephros, next to the liver, constitutes the largest organ of the abdomen at this stage. The enlarging mass of the mesonephros projects into the coelom on either side as a ridge beside the mesentery. Between the mesonephros and the dorsal mesentery, at the level of the upper third of the mesonephros, can be seen a thickening of the coelomic epithelium, the anlage of the fetal adrenal cortex.

THYROID

DEVELOPMENT. The first endocrine gland to appear in development is the thyroid. The first evagination of the floor of the pharynx can be seen about two and a half weeks after conception. This evagination elongates and extends anteriorly and caudally. While it is still attached to the pharynx by the thyroglossal duct, the evagination becomes bilobed. As early as the fourth week, the thyroid gland can synthesize thyroglobulin. By the seventh week, the thyroid has reached its definitive location, and the thyroglossal duct has atrophied. The endodermal cells multiply rapidly. The masses of cells are broken up by the ingrowth of the mesenchyme carrying blood vessels. The mesenchyme forms the capsule and stroma of the gland; the epithelial cells line the follicles. Thyroid follicles begin to form at about 8 weeks of age. After 10 weeks, the thyroid can accumulate iodine. Organification of iodide begins shortly thereafter, first monoiodotyrosine and then diiodo-

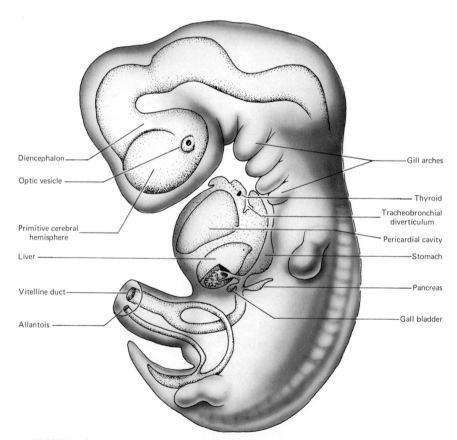

FIGURE 11–3. A lateral view of a 35-day-old human embryo, showing the development of the anterior portion of the nervous system and the gastrointestinal tract.

tyrosine being formed. Coupling of these iodinated tyrosines to form triiodothyronine and thyroxine is well established by 12 weeks' gestation.

Defective embryogenesis may result in complete or partial absence of the thyroid. This would result in impaired production of thyroid hormones, as would a hereditary deficiency in any of the enzymes involved in thyroxine synthesis. The latter might produce a goiter in the fetus due to stimulation by elevated levels of TSH. Overt cretinism may not occur, since both triiodothyronine and thyroxine can cross the placenta, though the passage is slow for thyroxine. This protection is not complete, for delay in osseous maturation is frequently found, and a high incidence of damage to the central nervous system is found despite early treatment of the newborn with thyroid hormone.

A large fraction of circulating thyroxine is bound to thyroxine-binding proteins: thyroxine binding globulin (TBG) and thyroxine binding prealbumin (TBPA). The concentrations of these binding proteins in fetal blood increase up to term. Associated with this rise in plasma-binding proteins is a progressive increase in fetal serum thyroxine (T_4) and triiodothyronine (T_3); however, T_3 values are low and in the adult hypothyroid range, whereas T_4 values are comparable to, or even exceed, maternal values. The concentration of free T_4 in cord blood at birth is slightly higher than the comparable value in maternal blood. The concentration of free T_3 in cord blood, in contrast, is significantly lower than the maternal free T_3 value.

Thus in the fetus the ratio of circulating T_4 to T_3 is higher than that in the normal adult. This high ratio is not reflected in the fetal thyroid gland. The mean T_4/T_3 content of fetal thyroid glands is similar to that of euthyroid adult glands. In the adult a significant fraction of circulating T_3 is derived from peripheral conversion of circulating T_4 to T_3, and it has been suggested that in the fetus the rate of this conversion may be decreased.

Despite the low concentrations of both total T_3 and free T_3 in fetal blood, the normal newborn is not hypothyroid. Throughout pregnancy the concentrations of free T_4 in maternal and fetal sera remain within normal limits, which may explain the euthyroid state of the normal newborn.

MATERNAL HYPOTHYROIDISM. Untreated maternal hypothyroidism is rarely a problem in pregnancy. Conception and the maintenance of pregnancy may be difficult in the hypothyroid state. Treatment of hypothyroid mothers with replacement amounts of thyroxine produces little effect on the fetus. However, the use of triiodothyronine may be hazardous, since this potent agent does not saturate the maternal TBG binding sites and may create a drain on the fetal thyroid hormones to fill these sites.

MATERNAL HYPERTHYROIDISM. In pregnancy total T_4 values are elevated because of increased maternal TBG (an estrogen effect). However, the values for free T_4 in the pregnant state are similar to those found in the non-pregnant state. Thus true thyrotoxicosis is not present in normal pregnancy.

Our present concepts of thyrotoxicosis suggest that with thyroid injury an antigenic protein is released from the gland into the circulation. This antigen stimulates the synthesis in peripheral lymphoid tissue of a compound which is apparently an antibody and also a stimulator of thyroid function. This substance has a prolonged reaction in bioassay systems and has been called the long-acting thyroid stimulator (LATS). LATS is a 7S globulin which can cross the placenta readily and is known to stimulate increased synthesis of thyroid hormone in the thyroid gland. The fetal thyroid responds to LATS. The result is enlargement of the gland and neonatal thyrotoxicosis. As LATS disappears from newborn blood (half-life of six days), the symptoms of thyrotoxicosis subside. Since the clinical disease is transient, treatment of the newborn with iodide is usually sufficient to control symptoms.

Mothers receiving thiouracils for hyperthyroidism may have babies with congenital goiter. The thiouracils can cross the placenta and interfere with the synthesis of thyroid hormones. This in turn stimulates the fetal pituitary to release more TSH. There may be intrauterine hypothyroidism. Iodides administered to the mother have also been shown to produce goiters in the fetus with or without hypothyroidism.

FETAL PANCREAS

The fetal pancreas arises as an outgrowth of the duodenal endoderm. A dorsal bud eventually becomes the body and tail of the pancreas. A smaller ventral bud grows in the angle between the duodenum and the hepatic diverticulum. This ventral bud eventually develops into the common bile duct and the head of the pancreas. Dorsal and ventral outgrowths fuse at about seven weeks in the human embryo. The endocrine cells of the islets originate from the duct epithelium and bud off from it. Acinar cells differentiate at about the third month, at a time when distinct islets of Langerhans are present. Precursors of the primary islets have been observed in embryos 39 to 54 mm. in length. Differentiation of the islets into α and β cells (with α outnumbering β) has been noted to begin at the 130 mm. stage. At full term, the pancreas has islets composed of 60 per cent β and 30 per cent α cells. Beta cells are known to produce insulin.

At about 24 weeks' gestation (> 1000 gm.), granules are seen in the acinar cells and proteolytic activity can be demonstrated chemically in the fetal pancreas. However, lipase and amylase are virtually absent at this stage. Trypsin begins to appear at 24 weeks and increases to adult values by birth. Amylase does not appear in pancreatic secretion until several months after birth. Lipase is present from the fourth month of gestation onward.

Insulin can be extracted from the pancreas of the human fetus after about the twelfth week of gestation. These extracts also contain proinsulin, which comprises 0.26 to 1.6 per cent of the total immunoreactive insulin in the pancreas regardless of the fetal age. After partial tryptic digestion of this fetal proinsulin the insulin immunoreactivity was enhanced by 40 per cent and the biological activity by over 800 per cent. Thus it seems likely that different portions of the insulin molecule are responsible for its immunoreactivity and its biological activity. The concentration of insulin in fetal blood, after the development of insulin in the pancreas, is comparable to that found in the newborn. The placenta is relatively impermeable to insulin, so that the fetus must supply its own hormone. The fetus near term can respond somewhat to a glucose load by increasing the concentration of insulin in his serum; however, the response is feeble, and for all practical purposes the normal fetus and newborn are considered to be unable to release insulin rapidly on demand. Offspring of diabetic mothers, in contrast, do release insulin when stimulated by a glucose load, possibly because of exposure *in utero* to hyperglycemia.

The conceptus also possesses enzymatic mechanisms for the proteolytic inactivation of insulin. These insulin degrading systems in the placenta may perhaps contribute to the increased turnover of maternal insulin.

FETAL PITUITARY

ACTH has been detected in fetal pituitaries by the tenth week of gestation. In anencephaly, if the pituitary is poorly developed (due to failure of the hypothalamus), the fetal adrenals are atrophic. It seems likely that this hypoplasia of the adrenal cortex results from insufficient release of ACTH from the fetal pituitary. Indeed the concentration of ACTH in the blood of anencephalic fetuses is very low. There is little or no transplacental passage of ACTH. Winters (1974) found marked differences in the concentrations of ACTH in fetal plasma before and after the thirty-fourth week of gestation. The highest levels occurred before the thirty-fourth week and the maximum rate of adrenal growth in the fetus (after the thirty-fourth week) occurs during a time when the concentration of ACTH in the plasma is low. This suggests that ACTH may not be the sole tropic stimulus, especially of the fetal zone. Although the concentration of cortisol in fetal blood rises with the spontaneous onset of labor, there is no evidence for an increase in ACTH in the blood. Thus an increased concentration of ACTH in plasma is not an obligatory prelude to the initiation of labor in the human.

Human growth hormone has been found in fetal pituitary glands by both bioassays and immunoassays. Higher concentrations are found in the umbilical arteries than in umbilical and maternal venous blood, indicating that the fetal anterior pituitary gland is an actively functioning one. The high concentrations of growth hormone in umbilical cord blood persist during the first 48 hours of neonatal life. The mean half-life of growth hormone in infants is 13 minutes, compared with 20 to 30 minutes in adults. It is not known what role growth hormone may have in the fetus.

Cytologic differentiation of the pituitary begins at about eight weeks' gestation and the first PAS-positive basophils appear at about nine weeks. Although histochemical evidence of thyrotroph formation does not occur until 13 weeks, TSH is found in fetal serum at 11 weeks. Extracts of fetal pituitary contained biologically active hormone as early as 12 weeks, and incorporation of labeled amino acids into labeled TSH occurred in cell cultures of pituitary glands from fetuses 14 weeks old.

There appears to be no correlation between maternal values of TSH and any of the parameters of fetal thyroid function. Therefore, it is unlikely that maternal TSH crosses the placenta to any appreciable extent. Anencephalics lacking a hypothalamus have active thyrotrophs in their pituitaries, and appear to have fairly normal thyroid differentiation and some thyroxine synthesis. It would seem that release of TSH by the pituitary is not completely under hypothalamic control.

Both vasopressor and oxytocic activities can be demonstrated in human fetal pituitaries as early as 10 weeks' gestation and increase in amount throughout gestation. The ratio of activities at term is 1:1. The hypophyseal-portal system is fully developed in fetuses at term.

Studies by Gitlin and his associates have shown that the fetal pituitary is capable of forming gonadotropins *in vitro*. The amounts of FSH and LH in pituitaries of fetuses from 12 to 20 weeks' gestation range from 3800 to 44,000 m.I.U./gland for LH, and 56 to 260 m.I.U./gland for FSH, with no obvious sex differences. Gonadotropic cells account for about 11 per cent of all granular cells in the pituitary at 12 weeks' gestation. It is not at all certain how much of gonadal development in the human requires gonadotropins from fetal pituitary. The possibility that hCG from the placenta may serve in this capacity has been suggested.

Recent studies have shown a clear sex difference in the secretion of FSH during fetal life. The concentration of FSH in the serum of 20-week female fetuses approaches the adult castrate level. Male fetuses have little or no circulating FSH at this stage of gestation. Since the circulating estradiol concentrations are equally high in both male and female fetuses it would appear that FSH secretion in the female fetus is not suppressed by estrogens during the first half of gestation.

During the latter half of pregnancy the gonadotropin-secreting cells of the pituitary gradually regress. In contrast, prolactin-secreting cells increase steadily until the end of gestation, and concentrations of prolactin in fetal sera rise throughout pregnancy. A few days after birth prolactin cells regress, while gonadotropes again become more developed. Since estrogen stimulates prolactin production and inhibits gonadotropin release, it is possible that the rising levels of estrogen, particularly in the latter half of pregnancy, and the abrupt decline in estrogen after birth might be responsible for the differential growth and function of these two pituitary cell types.

FETAL GONADS

The morphological differentiation of the gonads has been considered in Chapter 2. Only the testes play any significant role in steroid metabolism in the fetus. Morphologic differentiation of Leydig cells is first evident in interstitial cells at about eight weeks' gestation. Since hCG is present in small amounts in fetal tissues, and since the increase in hCG secretion coincides with the increase in Leydig cells, it is possible that the development of fetal Leydig cells is initiated by hCG.

Also at about eight weeks' gestation, the fetal testis first gives histochemical evidence for the presence of 3β-hydroxysteroid dehydrogenase. Maximal activity is attained at about 12 weeks' gestation. Human fetal testes can synthesize testosterone *in vitro* from a variety of precursors. The yield of testosterone tends to increase from the ninth to the fifteenth week. Most investigators agree that it is probably the testosterone secreted by the fetal testis which is responsible for the differentiation of the Wolffian-duct structures and the male external genitalia. Another testicular hormone (see Chapter 2) inhibits the development of the müllerian structures.

In contrast to the fetal testis, the fetal ovary is involved in only minimal steroid metabolism. The reduction of progesterone to 20α-dihydroprogesterone occurs, and possibly a small conversion of acetate to cholesterol, pregnenolone and progesterone. There is no evidence that the fetal ovary forms estrogens.

ADRENAL GLAND

The primordial cortical cells proliferating at the juncture of the mesonephros and the dorsal mesentery form invasive cords of cells, which soon separate from the coelomic epithelium and condense to form two large cell masses on either side of the aorta. Adjacent to these cortical cells are the medullary cells which had migrated from the neural crest. They begin to migrate into the interior of the adrenal cortical masses at about seven weeks' gestation. As the adrenal gland develops, an outer cortical layer can be distinguished. This layer encircles the original mass of cells. The former will be the definitive cortex of the adult; the latter forms the fetal zone of the adrenal cortex. This zone comprises about 80 per cent of the gland in fetal life. After birth these cells undergo involution and disappear by about 6 months of age.

The adrenal gland is one of the largest of the fetal organs. At four months' gestation it is larger than the kidney, but thereafter its relative size decreases, until at birth it is one-third the size of the kidney. With the rapid involution of the fetal zone during the neonatal period, the adrenal glands are reduced in size to about half that seen in the newborn infant.

The role of the fetal zone of the adrenal cortex has puzzled many investigators. These cells develop morphological changes indicative of functional differentiation during the first trimester. Comparable development of the endoplasmic reticulum, mitochondria, and Golgi complex is not seen in the definitive zone until the last part of the second trimester or the beginning of the third trimester. Adrenal glands of anencephalic fetuses appear to develop normally during the first third of gestation, but after the twentieth week, the fetal zone undergoes involution comparable to that normally seen after birth. It has been suggested that hCG produced during the first months of pregnancy stimulates adrenal growth. Maintenance of the adrenal cortex after the fourth or fifth month of gestation would depend on other trophic influences, such as ACTH from the fetal pituitary.

There is abundant histochemical and biochemical evidence for the presence in the adrenal cortex of enzymes involved in steroid synthesis and metabolism. The steroid nucleus can be synthesized from acetate, but interestingly enough, there appears to be a relative inactivity in the enzyme 3β-hydroxysteroid dehydrogenase. Both *in vivo* and *in vitro* studies have shown that the conversion of Δ5,3β-hydroxysteroids to Δ4,3-ketosteroids is rather limited (Figure 11–4). Acetate or cholesterol is readily converted to pregnenolone, 17-hydroxypregnenolone, dehydroepiandrosterone, 16α-hydroxydehydroepiandrosterone and their respective sulfates. If one bypasses the block, providing progesterone as a substrate, 17-hydroxyprogesterone, 16α-hydroxyprogesterone, and androstenedione are formed early in gestation. After eight weeks' gestation, both 21-hydroxylase and 11β-hydroxylase are active in the adrenal cortex. Thus, progesterone from the placenta can serve in the adrenal as the precursor of corticosterone and cor-

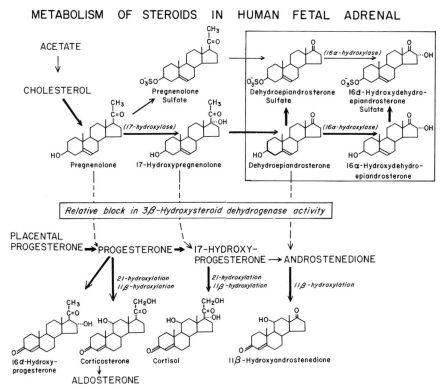

METABOLISM OF STEROIDS IN HUMAN FETAL ADRENAL

FIGURE 11–4. Pathways of steroid metabolism in the human fetal adrenal. The relative block in 3β-hydroxysteroid dehydrogenase activity is indicated by the dotted lines connecting the Δ5,3β-hydroxysteroids and the Δ4,3-ketosteroids.

tisol as well as the androgens (androstenedione and 11β-hydroxyandrostene-dione) and 16α-hydroxylated steroids. Aldosterone can also be synthesized from progesterone by the adrenal.

Perfusion studies as well as *in vitro* incubations with estrogens and Δ5–3β-hydroxysteroid substrates showed that a variety of fetal tissues, including the adrenal cortex, are capable of sulfurylating these compounds. The adrenal sulfokinases bring about a rapid and nearly complete conjugation of steroids, so it is not surprising that pregnenolone sulfate, dehydroepiandrosterone sulfate, and 16α-hydroxydehydroepiandrosterone sulfate as well as estrogen sulfates are found in fetal blood.

In summary, the human fetal adrenal cortex is an active endocrine gland producing large quantities of dehydroepiandrosterone and 16α-hydroxydehydroepiandrosterone. These products are rapidly sulfurylated and passed in the blood stream to either the liver or the placenta, where further transformations may occur. Because of a relative lack of 3β-hydroxysteroid dehydrogenase, the fetal adrenal cortex does not form glucocorticoids and mineralocorticoids from cholesterol or pregnenolone. A more likely substrate is progesterone, which is formed in large amounts in the placenta, and which can readily cross the placenta to the fetus. The fetal zone of the adrenal cortex is a likely source of the Δ5,3β-hydroxysteroids and the 16α-hydroxylated steroids found in fetal and newborn blood, since these steroids tend to disappear as the fetal zone involutes.

Other tissues of the fetus are intimately involved in the metabolism of steroids. In particular, the fetal liver is important in the reduction of progesterone (primarily to pregnanediol) and the 16α-hydroxylation of C19 and C18 steroids. 15α-Hydroxylation and 14α-hydroxylation of steroids also occur in the fetal liver. Both sulfates and glucosiduronates of steroids are formed readily in the liver. Thus a collaboration between the fetal adrenal and fetal liver yields large amounts of dehydroepiandrosterone sulfate and 16α-hydroxydehydroepiandrosterone sulfate, which circulate in the blood and reach the placenta for conversion to estrogens. In fetuses in whom either adrenal or liver function is compromised, estrogen excretion in maternal urine is reduced.

THE PLACENTA

The most obvious and immediate need for the developing embryo is the maintenance of a nutritive and highly vascular endometrium into which the trophoblast can insinuate itself. This is accomplished by the trophoblast's elaboration of several protein tropic hormones and two types of steroid hormones, estrogens and progestins. Although the maternal ovary by its secretion of estrogens and progesterone is responsible for the initial preparation of the endometrium for implantation, the role of the corpus luteum of pregnancy becomes minimal after the first few weeks. The trophoblast takes over.

Human Chorionic Gonadotropin (hCG)

A glycoprotein tropic hormone, similar to LH, is secreted by the tro-phoblast within the first few days after implantation. This hormone, hCG, formed by the syncytiotrophoblast, is found only in the presence of tropho-blastic tissue. Like the other glycoprotein hormones (FSH, LH, TSH) hCG consists of two nonidentical subunits. These subunits can be dissociated by incubation either with 10M urea or with 1M propionic acid. The α-subunit of hCG has been found to be a glycoprotein consisting of 92 amino acids and having two sites for carbohydrate attachment; its amino acid sequence has been determined. When the polypeptide chain of the α-subunit of hCG is aligned with the cysteine residues of the α-subunits of ovine LH and bovine TSH, 72 per cent of the amino acid residues of the three hormones are identical.

The larger β-subunit of hCG is the hormone-specific unit. Unlike the entire hCG molecule, the individual subunits are not concentrated by the ovary and possess no biological activity. A sizeable portion (30 per cent) of the hCG molecule is carbohydrate, and the sialic acid moiety is essential for biological activity. The molecular weight of the entire molecule is about 36,000.

The concentration of hCG in both maternal blood and urine rises to a maximum during the first trimester and declines thereafter to a low level for the latter portion of pregnancy (Figure 11–5). Very little hCG can be de-tected in fetal fluids, suggesting that the placenta secretes the hormone primarily to the mother. This does not rule out the possible influence of this hormone on fetal development, particularly at the period of peak production.

Many of the biological effects of hCG are similar but not identical to those of LH. It has been shown that hCG is capable of inducing ovulation of follicles previously primed by FSH and LH. Follicular cells are luteinized under the influence of hCG, and the corpus luteum formed during the men-strual cycle or in pregnancy is stimulated to produce more progesterone when hCG is added *in vitro* or *in vivo*. The actual lifespan of the corpus luteum is extended beyond its usual duration by hCG. In the testis, a com-parable LH-like stimulation of interstitial cells occurs under the influence of hCG. This latter effect is of importance when considering the role of hCG in the development of the fetal testis.

The molecular basis of these gonadal effects is not fully known. Some effects of hCG are abolished by inhibitors of RNA and protein synthesis, suggesting a locus of action involving protein (enzyme) synthesis. Cyclic AMP, LH, and hCG raise luteal progesterone production, and this increase is associated with expanded synthesis of RNA and protein. Thus cyclic AMP may be an intracellular messenger for LH and hCG. Certainly a major consequence of the effect of hCG on gonads is increased steroid production, but the phase of the biosynthetic pathway at which this effect occurs is still unknown. Some evidence suggests an effect of hCG on one of the steps in the conversion of cholesterol to pregnenolone.

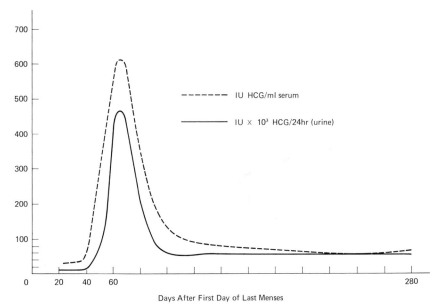

FIGURE 11–5.   The concentrations of human chorionic gonadotropin (hCG) in serum and in urine as a function of the length of gestation. (Redrawn from data presented by Albert and Berkson, 1951.)

Cedard has found a stimulation of estrogen synthesis in the perfused placenta as a result of the addition of hCG (or LH) to the perfusion fluid. She postulated that LH or hCG stimulates adenyl cyclase, increasing the intracellular cyclic AMP, which in turn increases the amount of active phosphorylase kinase. Such an effect would increase the breakdown of glycogen, yielding more glucose-6-phosphate. The latter, by its metabolism via the pentose pathway, produces increased amounts of NADPH. The hydrogen from NADPH would be available for steroid hydroxylation and aromatization. In the placental perfusion system of Cedard, cyclic AMP mimics the effects of LH or hCG on estrogen synthesis.

Recently hCG was shown to stimulate the incorporation of ^{14}C-adenine into ^{14}C-cyclic AMP in the term placenta, further implicating adenyl cyclase as a site of action. Such a stimulation of adenyl cyclase by hCG is in agreement with the findings of Demers. He determined that hCG (50 to 250 I.U./ml) caused an acute glycogenolytic response in placental tissue *in vitro*, coupled with a decrease in placental glycogen concentration and an increase in placental cyclic AMP concentration. Demers studied the inactive and active glycogen phosphorylases and noted a shift from the inactive to active form of the enzyme with hCG stimulation. This would be compatible with Cedard's theory of an effect of cyclic AMP on the phosphorylase phosphokinase.

Thus the placental hormone hCG, perhaps via adenyl cyclase, enhances the synthesis of steroids in the gonads as well as in its own cells. The possibility that the placenta controls its own steroid synthesis is intriguing, but evidence is not yet available to accept or reject it. By means of its gonadal effects, hCG prolongs the life span of the corpus luteum and possibly exerts an influence on the developing fetal gonads.

Present-day assays for hCG are either biological or immunological. Some of the bioassays depend upon the direct action of hCG on the ovary. One of the most sensitive involves the end point of ovarian hyperemia in the immature rat. Other biological assays utilize effects secondary to the stimulation of gonadal hormone secretion by hCG. Among the most sensitive in this group are the quantitative measurement of increase in uterine weight and the presence of vaginal cornification as a result of administration of hCG to the immature rat.

A wide variety of immunoassays for hCG are available. These tests include the techniques of hemagglutination inhibition, complement fixation and radioimmunoassay. Agglutination inhibition tests, utilizing the ability of tannic acid-treated erythrocytes to absorb antigens and their subsequent agglutination in the presence of antibody, are now used widely in the diagnosis of pregnancy. The known ability of LH to cross-react with hCG immunologically has complicated the immunoassays of hCG. Many laboratories now have antibodies to the hCG β-subunit, thus making the radioimmunoassay of hCG far more specific.

Because hCG is cleared rapidly, approximately equal amounts are found in serum and urine. Thus assays may be performed on either, and the patterns of hCG concentration in both during pregnancy are very similar (Figure 11–5). Since hCG is found only in the presence of viable trophoblast, its assay serves as a diagnostic test of pregnancy, or more rarely, in patients with trophoblastic disease. There is a rough correlation between the concentration of hCG and the amount of trophoblastic tissue present. Values of hCG below normal should suggest death of the conceptus, whereas values above normal may indicate multiple pregnancy or trophoblastic disease. Serial values are helpful in doubtful cases and, of course, the measurement of other hormones provides further evidence regarding the viability of the placenta and fetus.

HUMAN PLACENTAL LACTOGEN (hPL)

Another protein hormone formed by the trophoblast is human placental lactogen (hPL), or human chorionic somatomammotropin (hCS). In contrast to hCG, it does not appear to have a carbohydrate component. It is a polypeptide of 191 amino acids with a molecular weight of about 20,000. Its amino acid composition and primary structure are very similar to those of growth hormone and prolactin. It is not surprising therefore that hPL has both growth-promoting and lactogenic properties. Preparations of the hormone can cause lactation in pseudopregnant rats and positive reactions in assays of the pigeon crop-sac (an analogue of the mammalian breast). The experiments of Turkington have shown that hPL (in combination with insulin and cortisol) stimulates the synthesis of casein and other proteins in cultures of mammary epithelial cells.

Some of the earliest studies with hPL demonstrated its somatotropic action. In hypophysectomized rats, the width of the tibial epiphyseal cartilage increases after administration of hPL. The growth-promoting action of hPL is far weaker than that evoked by comparable doses of pituitary growth hormone. Josimovich has suggested that hPL potentiates the action of growth hormone, rather than having a unique somatotropic action of its own. Mammotropic and luteotropic actions of hPL have also been documented.

More recently, a variety of metabolic changes in pregnancy have been ascribed to the action of hPL. Among these are insulinogenic and diabetogenic effects. This aspect of hPL action will be discussed later, when we consider the metabolic alterations of pregnancy.

As with hCG, there seems to be no doubt that hPL is synthesized solely by trophoblastic tissue. *In vitro* cultures of placental fragments can form labeled hormone from ^{14}C-amino acids. By immunofluorescent techniques, the syncytiotrophoblast has been implicated in its synthesis.

In the serum of pregnant women hPL can be detected as early as the sixth week of gestation, and its amount rises steadily during the entire period of gestation (Figure 11–6). The blood levels show little variation and are not affected by blood glucose concentrations. Low values of hPL have been reported with chronic fetal distress. HPL disappears promptly from the blood after delivery, with a half-time in the serum of about 30 minutes.

Very little is known about the factors which regulate the synthesis and secretion of hPL. In several studies the concentration of hPL in maternal serum could be correlated with placental weight, whereas in other studies no correlation was found. As there is considerable variation in different placental fragments in the synthesis of hPL *in vitro,* factors within the trophoblastic cells may govern hPL production. Recently, Tyson and his associates showed that the concentration of hPL in maternal serum was consistently higher in mothers undergoing prolonged food deprivation than that in mothers on complete diets. Prolonged fasting can be mimicked *in vitro* by adjusting the culture medium. Friesen and his colleagues found that under such conditions the synthesis and secretion of hPL increased. This effect of starvation on hPL concentration would lead to enhanced fat mobilization in the calorically deprived mother.

In addition to hCG and hPL the placenta forms a thyrotropin, human chorionic thyrotropin (hCT). Recent evidence gathered by Gibbons and colleagues (1975) suggests that placental fragments are capable of incorporating labeled proline, glutamic acid or histidine into thyrotropin releasing hormone (TRH) and into luteinizing releasing hormone (LRH). Thus the placenta has an enormous potential for the synthesis of biologically active peptide and protein hormones.

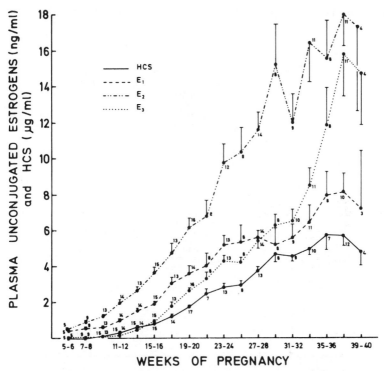

FIGURE 11–6. Concentrations of unconjugated estrogens and placental lactogen in the plasmas of 22 normal women during the course of pregnancy. The graph depicts the means and the vertical bars the standard error of the means of the values for estrone (E_1), estradiol (E_2), estriol (E_3) and human placental lactogen (human chorionic somatomammotropin, hCS). The figures indicate the number of subjects included in each two-week period. (Reproduced, with permission, from DeHertogh, R., Thomas, K., Bietlot, Y., Vanderheyden, I, and Ferin, J. J. Clin. Endocrinol. Metab. *40*:93–101, 1975.)

STEROID SYNTHESIS IN THE PLACENTA

It has been documented that the human placenta possesses the enzymes necessary for the conversion *in vitro* of acetate to cholesterol (see Chapter 3); however, this transformation is very small, and it has been shown that *in vivo* maternal blood cholesterol is used in preference to endogenous cholesterol for the synthesis of progesterone. The steps involved in the conversion of cholesterol to progesterone are outlined in Figure 11–7. Cholesterol, pregnenolone sulfate and pregnenolone are very efficient precursors of progesterone. Once formed, most of the progesterone is stored as such or secreted by the placenta. A small amount is converted to 20α-dihydroprogesterone, and even smaller amounts to 6α- and 6β-hydroxyprogesterone.

Although very small amounts of 17-hydroxylase, 16α-hydroxylase, and 17-20 desmolase have been reported in the placenta by some investi-

SYNTHESIS AND METABOLISM OF PROGESTERONE IN HUMAN PLACENTA

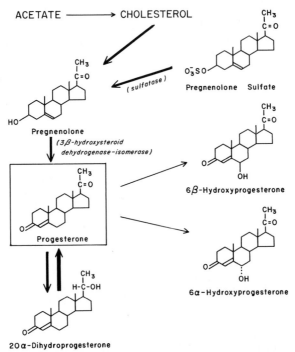

FIGURE 11–7.   The synthesis and metabolism of progesterone in the human placenta.

gators, it seems unlikely that these enzymes are quantitatively significant in the overall pattern of steroid metabolism. Thus, with the virtual absence of 17-hydroxylase and 17-20 desmolase, the conversion of C21 steroids to C19 steroids is impossible. The placenta, therefore, cannot convert any appreciable amount of progesterone to androstenedione (or pregnenolone to dehydroepiandrosterone). It must be remembered that all other steroid-producing glands are able to perform this elemental conversion (even the adrenal, the primary concern of which is the synthesis of C21 steroid hormones). We may think of the placenta as an incomplete steroid-synthesizing gland.

Androgens and estrogens are formed from either pregnenolone or progesterone (see Chapter 3). If the 17-hydroxylation and subsequent side-chain cleavage of these precursors cannot occur in the placenta to any significant extent, how are androgens and their estrogen offspring produced?

Estrogens can be isolated in large quantities from placenta, and a considerable body of evidence indicates that they are actually synthesized there. The precursors of these estrogens are C19 steroids such as dehydroepiandrosterone   and   16α-hydroxydehydroepiandrosterone   (Figure

11–8). They are brought to the placenta in the blood from either fetal or maternal sources. Usually they arrive in the sulfurylated form, but the high activity of placental sulfatase ensures an essentially complete conversion to the free steroids. The conversion of dehydroepiandrosterone to androstenedione is catalyzed by 3β-hydroxysteroid dehydrogenase (similar to or the same as the enzyme used in the conversion of pregnenolone to progesterone). This enzyme and the aromatizing system for the conversion of androgens to estrogens are very active in the human placenta. If sufficient C19 precursors can be provided, the placenta is capable of forming large amounts of estrogen. The almost complete lack of 16α-hydroxylase requires that the placenta be provided with a 16α-hydroxylated C19 steroid in order to form estriol.

THE MATERNO-FETO-PLACENTAL UNIT

The attractive complementarity of the enzymes of the fetal adrenal cortex and those of the placenta have encouraged the adoption of the con-

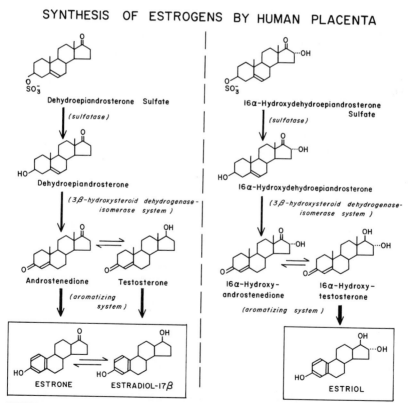

FIGURE 11–8. Diagram of the reactions by which estrogens are synthesized in the human placenta. Note the absence of 16α-hydroxylase in the human placenta.

cept that the fetus and placenta collaborate in the synthesis of estrogens. The term feto-placental unit has been largely accepted now; however, it is important to remember two things: (1) this collaboration is applicable only to steroid metabolism as far as we know, and (2) it should not in any way detract from the important maternal contributions to steroid metabolism.

It would seem beneficial at this point to review what we know about the formation and excretion in pregnancy of each of the major groups of steroid hormones. Then we can go on to a discussion of other endocrine and metabolic changes of pregnancy. The feto-placental unit will be referred to when appropriate, but the emphasis will be on the entire materno-feto-placental unit.

ESTROGENS AND ANDROGENS

Concentrations of the three principal estrogens (estrone, 17β-estradiol and estriol) are markedly elevated during pregnancy (Table 11–2). Excretion of estrone in the urine in pregnancy is increased one hundred-fold over that in the non-pregnant state, and estriol one thousand-fold. The only endocrine gland that can alone and unaided synthesize estrogens from acetate or cholesterol in significant quantities is the maternal ovary (see Chapter 3). However, removal of the maternal ovary after about six weeks of pregnancy appears to have little effect on the excretion of estrogens in maternal urine.

The hallmark of the estrogens is the aromatic or phenolic A ring. As was discussed previously, the placenta has a powerful aromatizing system and can convert a variety of C19 precursors to estrogen. The concentrations of dehydroepiandrosterone sulfate and 16α-hydroxydehydroepiandrosterone sulfate are higher in umbilical arterial plasma than in venous plasma, suggesting production of these precursors by the fetus. There seems to be little doubt that most of this dehydroepiandrosterone is synthesized in the fetal adrenal cortex, where it may also be 16α-hydroxylated or sulfurylated, or

Table 11–2.   **EXCRETION OF ESTROGENS AND PROGESTINS IN URINE (mg/24 hr.)**

	Non-Pregnant (Midcycle)	Pregnant (Term)	Pregnant with Anencephalic
Estrone	0.021[1]	0.93 to 2.1[2]	0.005 to 1.26[6]
17β-estradiol	0.008[1]	0.33 to 0.90[2]	0.014 to 0.290[6]
Estriol	0.029[1]	14 to 43[4]	0.80 to 4.7[6]
Pregnanediol	2 to 4[3]	15 to 70[5]	5.7 to 56.8[6]

[1]Brown, J. B., and Matthew, G. D. Rec. Prog. Horm. Res., 18:337, 1962.
[2]Brown, J. B. Lancet 1:704, 1956.
[3]Klopper, A., Michie, E. A., and Brown, J. B. J. Endocrinol., 12:209, 1955.
[4]Coyle, M., and Brown, J. B. J. Obstet. Gynaec. Brit. Cwlth., 70:225, 1963.
[5]Shearman, R. J. J. Obstet. Gynaec. Brit. Cwlth., 66:1, 1959.
[6]Frandsen, V. A., and Stakemann, G. Acta Endocrinol., 47:265, 1964.

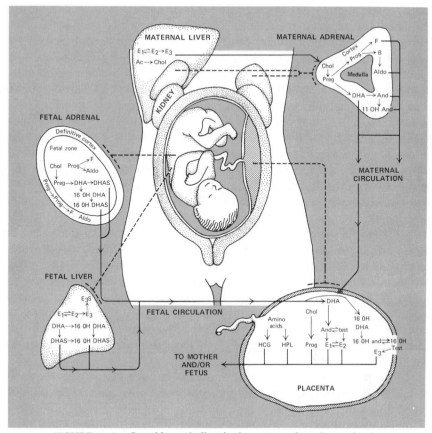

FIGURE 11–9. Steroid metabolism in the materno-feto-placental unit.

both. The fetal liver contributes to estriol synthesis, since it is capable of 16α-hydroxylating any circulating dehydroepiandrosterone. The conjugation of these precursors with phosphoadenosyl phosphosulfate ensures a greater water solubility to aid in transport. Within the placenta, highly active sulfatases remove these groups to release the free steroid (Figure 11–9).

The disproportionate rise in estriol concentration in pregnancy as compared to the rise in estrone and 17β-estradiol is a reflection of a remarkably active 16α-hydroxylase in the fetal adrenal and liver. Thus, measurement of this estrogen in urine or amniotic fluid can serve as an index of fetal well-being. Usually, serial determinations of estriol are used in conjunction with other criteria to aid in deciding the optimal time for delivery in pregnancies in which complications such as diabetes, toxemia, intrauterine growth retardation, Rh incompatibility, and so forth exist. Fortunately, the techniques for measuring estriol are accurate and fairly rapid.

In anencephaly, if the fetal adrenal is atrophic, estriol excretion in the maternal urine is markedly reduced (Table 11–2). In the event of fetal death, estrogen excretion is also reduced. It would, however, be unwise to consider the fetus as the sole supplier of the precursors for placental synthesis of estrogens. A variety of studies of urinary steroids in pregnant women with Addison's disease and in pregnant women who have undergone adrenalectomy for Cushing's disease have shown clearly that estrogen excretion is reduced, in some cases to half the normal pregnant level. This would suggest that the maternal adrenal provides about one-half the estrogen precursors in pregnancy. It would seem that the very high estrogen concentrations in the blood and urine of a pregnant woman are not required for the well-being of the offspring, since in instances of reduced estrogens because of maternal adrenal insufficiency, no abnormalities have been noted in the newborn. This leaves us with the question of why so much excess estrogen is synthesized. Presumably the "excess estrogen" is largely estriol, a biologically rather inactive steroid. The further hydroxylation of estriol to form 15α-hydroxyestriol yields an even less estrogenic compound. The fetal liver has very active 15α-hydroxylases to inactivate the estrogens. Indeed, it seems quite likely that hydroxylation at several carbons (C14, for example) occurs to a significant degree and may serve further to protect the fetus against biologically active estrogens synthesized in the placenta.

This tendency to hydroxylate and conjugate estrogens (primarily sulfates) is found in the newborn also. Estrogens administered to either males or females are rapidly metabolized to highly polar (hydroxylated) estrogen sulfates, and are thus rendered innocuous.

The immediate precursors of the estrogens are the androgens, androstenedione and testosterone. They too are increased in concentration in the blood and urine in pregnancy. The increase in total circulating testosterone is undoubtedly due to the increased amount of the testosterone binding protein in the blood of pregnant women. The actual amount of free testosterone is not elevated.

PROGESTINS

It has been estimated that from 250 to 450 mg. per day of progesterone are formed by the placenta in the last trimester of pregnancy. These enormous amounts of progesterone must be metabolized and excreted. The concentration of progesterone in maternal peripheral plasma is increased during pregnancy, and this increase is paralleled by a rise in urinary pregnanediol glucosiduronate. The conversion of progesterone to pregnanediol and the glucosiduronization of the latter takes place primarily in the maternal liver. Based upon differences in concentration of progesterone in umbilical artery and vein, it seems quite likely that large amounts of progesterone are metabolized by the fetus. It is estimated that about one-half of the placental progesterone reaches the fetus. There it can be metabolized by the fetal liver (forming 20α-dihydroprogesterone and pregnanediol) or by the fetal adrenal cortex, where it may serve as a precursor of glucocorticoids, mineralocorticoids or 16α-hydroxyprogesterone (Figure 11–4). The 20α-dihydroprogesterone formed in the fetal liver may reach the placenta and there be converted back to progesterone. This would permit the reutilization of progesterone, which is indispensable for the maintenance of pregnancy.

Unlike the estrogens, progesterone can be synthesized in the placenta from acetate or cholesterol, and therefore its concentration may be a reflection of placental function independent of the fetus. Obviously, though, a severely compromised placenta will be unable to support the fetus, and falling pregnanediol values may be just as ominous for the fetus as falling estriol levels.

Following delivery of the placenta, there is a very rapid fall in the concentration of progesterone in maternal plasma. At one time, this sharp decline in progesterone was believed to precede delivery and to be instrumental in the initiation of labor. At present, this view is not widely held, and it appears that the fall in progesterone is secondary to placental separation.

We can summarize the present concepts of estrogen and progesterone metabolism in pregnancy as follows. Although the corpus luteum of pregnancy undoubtedly furnishes most, if not all, of the progesterone in the first few weeks of gestation, as soon as the trophoblast cells become established they proceed to form increasing amounts of this steroid. Maternal (or fetal) cholesterol serves as the primary precursor, and the progesterone formed is distributed to both fetus and mother. Its biological action in pregnancy has been discussed in Chapter 3. Rapid metabolism of progesterone to the weaker gestagen, 20α-dihydroprogesterone, is ensured by active 20α-reductases in maternal and fetal liver. 20α-Dihydroprogesterone can be reclaimed by the placenta to be reutilized as progesterone (Figure 11–7). Further reduction of the biological activity of the gestagens is achieved by the reduction of ring A. Both maternal and fetal livers are capable of converting 20α-dihydroprogesterone to pregnanediol. The latter is conjugated to the glucosiduronate by the maternal liver and excreted in the maternal urine. Pregnanediol is readily measured and may serve as a useful index of placental function.

Estrone and 17β-estradiol are formed in the placenta from androstenedione and from testosterone, respectively. These androgens are formed from dehydroepiandrosterone supplied by either the maternal or the fetal adrenal. The dehydroepiandrosterone is transported to the placenta primarily as the sulfate; however, the very active sulfatases in this tissue can rapidly remove the sulfate group, providing the free steroid for conversion to androstenedione. Estriol is synthesized from 16α-hydroxydehydroepiandrosterone by identical pathways. 16α-Hydroxylation occurs primarily in fetal liver and fetal adrenal cortex; however, the maternal adrenal can also provide a certain amount of 16α-hydroxydehydroepiandrosterone for estriol synthesis. Disorders of the fetus which impair either fetal adrenal or fetal liver function should be reflected in diminished estrogen excretion, which can be measured in either maternal urine or amniotic fluid. Disease or absence of the maternal adrenals will usually reduce estrogen production and excretion also.

Anencephaly, intrauterine growth retardation and chronic fetal distress may be frequently diagnosed and followed by serial estriol determinations in conjunction with other measures of fetal function. If proper facilities and experienced personnel are available, amniocentesis should be performed, and it is preferable to measure the concentration of estriol in amniotic fluid rather than in maternal urine. Particularly in Rh disease, amniotic fluid estriol is probably a more reliable indicator of fetal well-being. These determinations are used in conjunction with intrauterine fetal transfusions to follow the condition of the fetus. Berman et al. have found that in suspected placental insufficiency, values of less than 100 μg. of estriol per liter of amniotic fluid were consistently associated with severe fetal distress and seemed to reflect more accurately the condition of the fetus than did the concentration of estriol in maternal urine. In anencephaly, not only are estrogens reduced, but the concentrations of both dehydroepiandrosterone sulfate and 16α-hydroxydehydroepiandrosterone sulfate in cord plasma are also markedly diminished.

CORTICOSTEROIDS

There is little or no evidence for the synthesis of glucocorticoids or mineralocorticoids by the placenta. In pregnant women who have Addison's disease or who have been adrenalectomized for Cushing's disease, if corticoid therapy is interrupted, the excretion of 17-hydroxycorticosteroids in the urine disappears. Thus the maternal adrenal is the major if not the sole source of the 17-hydroxycorticosteroids excreted in the urine in pregnancy. The concentration of cortisol in maternal blood during pregnancy is increased two- to three-fold over that in the non-pregnant state. However, this increase can be explained by the increase in transcortin in the blood. The capacity for binding cortisol in maternal blood increases during pregnancy and is greater than that of the fetus. This may explain the two to five times greater concentration of 17-hydroxycorticosteroids in maternal blood than in fetal blood. In addition, the production rates of cortisol are increased two- to three-fold in pregnancy and range from 20 to 40 mg. per day.

There is no doubt that both cortisol and progesterone can cross the placenta and reach the fetus. It is unknown, however, how much fetal cortisol is synthesized by the fetal adrenal from progesterone and how much is derived from the maternal blood. The cortisol production rate in the newborn, when corrected for surface area, is within the normal adult range.

In addition to cortisol, the production rate of aldosterone in pregnancy may be increased two- to five-fold over that in the non-pregnant state. This may simply be a reflection of the increased renin-angiotensin which is observed in the pregnant state.

METABOLIC CHANGES IN PREGNANCY

Carbohydrate Metabolism

Intermediary metabolism is geared for the storage of excess energy after eating, to provide for periods of fasting. Glucose absorbed from the gastrointestinal tract, for example, is distributed throughout the extracellular fluid. A certain proportion is used for immediate metabolic needs; the rest can be hoarded in the liver or muscle as glycogen or in adipose tissue as triglyceride. Insulin is released during the process of absorption of glucose and amino acids, and high levels in the blood are sustained by the insulinogenic effects of the rising concentrations of glucose and amino acids in the blood. Insulin facilitates the entrance of glucose and amino acids into the cells of muscle and adipose tissue. Any excess carbohydrate may be stored as glycogen in muscle or as glycogen or triglyceride in adipose tissue. Amino acids are stored as protein in muscle. Insulin also promotes the activity of the enzyme that facilitates the cleavage of exogenous esterified fats (lipoprotein lipase) so that dietary fatty acids can be stored. Thus, insulin is a true anabolic hormone.

The major metabolic consideration is to provide continued access of the central nervous system to glucose. It is estimated that approximately 140 gm. of glucose are oxidized by the adult central nervous system daily.

The picture is considerably complicated in pregnancy. Another structure has been added, the conceptus, which is also a glucose consumer. The fetus also requires amino acids for growth. Thus, the mother is subjected to a constant drain on her glucose and amino acid supplies. There is at present a considerable body of evidence that suggests that the conceptus continues to abstract glucose and amino acids at much the same rate in the face of maternal starvation as it does in the fed state. Thus, the fetus is a true parasite which will manage to thrive regardless of peril to the maternal organism. In prolonged malnutrition, the fetal organs have different priorities for utilizing available nutrients. Under such conditions, brain weight is maintained in preference to other organ weights.

In order to provide a constant supply of glucose and amino acids to the fetus, the mother must make certain adjustments. No longer can she use glucose as indiscriminantly as before. As gestation proceeds, there is a diminished peripheral utilization of glucose in the mother. The usual hypoglycemic response to insulin therapy is reduced in pregnancy. This progressive biologic resistance to insulin which is observed in pregnancy is compensated for by increased plasma concentrations of the hormone. In addition, there is an increase in insulin response to a glucose load and islet cell hyperplasia of the maternal pancreas.

Fat Metabolism

The metabolism of the pregnant woman is altered in order to conserve carbohydrate for the conceptus and for her own nervous system. Control mechanisms are shifted so that fat is consumed whenever possible. Fat stores are mobilized and the concentration of free fatty acids (FFA) in the plasma is increased. One might anticipate that such changes would be mediated by the lipolytic hormones, such as epinephrine, growth hormone and glucagon. There is no evidence to support this, however. Excretion of catecholamines is increased in the 24 hours following delivery, but there is no persistent elevation during pregnancy. Growth hormone values in maternal blood are low, in marked contrast to the high concentrations found in fetal blood.

HPL, which is secreted by the placenta, is also a lipolytic hormone in the sense that its administration increases the rate of lipolysis and the concentration of FFA in the plasma. There are no known controls for the secretion of hPL, and it is quite possible that the autonomous release of large quantities of the hormone daily is responsible for many of the diabetogenic changes noted in pregnancy. Grumbach has theorized that the role of hPL in lipid metabolism in the mother spares carbohydrate by providing an alternate source of energy, FFA, which in turn, in high concentration in the cell, inhibits glucose utilization. The triglyceride and glycogen stores are replenished, of course, with each feeding of the mother, by the usual mechanisms. The anti-insulin effect of hPL (which is chronically elevated) is probably responsible for the high insulin values of pregnancy.

Free fatty acids cross the human placenta from mother to fetus. They are derived primarily from circulating free fatty acids bound to albumin in maternal serum. To a lesser extent, the fetal fatty acids may originate from maternal serum triglyceride which is cleaved to free fatty acids and glycerol by a placental lipoprotein lipase. There is some evidence that increased quantities of free fatty acids cross to the fetus from the mother in diabetic pregnancies. The concentration of free fatty acids in maternal blood increases significantly in pregnancy, and diabetic mothers have even higher levels of free fatty acids. The arteriovenous difference in the concentration of free fatty acids in umbilical blood suggests that the fetus extracts significant quantities of free fatty acids. Nevertheless, the concentrations of free fatty acids in fetal blood are very low compared with those in the adult.

Protein Metabolism

The continued growth of the fetus, and for that matter of the adolescent mother, is dependent upon ready availability of all the amino acids and the energy to run the machinery of protein synthesis. Insulin plays an important role in protein synthesis by its facilitation of transport of amino acids into the cell. The continued secretion of insulin at a higher than normal rate in pregnancy ensures adequate protein synthesis for the mother. It is still not certain whether or not a similar role for insulin exists in the fetus.

Concentrations of all the amino acids are higher in the fetus than in the mother; the feto-maternal α-amino nitrogen ratio ranges from 1.03 to 3.00.

During pregnancy, the levels of amino acids in maternal plasma are lower (3.2 mg./100 ml.) than they are in the non-pregnant normal woman; however, these values return to normal (4.3 mg./100 ml.) a few days after delivery. The levels in the fetus appear to be comparable to levels in the adult non-pregnant female.

Phenylalanine concentrations in the pregnant female are markedly lower than those in the non-pregnant female, whereas the fetal level of phenylalanine is increased over that in the mother. Kang and Paine have shown that in a pregnant woman who is heterozygous for phenylketonuria, the plasma phenylalanine level was actually increased. Since the feto-maternal ratio remains nearly the same even when phenylalanine concentrations are increased, the fetal level would be tremendously increased over normal. It has been suggested that patients with phenylketonuria who become pregnant should have adequate but not high-protein diets during pregnancy, to avoid damage to the fetus.

The concentration of serum albumin falls markedly as pregnancy advances, and there is a minor drop in gamma globulin. Both beta globulin and fibrinogen increase; however, the total plasma proteins in pregnancy are reduced. Some serum enzymes, such as alkaline phosphatase, diamine oxidase and oxytocinase, are markedly elevated. The changes which occur in the blood coagulation system are reviewed in the next chapter.

Minerals and Vitamins

Probably no single class of compounds has been the target of so much misrepresentation and misuse as the vitamins, and minerals are not far behind in this respect. There is a general tendency in obstetric practice to supplement the dietary intake of most of the minerals and vitamins during pregnancy, particularly during the second half. The rationale for this procedure is discussed in Chapter 13.

CALCIUM. We know very little about alterations to calcium metabolism in pregnancy. The conceptus requires a total of about 30 gm. of calcium, most of this during the second half of pregnancy. Osteomalacia has occasionally been noted in pregnant women, but in each case it has been under unusual social conditions. The large stores of calcium in maternal bone make it unlikely that depletion would occur with an adequate diet. Since the efficiency of the absorption of calcium in the gastrointestinal tract is geared to the need for it, it is quite likely that calcium is absorbed more efficiently in pregnancy.

The concentration of PTH in blood is in the normal nonpregnant range throughout the first half of pregnancy. It then declines from 20 to 24 weeks and subsequently rises progressively to term, reaching a value 2.4 times greater than in the non-pregnant woman. There is some evidence that the concentration of calcitonin in maternal serum doubles late in pregnancy. Such an increase could serve as a compensatory mechanism to limit the osteolytic activity of the increased PTH while permitting the effects of PTH on the gut and kidney to provide for the additional calcium requirements of pregnancy.

The total calcium concentration in maternal serum falls during pregnancy. This decrease begins shortly after conception and progresses until the middle of the third trimester. Serum inorganic phosphorus falls progressively until about 30 weeks, then rises and reaches normal values at term. The decrease in concentration of calcium in the serum may be secondary to a decrease in serum protein. The patterns of calcium and albumin concentrations in the serum during pregnancy are identical. Thus the concentration of ionic calcium may not change during pregnancy. Calcium ions are transferred from maternal to fetal blood against a concentration gradient; the concentrations of serum calcium in the fetus exceed those in the mother by 1 to 2 mg./100 ml. by term.

The current recommended dietary allowance for calcium during pregnancy is 1200 mg. daily, which represents an increase of 400 mg. daily over the allowance for the non-pregnant adult. The requirement for calcium during lactation represents an increase of from 150 to 300 mg. daily over the allowances for the non-pregnant female, and depends upon the volume of milk produced. The consumption of one quart of whole milk daily will provide the amount of calcium required for both pregnancy and lactation.

IRON.   In contrast to calcium, there is no large storage depot for iron, and deficiency may develop, particularly in women, since they lose approximately 50 ml. of blood with each menstrual flow. The replacement of this amount of blood alone requires 0.9 mg. of iron per day. The average unsupplemented diet contains barely enough iron to meet these requirements.

Despite the fact that there is little or no storage iron in the average woman, the iron requirements of pregnancy are considerable. The average full term infant contains about 250 to 300 mg. of iron. The fetal blood in the placenta and cord accounts for another 50 mg. The maternal red cell volume increases during pregnancy, and another 500 to 600 mg. of iron is needed for the extra 450 to 550 ml. of red cells present. Thus, the total maternal requirement for iron on the average is between 750 and 900 mg., or about 5 mg. per day in the second and third trimester.

Although it is relatively easy to calculate the physiological iron requirement, precise calculation of the nutritional iron requirement is not feasible, since ingested iron is not quantitatively absorbed from the intestine. The ferrous form is more readily absorbed than the ferric form; thus, the presence or absence of reducing agents, such as ascorbic acid, will influence the amount of iron absorbed. The recommended supplements for the pregnant woman are based upon the physiological requirements.

VITAMINS.   It is conventional to increase the "recommended allowances" of vitamins during pregnancy. Clear-cut evidence for vitamin deficiency during pregnancy is lacking for most of the vitamins; however, folic acid appears to be an exception. After delivery, in one series of patients, as many as one-third showed evidence of folic acid deficiency. Kitay believes that it is probably the most common deficiency of a water-soluble vitamin in America.

Defects in ingestion, absorption, utilization or metabolism of folate may be involved in the etiology of folate deficiency. Other possibilities are increased demands, as in multiple pregnancy or closely spaced pregnancies, or hemolytic disease. Willoughby and Jewell found that supplementation of dietary folate with 355 μg. of folic acid daily during the third trimester maintained the serum folate level of most women within the normal non-pregnant range.

SUMMARY

The demands on the maternal organism during pregnancy are seemingly endless. Whether we look at endocrine alterations, shifts in carbohydrate and fat metabolism, protein anabolism, or stores of minerals and vitamins, we see the development of a protective organism which envelops the growing offspring and provides the necessary nutrients for the development of the fetus.

We speak of the "drain" on the maternal organism by the fetus, and yet as one examines the pregnant state one is impressed by the marvelous adaptation of each to the other. The mother alters her hormonal status in order to conserve carbohydrate and amino acids for the fetus. HPL probably plays an important role in this shift, so that the mother can utilize energy from fat whenever possible. This hormone, perhaps fortunately, does not cross the placenta, so that the fetus is not under a diabetogenic influence. Rather, fetal insulin is found in abundant amounts, and may constitute the major fetal hormone involved in growth of cells and tissues. The role of fetal growth hormone is uncertain at present.

The interplay of steroid hormone synthesis between fetus and placenta provides a useful diagnostic tool: estriol in maternal urine or amniotic fluid. The fact that both the fetal adrenal and the fetal liver are involved in providing precursors to the placenta for estrogen synthesis gives us a means of measuring the viability of these two organs and thus, indirectly, of the fetus. The differentiation of estriol as a product of both fetus and placenta from hormones synthesized only or largely in the placenta (hCG, hPL, progesterone) provides a means of assessing placental and fetal function.

SUGGESTED SUPPLEMENTARY READING

Klopper, A., and Diczfalusy, E. (Eds.). Foetus and Placenta. Blackwell Scientific Publications, Oxford, 1969.
   A multiauthored compendium of facts about the fetus and the placenta, with the emphasis on placental morphology and function. Steroid metabolism is covered in detail.
Pecile, A., and Finzi, C. (Eds.). The Feto-placental Unit. Excerpta Medica Foundation, Amsterdam, 1969.
   This book represents the papers presented at an international symposium held at Milan

in September, 1968. Large segments deal with structural studies, endocrinology and clinical investigations.

Philipp, E. E., Barnes, J., and Newton, M. (Eds.). Scientific Foundations of Obstetrics and Gynaecology. F. A. Davis Co., Philadelphia, 1970.
A good sourcebook. See especially Section V, Metabolism and Nutrition, and Section X, Endocrinology.

Ryan, K. J. (Ed.): Obstetrical Endocrinology. Clin. Obstet. Gynecol. *16*: No. 3, 1973.
A series of papers on endocrine function in pregnancy. The following are particularly appropriate to this chapter: Thyroid function and dysfunction during pregnancy, by H. A. Selenkow, M.D. Birnbaum and C. S. Hollander; and Endocrine regulation of metabolic homeostasis, by S. S. C. Yen.

Villee, D. B.: Human Endocrinology: A Developmental Approach. W. B. Saunders Co., Philadelphia, 1975.
The development of endocrine function in the placenta and fetus is outlined. Special emphasis is placed on endocrine regulation of differentiation and maturation. Changes during the perinatal period are covered in detail.

Waisman, H. A., and Kerr, G. (Eds.). Fetal Growth and Development. McGraw-Hill Book Co., New York, 1970.
Procedures of a conference held in November, 1968. The 25 chapters cover a variety of subjects related to the perinatal period. See especially Chapter 9, Steroid Synthesis and Catabolism in the Human Fetoplacental Unit, and Chapter 16, Energy Metabolism and Fetal Development.

Chapter 12

PHYSIOLOGIC ADJUSTMENTS IN PREGNANCY

The complex and multiple physiological changes observed in pregnancy, even in the early weeks, are adjustments brought about, for the most part, by placental secretions. Some of these presumably occur in the interests of the growing embryo or fetus, but once a single alteration is accomplished, then the maternal body may react with a series of additional adjustments required for her own interests.

As an example, one of the earliest changes is simply an extension and exaggeration of a progesterone effect which regularly occurs during the luteal phase of the menstrual cycle. A steadily increasing plasma progesterone concentration results in a lower "setting" in the hypothalamus for plasma pCO_2. The partial pressure of CO_2 falls from about 40 to about 32 mm. Hg. In addition to this, the respiratory centers become more sensitive to small increases in carbon dioxide concentration. A rise of 1 mm. Hg pCO_2 results in an increased ventilation of some 6 liters per minute, which is several times the change noted in non-pregnant individuals.

There is a persistent increased ventilation beginning early in pregnancy and amounting to about a 42 per cent increase by term (Cugell et al.). At any period of gestation, the rise is greater than the increase in total oxygen requirement. With the fall in carbon dioxide tension, one might predict a rise in the oxygen tension. The pO_2 does increase from a non-pregnant mean of 95 mm. Hg to a mean of 106 mm. Hg (Andersen et al.) early in pregnancy, and remains at this level until term.

Presumably the reduced pCO_2 of maternal plasma makes the elimination of carbon dioxide by the fetus easier. But if such a fall occurred without other adjustments, it would alter the pH of the maternal blood, which would be detrimental. To compensate, the plasma bicarbonate is reduced through renal excretion, and with this there must be a corresponding reduction in plasma sodium concentration. The fact that the pH of the blood remains normal indicates that compensation is complete, leading to a state which may be referred to as a chronic respiratory alkalosis fully compensated for by a chronic metabolic acidosis.

The decreased sodium concentration leads to a drop in the osmolality of the plasma, from about 290 to about 280 mOsm./kg. of plasma water. This is even greater than the drop which occurs in a non-pregnant individual after a large water load, and is more than enough to "turn off" the secretion of the antidiuretic hormone from the neurohypophysis. This, in turn, leads to polyuria, to periods of increased thirst and to exaggerated diuresis following water loads, symptoms which are observed in most women during the first trimester.

Thus, what may have been a single adjustment of the respiratory center brought about by one hormone leads to a whole series of readjustments by the mother; so it is with many of the physiological changes to be discussed.

CIRCULATORY ALTERATIONS

From a clinical viewpoint, the cardiovascular changes of pregnancy are of paramount importance, particularly when one is dealing with patients who have some form of heart disease. Until recently, it was not realized that the major part of these changes takes place during the first trimester, before marked changes in the uterine circulation occur. The patterns of change in the last month or two have also been misunderstood, because many of the studies reported have been conducted with patients lying in the supine position, which may profoundly alter circulatory dynamics.

Blood Pressure

The mean arterial pressure (estimated by adding one-third of the pulse pressure to the diastolic pressure) reaches its lowest point in the middle trimester, then rises slowly until term (Figure 12–1). It has been shown that mean pressures of 90 or more in the middle three months are ominous, since they are associated with a marked increase in the incidence of gestational hypertension or preeclampsia in the third trimester. Similarly, mean arterial pressures exceeding 105 mm. Hg. in the third trimester should be regarded as hypertension.

Cardiac Output

By the end of the first trimester, the cardiac output has increased by 25 to 50 per cent. Thereafter, an additional 10 per cent increase occurs in the middle trimester, and so long as the patient lies in the lateral recumbent position, this augmented cardiac output persists until term. The previously reported declines in the last month or two were probably due to the fact

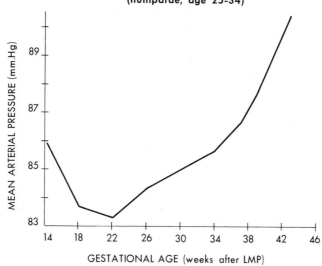

MEAN ARTERIAL PRESSURE BY GESTATIONAL AGE
FOR SINGLE WHITE TERM LIVE BIRTHS
(nulliparae, age 25-34)

GESTATIONAL AGE (weeks after LMP)

FIGURE 12–1. Mean arterial pressure in pregnancy.

that subjects were lying on their backs during measurements, and the large uterus compressed the inferior vena cava, interfering with return of the blood to the right atrium.

The basal heart rate increases by about 10 beats per minute, which accounts for a small part of the increased output. The major part is due to an increase in the stroke volume. The maximal increases in cardiac output (with subjects at rest and not in labor) vary from 1.2 to 3.1 l. per minute. Inasmuch as there is actually a fall in the mean blood pressure, especially during the middle trimester, there is obviously a significant fall in the total peripheral resistance. In the second half of pregnancy, the utero-placental circulation constitutes a low resistance shunt, which accounts perhaps for a major part of the reduced peripheral resistance, but there is also noted a reduced peripheral resistance in the forearm and the leg.

The cause of the early increase of cardiac output is not known, but it may be the hormonal changes of pregnancy. The circulatory changes of gestation may be mimicked in the non-gravid ewe by the intravenous infusion of estrogens.

During labor, the cardiac output increases about 15 to 20 per cent with each uterine contraction. This begins even before the contraction can be palpated, and returns to normal by the time the peak of the contraction is reached. At the same time, there is an increase in the mean blood pressure, so that the work of the heart is significantly elevated. Neither the increased cardiac output nor the rise of pressure is abolished by regional anesthesia, so the effects are not the result of pain. Apparently, about 500 ml. of blood are squeezed into the central blood volume very early in the uterine contraction phase, and this raises the right atrial pressure. At the height of the contraction, the utero-placental circulation, which represents an area of low resistance, is temporarily excluded, which could account for the rise in the mean blood pressure and the decline in cardiac output.

In addition to the circulatory changes with each contraction, there is an overall increase of some 40 per cent in the cardiac output by the end of the second stage of labor, but contrary to the changes with labor contractions, this general increase is largely abolished by continuous caudal analgesia, suggesting that pain and apprehension play an important role in its genesis. Immediately following delivery, the uterus contracts firmly, and again there is an auto-transfusion effect, but this is largely balanced by external blood loss, which averages 300 ml. before and after delivery of the placenta. The administration of ergometrine for its oxytocic effect increases the mean blood pressure in many women (especially those who are hypertensive), and this would further increase the work of the heart.

One might suppose that labor and vaginal delivery should be avoided in women with heart disease, but clinical experience has shown that the various postoperative complications which may follow cesarean section may be more hazardous than the increases of cardiac work associated with normal labor and delivery. From what has been said, it is apparent that labor in women with cardiac disease should be as free of pain and apprehension as possible. Voluntary or involuntary bearing down efforts in the second stage should be avoided, as should potentially pressor agents such as epinephrine (commonly mixed with local anesthetic agents) and ergometrine or Methergine.

Blood Volume

An increase in total plasma volume also occurs early in pregnancy (Figure 12–2) but lags behind the increase in cardiac output. About 75 to 85 per cent of the increase has occurred by the midpoint of pregnancy. With singleton pregnancies, the plasma volume increases up to the eighth month and then levels off until term. With twins, the increase is greater, and continues to rise until the time of delivery.

The magnitude of the increase in plasma volume may vary from 40 to 90 per cent, depending upon the size of the individual and the size of the fetus. Using the empiric prediction formula devised by Smith and Yarbrough, a small woman 155 cm. tall and weighing 55 kg. should have a peak increase of 88 per cent (from 2120 ml. to 4000 ml.), whereas a woman 175 cm. tall and weighing 77 kg. will have a peak increase of 60 per cent (from 3000 ml. to 4800 ml.).

The red cell volume also increases, but frequently lags behind the plasma volume, resulting in a reduced hematocrit and hemoglobin concentra-

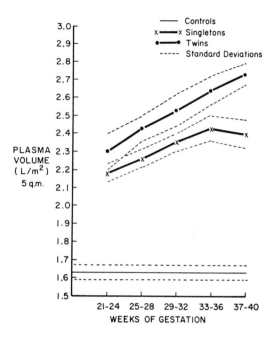

FIGURE 12-2. Plasma volume changes in pregnancy. (From Rovinsky and Jaffin, Amer. J. Obstet. Gynec., *93*:1, 1965. C. V. Mosby Co., St. Louis.)

tion. This has sometimes been referred to as the physiological anemia of pregnancy. In the majority of instances, however, when adequate amounts of iron are supplied, the drop in hemoglobin concentration is quite modest: approximately 1 gm./100 ml. of blood. Hemoglobin values below 11.5 gm./100 ml. suggest iron deficiency. The mean increase in total blood volume for a group of women is shown in Figure 12-3.

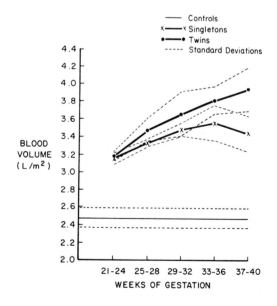

FIGURE 12-3. Blood volume changes in pregnancy. (From Rovinsky and Jaffin, Amer. J. Obstet. Gynec., *93*:1, 1965. C. V. Mosby Co., St. Louis.)

Several explanations have been advanced for the increased blood volume. The fact that the changes occur early implicates hormonal effects. There is a decrease of intrinsic vascular tone, which appears to be a direct hormonal effect upon the vessel wall. When this is coupled with the increasing vascular capacity of the uterus, the size of the intravascular compartment is considerably increased. In the second and third trimesters, the presence of a low resistance shunt in the utero-placental circulation is thought to lead to an increased blood volume in the same way that an arteriovenous fistula does. In the final analysis, there must be more space for blood, and whether this increased space is underfilled or overfilled is an interesting theoretical question. In many respects, the changes observed in late pregnancy, such as increased aldosterone secretion rates, sodium retention, the tendency for hypotension to develop when the patient assumes the upright posture and the presence of a high venomotor tonus (to compensate for the intrinsic hormonal atonic effect), resemble the changes noted with hypovolemia rather than with hypervolemia. Thus, it would appear that the volume receptors in pregnancy do not respond in the same way that they do when hypervolemia occurs in a non-pregnant subject, suggesting that the potential intravascular compartment may be even larger than the increased blood volume.

Regional Blood Flows

Measurements of hepatic and of cerebral blood flows do not reveal any changes as the result of pregnancy. Renal blood flow, however, increases by about 400 ml. per minute (almost a 50 per cent increase), and this occurs during the first trimester. Thereafter, the renal blood flow maintains a plateau, and tends to decline in the last month or so. Once again, those studies showing the greatest decline toward the end of pregnancy were conducted with the patient in the supine position. Studies of flow in the forearm (primarily flow to the muscles) show a considerable increase after the twenty-fourth week (Figure 12–4), and this is accompanied by a marked decrease in the peripheral resistance within the forearm (Figure 12–5).

Although one cannot state with precision the distribution of the additional 1200 to 3100 ml. per minute of cardiac output, it is apparent that the uterus, the kidneys and the extremities account for the major share.

Supine Hypotension Syndrome. In discussing the cardiovascular adjustments in pregnancy, repeated references have been made to the effects of the supine position. Some degree of inferior vena caval compression occurs in all women who lie on their backs in the second half of pregnancy. The venous pressure below the uterus increases to 20 to 30 mm. Hg., and this pressure, of course, is transmitted to the veins of the lower extremities. To some extent this is true in the quiet standing position, resulting in a pooling of blood volume in the legs and thighs and causing a tendency to

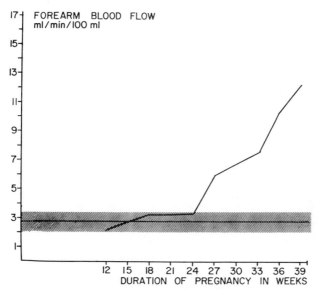

FIGURE 12–4. Blood flow through the forearm during pregnancy. Range of control values is within shaded area. (From Spetz, Acta Obstet. Gynec. Scandinav., *43*:12, 1964.)

hypotension and fainting. The increased venous and capillary pressures in the legs with quiet standing leads to edema of the legs and an increased tendency to develop varicosities.

About 8 per cent of women who lie on their backs experience a fall of

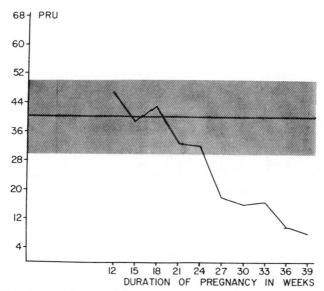

FIGURE 12–5. Peripheral resistance in the forearm during pregnancy. The 95 per cent confidence limit for the mean of controls is shown by the shaded area. (From Spetz, Acta Obstet. Gynec. Scandinav., *43*:12, 1964.)

30 per cent or more in the systolic blood pressure. After four or five minutes, a bradycardia ensues, indicative of a vaso-vagal reflex. The combination of hypotension and bradycardia (the supine hypotensive syndrome) reduces the cardiac output by about half, and at this point many women faint and appear shocky. The distress ordinarily induces women to turn on their sides, but they are unable to do so while they are in the lithotomy position for delivery or strapped on the operating table for cesarean section. Under these circumstances, lifting the uterus and displacing it to the left side is essential for correcting the syndrome.

The typical changes which occur are illustrated in Figure 12-6. In addition to these maternal changes, evidence of fetal distress may be elicited. It is important for the physician to be thoroughly familiar with these cardiovascular events in order to recognize and treat them.

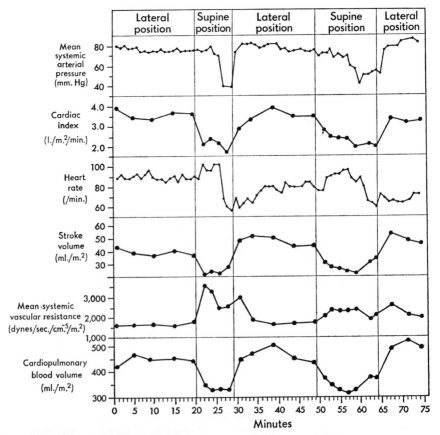

FIGURE 12–6. Cardiovascular changes induced by the supine position in late pregnancy. (From Kerr, Brit. Med. Bull., 24:19, 1968.)

WATER AND ELECTROLYTES

The changes in body water and electrolytes during gestation are of importance if we are to understand the significance of generalized edema, which occurs at some time in a large number of women during late pregnancy. Total body water in pregnant women at term and without edema increases by some 7.5 l. Estimates by Hytten and Leitch of the distribution of this water are shown in Table 12-1.

In this table, the extracellular, extravascular volume of fluid added represents the difference between the increases in total water, as measured by the deuterium oxide method, and the subtotals in the first part of the table, which account for the increases in specific sites. These volumes are primarily in the interstitial space, where water enters loosely into combination with the ground substance, a material quite like the Wharton's jelly of the umbilical cord. A direct, absolute measurement of the total extracellular space is not possible because tracers such as sodium, chloride, thiocyanate or sulfate may enter the cells to a variable degree, whereas large non-electrolytes such as inulin or mannitol may fail to penetrate the spaces completely. The true value lies somewhere between the sodium space and the inulin space, and the estimates of gain during pregnancy shown in Table 12-1 are based upon this assumption.

Assuming that these estimates for a standard woman (in this case, the average size and weight of 34 primigravidas studied in Aberdeen) are correct, it will be noted that even in the absence of clinical edema, the

Table 12-1.   ESTIMATE OF EXTRACELLULAR AND INTRACELLULAR WATER ADDED DURING PREGNANCY (in ml.) *

Site	Total Water	Extracellular	Intracellular
Fetus	2343	1360	983
Placenta	540	260	280
Amniotic fluid	792	792	0
Uterus	743	490	253
Breasts	304	148	156
Plasma	920	920	0
Red cells	163	0	163
Subtotals	5805	3970	1835
Measured increase:			
No edema	7500		
Leg edema	7880		
General edema	10,830		
Extracellular, extravascular volume:			
No clinical edema		1695	
Leg edema		2075	
General edema		5025	

*From Hytten, F. E., and Leitch, I. The Physiology of Human Pregnancy. Blackwell Scientific Publications, Oxford, 1964.

tissues are more hydrated to the extent of an additional 1.7 l. of water, and that when generalized edema is clinically evident, an additional 3.3 l. has been added. Of course, each liter of added extracellular water requires the retention of sufficient sodium chloride to maintain normal osmolality.

Demonstrable edema limited to the ankles is due to the increased capillary filtration pressure which occurs in the upright position. It is not, therefore, related to the renal tubular handling of salt and water and is of little physiological significance. Treatment (if indicated at all) would be to correct posture to reduce capillary pressure, or to wear elastic hose to increase the tissue pressure. It is patently irrational to utilize a diuretic drug for the treatment of localized ankle edema.

Generalized edema, manifested by a rapid weight gain, swelling of the fingers and other evidence of edema in the upper half of the body, is recorded in the prenatal charts of about one-quarter of the obstetrical population at some time. When accompanied by a rise of systolic and diastolic blood pressure and the appearance of proteinuria (a combination occurring in about 4 per cent of primigravidous women and less often in multiparas), it is called preeclampsia, a distinctly abnormal disease process (see Chapter 16). It is because of this association that generalized edema is viewed with alarm by obstetricians, and efforts are made to prevent its occurrence by routine restriction of sodium chloride or treatment of the edema with diuretic drugs.

The problem, however, is not that simple. Whether generalized edema in the absence of hypertension or proteinuria is abnormal or should even be treated is an interesting question which may have no logical answer at this time. Certainly the majority of women with generalized edema do not develop preeclampsia, and Hytten has noted that such women with edema produce infants of slightly greater birth weight with reduced perinatal mortality. It is possible that general edema without rises in blood pressure or proteinuria is due to the reduced colloid osmotic (oncotic) pressure of the plasma and is therefore a part of the physiological adjustments to pregnancy, whereas the general edema associated with preeclampsia is the result of disturbed renal function, resulting in a primary retention of sodium chloride and a secondary gain of water. Unfortunately, the various causes of general edema in pregnancy cannot readily be differentiated.

HEMATOLOGIC CHANGES IN PREGNANCY

In discussing the changes in blood volume, it was noted that the concentration of hemoglobin ordinarily falls during pregnancy by about 1 gm./100 ml. even when iron supplementation is given. Despite this fact, there is an addition of about 450 ml. of red blood cells during pregnancy, which results from an accelerated red cell production. This volume of red blood cells contains 500 mg. of iron, which is equal to or greater than the total body stores of iron in non-pregnant women, according to Pritchard. In addition to this, the fetus contains 300 mg. of iron, and the total iron requirement frequently exceeds the amounts contained in the average American diet. For this reason, routine iron supplements are indicated, especially in the second half of pregnancy.

Megaloblastic anemia due to folic acid deficiency is not uncommon during pregnancy, especially among populations whose diets are low in animal protein and fresh or uncooked vegetables. The requirement for folic acid is about five times as great during pregnancy, and approximates 0.5 mg./day. Routine supplementation of the diet with about 0.3 mg./day would probably prevent megaloblastic anemias in essentially all pregnant patients.

The leukocyte count is modestly elevated during gestation to values ranging between 6000 and 12,000 per cu. mm., and may exceed 20,000 during labor. The cells contain increased quantities of alkaline phosphatase.

During pregnancy, there are changes in the concentrations of a number of the components of the coagulation system of the plasma. Fibrinogen levels increase by about 50 per cent, and certain other factors are appreciably increased (Table 12–2). No doubt these are brought about mainly by the influence of estrogen and progesterone.

In addition to the changes noted in Table 12–2, there is a modest increase in the platelet count, a slight decrease in platelet adhesiveness, an increase of plasminogen concentration, and a decrease in the fibrinolytic activity of plasma. From a qualitative standpoint, most of the changes in the coagulation system of plasma are also found in women who have been on the combined forms of oral contraceptives for several months, suggesting that the causes of the various adjustments are hormonal.

Studies of whole blood coagulation kinetics with the thromboelastograph (Markarian and Jackson) reveal a significant hypercoagulability. The overall coagulability (reaction time plus fibrin formation time) is reduced from 12 to eight minutes, and in addition to this there is a significant reduction in the fibrinolysis time. Although normal pregnancy does not appear to be complicated by an increased incidence of spontaneous thrombosis before

Table 12–2.   COMPONENTS OF THE COAGULATION SYSTEM

Factor	Name	Effect of Pregnancy
I	Fibrinogen	Moderate increase
II	Prothrombin	No change
III	Thromboplastin	May be liberated in abruptio placentae
IV	Calcium	Slight reduction
V	Proaccelerin	No change
VII	Proconvertin	Marked increase
VIII	Antihemophilic	Increased in some
IX	Plasma thromboplastin component	Increased in some
X	Stuart factor	Marked increase
XI	Plasma thromboplastin antecedent	No change
XII	Hageman factor	– – –
XIII	Fibrin stabilizing factor	Decreased

the time of delivery, pregnant women are more subject to episodes of disseminated intravascular coagulation when such complications as infection, abruption of the placenta, eclampsia, vascular injury and amniotic fluid embolism occur.

CHANGES IN THE URINARY TRACT

Morphologic Alterations

About 80 per cent of pregnant women have significant dilatation of both ureters and renal pelves, as demonstrated by intravenous pyelography. The change begins as early as the tenth week, and for this reason is thought to be hormonal in origin. Kymographic studies show that the dilatation is related to the degree of atony. Later in pregnancy, however, mechanical factors play a role, because the dilatation is almost always greater on the right side, especially above the pelvic brim. This is attributed by some to the usual dextro-rotation of the gravid uterus. The volume of urine in the ureters and pelves, the so-called dead space, is increased from a normal 6 to 15 ml. to 20 to 60 ml. (Crabtree), a fact which physiologists must consider when performing renal function studies. The delayed flow time also contributes to the increased susceptibility of pregnant women to urinary tract infection.

Renal Blood Flow

The amount of plasma flowing through the kidneys is commonly measured by the clearance of a compound such as para-aminohippurate, 92 per cent of which is extracted during a single circuit. Its clearance (urinary excretion per minute divided by the plasma concentration) is therefore a measure of the effective renal plasma flow, which may be converted to whole blood flow by use of the hematocrit.

Serial measurements during pregnancy have been made by a number of workers. Perhaps the most reliable are those made in 1958 by Sims and Krantz, which show an increase of renal plasma flow from a non-pregnant mean of 600 ml./min. to 836 ml./min. in the first trimester, a drop to 750 ml./min. in the middle trimester and a return to the non-gravid flow at term.

The fact that the greatest rises are noted early in pregnancy suggests that the causes are hormonal. Some have attributed the increase to the action of human chorionic somatomammotropin (hCS or hPL).

Glomerular Filtration Rate (GFR)

The volume of glomerular filtrate formed per minute is determined by measuring the clearance of a substance which is not reabsorbed by the renal tubules, such as inulin. For non-pregnant women, the normal values are between 100 and 120 ml./min. per 1.73 square meters of body surface. From the fourth through the eighth month of pregnancy, the GFR ranges between 150 and 180 ml./min., an increase of about 60 per cent. The serial changes noted by Sims and Krantz are shown in Figure 12–7.

Inasmuch as the increase of GFR is proportionately greater than the increase in renal plasma flow, the proportion of plasma filtered, i.e., the filtration fraction, rises from 0.18 to about 0.22.

The clearance of creatinine roughly parallels the GFR and in most hospital laboratories is the method commonly used to follow changes in the renal function of filtration.

Significance of the Increased GFR

The fact that the renal excretion of creatinine and of urea fluctuates with their filtration rates accounts for the fact that their plasma concentrations are considerably reduced in pregnancy. Creatinine concentration decreases from 0.67 ± 0.14 to 0.46 ± 0.13 mg./100 ml. of serum. Blood urea nitrogen (BUN) decreases from a non-pregnant level of 13 ± 3 to 8 ± 1.5 mg./100 ml.

It must be remembered that if the amount of water filtered by the

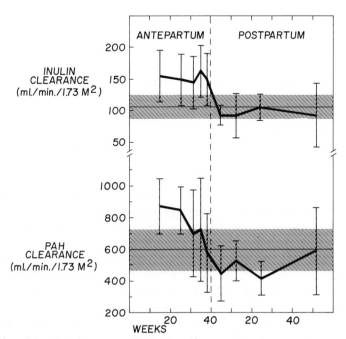

FIGURE 12–7.   The glomerular filtration rate and renal blood flow during pregnancy and the postpartum period. (Redrawn from Sims and Krantz, J. Clin. Invest., *37*:1764, 1958.)

glomeruli is increased some 60 per cent, so are all of the solutes which are contained in the plasma water, and this considerably increases the load of substances presented to the renal tubules for reabsorption. If, for example, the amount of filtered glucose is increased by half and the maximum ability of the kidney to reabsorb glucose (called Tm_g) is not increased, then it would be expected that clinical glycosuria would occur more often in pregnancy, even with modest levels of blood glucose. This is referred to as a lowered renal threshold for glucose, and knowledge of this threshold for any individual patient is particularly important when diabetes mellitus and pregnancy coexist.

Pregnancy is also characterized by a considerable amino aciduria, much of which is attributable to the increased GFR. There is considerable variation in the rates of excretion of the individual amino acids. The plasma concentrations of histidine, arginine and threonine are significantly increased in pregnancy despite increased renal clearances. Prior to the advent of accurate and rapid pregnancy tests, histidinuria was used in pregnancy diagnosis. The study of histidinuria by Page and his coworkers suggested that about half the increased excretion of histidine could be attributed to the increased GFR, about one-quarter to an actual lowering of the Tm (the reabsorptive capacity) and the remaining quarter to a reduction in the maternal metabolism of histidine. By analogy, one might surmise that there is a similar reduction in the metabolism of arginine and threonine.

The excretion of sodium is so complex that it defies analysis. For one thing, minute changes in the rate of sodium reabsorption cannot be measured accurately, yet might produce profound effects. If the percentage of filtered sodium which was reabsorbed by the kidneys increased from 99.4 per cent to 99.5 per cent in pregnancy this would lead to a positive balance of 34 mEq. per day, resulting in an increase in body weight of over 250 gm. daily. Those factors which tend to retain sodium in pregnancy are (1) an increased secretion of estrogen, (2) a slight increase in free plasma cortisol, (3) a marked increase in the secretion rate of aldosterone, and (4) the upright or supine posture, both of which raise the venous pressure in the lower half of the body and cause a pooling of sodium chloride and water.

Factors favoring sodium excretion in pregnancy are (1) the increased filtration fraction resulting from the high GFR, (2) the steadily increasing plasma concentrations of progesterone, which is natriuretic and antagonizes the action of aldosterone on the renal tubule, and (3) possibly some alteration in the "third factor," a poorly defined natriuretic substance which comes into play with expanding plasma volumes.

It is little wonder that obstetrical physiologists have a difficult time explaining the vagaries of sodium and water excretion in pregnancy, because it is apparent that virtually every factor controlling the retention or excretion of both is altered by pregnancy. About the only change which is essentially agreed to by most observers is that in the presence of preeclampsia there are a reduction in renal blood flow and glomerular filtration and a retention of sodium and water.

The Renin-Angiotensin-Aldosterone System

Normal pregnancy is characterized by a considerable increase in the renin activity of peripheral blood and an increase in the renin substrate concentration. Inasmuch as renin is an enzyme which acts upon the renin substrate to form angiotensin, the absolute concentrations of angiotensin are also increased in pregnancy. Despite this, the mean blood pressure in pregnancy is reduced, and it has been shown that the vascular reactivity (blood pressure response) to angiotensin is considerably reduced in pregnancy. In preeclampsia, this reduction of vascular reactivity is lost, which may account for the hypertension.

Inasmuch as angiotensin is a stimulant to aldosterone secretion by the adrenal cortex, there is a progressive rise in plasma aldosterone concentration during pregnancy. The clinical significance of these changes in the renin-angiotensin-aldosterone system in normal pregnancy is poorly understood.

CHANGES IN OTHER ORGANS AND SYSTEMS

Pulmonary System

The distinct changes in lung volumes and capacities in late pregnancy are summarized in Figure 12–8. The data are derived from the longitudinal studies of Cugell et al. Although there is no change in the basal respiratory rate or vital capacity, there is a considerable increase in the volume of air breathed per minute due to the increase in tidal volume. The residual volume and total lung volume are reduced, but the inspiratory capacity is increased. There is an increased alveolar ventilation, so that the distribution and mixing of gas is more efficient during late pregnancy.

The reported increases in total oxygen consumption in late pregnancy are between 40 and 60 ml./minute, an increase of about 20 per cent. Inasmuch as the increase of alveolar respiration exceeds 60 per cent, it can be seen that there is a persistent hyperventilation which accounts for the lowered pCO_2 of the blood we referred to at the beginning of this chapter. A further increase in ventilation and resultant lowering of the pCO_2 occur at higher altitudes, and upon this is superimposed the effect of pregnancy.

The symptom of dyspnea is a common complaint in pregnancy. The patient may say, "I feel short of breath sometimes, even when I am sitting quietly." Obviously this is not due to limitations of pulmonary capacity. One theory states that the sensation of dyspnea occurs when the ventilation response is inappropriate to the demand. Perhaps the hyperventilation of pregnancy in the face of reduced alveolar pCO_2 pressures is "inappropriate" and leads to the sensation of dyspnea.

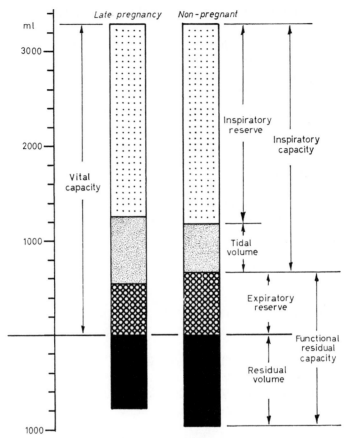

FIGURE 12–8. Respiratory changes in late pregnancy compared to the non-pregnant woman. (From Hytten, The Physiology of Human Pregnancy, Blackwell Scientific Pub., Oxford, 1964.)

Gastrointestinal-Hepatic System

Vagaries of appetite are very common in pregnancy, especially in the first trimester. The common symptom of nausea leads to anorexia, and yet the majority of women report increased appetites and a large number admit "cravings" for certain foods. Given a choice, the laboratory rat will select increased quantities of those foods required by pregnancy, but women have lost this instinctive response, and the cravings or dietary habits are at times bizarre, with little relationship to need. In general, there is an increased food intake which, in the first half of pregnancy, is larger than the requirements, as though the pregnant organism anticipates the need for the storage of nutrients.

The physiological basis for the nausea, which occurs most often upon arising in the morning, is poorly understood. It may be related to the rapidly rising estrogen levels in the blood. There is no doubt that most instances of excessive persistent vomiting (hyperemesis gravidarum) are due to deep-rooted psychological factors which result from a subconscious rejection of the pregnancy.

The gums are frequently swollen and hyperemic and may bleed easily, because of some hormonal effect. Pregnancy does not increase the frequency of dental caries, and any needed dental reparative work may be done at any stage of gestation without increasing the risk of complications.

During the first two trimesters there is a reduced gastric secretion of acid and a reduced secretory response to histamine. Concomitant with this is a reduced tendency for peptic ulcer formation.

Throughout pregnancy there is a relative atony of the gastrointestinal tract, and this applies to the esophagus, stomach, gallbladder, small intestine and colon. In all likelihood, this is due to the same hormonal effects which relax the smooth muscle components of the arteries, veins and ureters. The combination of relaxed cardiac sphincter and upward pressure of the expanding uterus upon the stomach leads to a high incidence of heartburn. This is obviously not due to hyperacidity, although the symptom may be relieved by antacid preparations or by an injection of neostigmine, which increases the tonus. The relative atony of the colon leads to an increased frequency of constipation, another common complaint in pregnancy.

The liver is undoubtedly performing more work in pregnancy, but data for women are lacking. In the rat, pregnancy produces an increased liver size, a higher protein content and a higher rate of oxygen utilization. Liver function tests in pregnant women are within normal limits except for the hippuric acid excretion after the administration of sodium benzoate. This is probably not a reflection of impaired function, but more likely represents an increased competition by other substances for conjugation. Liver biopsies fail to show any characteristic change in the histologic picture.

A few women may develop cholestasis during pregnancy, resulting in a generalized itching of the skin. The pruritus is due to the retention of bile salts, which are measurably increased even though jaundice is absent. The condition may be successfully treated by the administration of 30 to 60 mg. of phenobarbital three times daily for the purpose of activating certain hepatic enzymes concerned with bile salt metabolism. The cholestasis itself results from the placental steroids, and the same women who are susceptible are known to develop generalized pruritus later if placed upon the oral contraceptives.

Skin

About half of pregnant women develop red, slightly depressed streaks over the abdomen in the last trimester, and these sometimes appear over the breasts. These "stretch marks," or striae gravidarum, become silvery several months after delivery and persist indefinitely.

There is an increased pigmentation of the linea alba of the lower abdomen, of the vulva, of the areolar areas of the breasts and of the face. Typical chloasma on the forehead and cheeks, most marked in Caucasian brunettes and exaggerated by exposure to the sun, form the "mask of pregnancy." In addition, Caucasian women in particular develop vascular "spiders" or angiomata. They consist of dilated precapillary vessels radiating out from a central dilated arteriole, which appears as a bright red circle, and these are believed to be secondary to high estrogen concentrations. For some obscure reason, multiple filiform fibromas often develop around the neck and shoulder girdle.

SUMMARY

The physiological adjustments observed in pregnancy are brought about by a combination of hormonal and mechanical changes. In general, the hemodynamic alterations reach their peak in the seventh or eighth month and tend to decline as term is approached, whereas the endocrine and biochemical changes (except for hCG) tend to increase progressively until term. The most significant hormonal events which produce physiological adjustments are the steadily increasing concentrations of estrogen, progesterone and human chorionic somatomammotropin (hCS), which exerts a growth hormone–like effect. The most significant mechanical factors are the expanding size of the uterus and the progressive development of the utero-placental circulation, an area of low resistance.

Among the significant changes are the increased blood volume and cardiac output, the increased renal blood flow and glomerular filtration rate, a persistent hyperventilation, an increased hydration of the interstitial ground substance, a reduction of the plasma pCO_2 and osmolality, a reduced tonus of smooth muscle generally and a reduced rate of acid secretion by the stomach.

The physiological adjustments described may be considered normal for the state of pregnancy, but they frequently are the basis for minor symptoms, which include dependent or generalized edema, sensations of dyspnea, changes of pigmentation and of the fine blood vessels of the skin, heartburn, constipation, nausea, periods of thirst, orthostatic vertigo, hemorrhoids, varicose veins of the lower extremities and episodes of acute hypotension while in the supine position. The ability to explain such symptoms to the obstetrical patient in simple physiological terms may offer her gratifying reassurance.

SUGGESTED SUPPLEMENTARY READING

Hytten, F. E., and Leitch, I. The Physiology of Human Pregnancy. 2nd ed. F. A. Davis Company, Philadelphia, 1971.
A good reference monograph; 614 pages of text and references.
Kerr, M. G. Cardiovascular dynamics in pregnancy and labour. Brit. Med. Bull., 24:19, 1968.
A concise summary of the hemodynamic changes.
Rizza, C. R. Blood coagulation and haemostasis. In Philipp, E. E., et al. (Eds). Scientific Foundations of Obstetrics and Gynecology. F. A. Davis Company, Philadelphia, 1970.
A good summary of a highly complex subject.
Sims, E. A. H. Renal function in normal pregnancy. Clin. Obstet. Gynec., 11:461, 1968.
Contains the best data on changes of glomerular filtration and renal blood flow.

Chapter 13

PRINCIPLES OF ANTEPARTUM CARE

DIAGNOSIS OF PREGNANCY

The tradition of grouping the various signs and symptoms of pregnancy as presumptive, probable and positive is worthy of perpetuation in order to indicate degrees of certainty.

Presumptive symptoms are (1) absence of menses 10 or more days after the usual length of the cycle, (2) morning nausea or change of appetite, (3) frequency of urination, and (4) soreness of the nipples and breasts. If the missed period is accompanied by two of the three symptoms mentioned, the odds are about two to one that the woman is pregnant.

Probable signs of pregnancy are (1) an enlargement of the uterus, as compared with some prior examination, (2) a softening of the uterine isthmus (Hegar's sign), (3) a bluish or cyanotic color to the cervix and upper vagina due to hyperemia (Chadwick's sign), (4) an asymmetrical, softened enlargement of one uterine cornu, due to placental development in that area (Piskacek's sign), and (5) a positive test for hCG in the urine or serum. These signs are present during the first trimester, and each has a degree of certainty which varies from 80 to 95 per cent, the hCG test being the most accurate. Almost equally accurate is the continuous recording of the basal body temperature, but of course this must have been instituted prior to conception, which is frequently the case with infertility patients or with some women who employ the "rhythm" method of birth control. A distinct rise of temperature which is maintained for three weeks or longer is almost certainly the result of pregnancy.

Positive signs, diagnosing pregnancy with absolute certainty, include (1) the demonstration of a fetal heart beat by auscultation or electrocardiography or by the Doppler effect instrument, (2) the objective detection of fetal movements, and (3) the radiologic or ultrasonic demonstration of fetal parts. Most of these are not certain until the second trimester, but they may be the first indications of the presence or absence of pregnancy in rare instances of extreme obesity, multiple myomata uteri or secondary amenorrhea with spurious pregnancy (pseudocyesis).

Pharmacologic Tests

Pharmacologic tests for pregnancy have been in use for many years, but are not very reliable. For example, the administration of an oral progestin (or one of the combination oral contraceptives) for five days should produce uterine bleeding within four days after discontinuance, and the absence of a flow is a presumptive diagnosis of pregnancy.

TESTS FOR hCG. Tests for human chorionic gonadotropin (hCG) are legion, and a large number of animals have been utilized, such as mice, rats, rabbits, frogs and toads. In general, all these biologic tests utilizing animals have been displaced by *immunochemical* tests for hCG. Properly used, these tests have the advantage of rapidity, economy, reasonable accuracy and a level of sensitivity which is practical for ordinary usage. The original tests, using anti-hCG rabbit sera, involved either a precipitin reaction (McKean) or tanned red blood cells coated with hCG (Wide and Gemzell). The most common commercial tests use standardized latex particles coated with hCG and a standardized anti-hCG rabbit serum. The antibody is first mixed with the urine, then added to the hCG particles. The failure to agglutinate constitutes a positive test, because urine containing hCG has already neutralized the hCG antibodies, whereas urine devoid of hCG permits the antibodies to agglutinate the particles.

Immunochemical tests of exquisite sensitivity may, of course, be designed, but because hCG and LH cross-react, the sensitivity of commercial tests has been adjusted to a variable degree in an attempt to eliminate false positives. For example, the sensitivity of some currently available reagents in I.U. (international units) per liter, according to their manufacturers, are as follows:

Gravindex (Ortho): 3500
DAP (Wampole): 2000 to 3000
hCG test (Hyland): Product varies from 2000 to 8000
UCG (Wampole): 1000
Pregnosticon (Organon): Slide test, 1000 to 2000; tube test, 700 to 750; Accuspheres, 750 to 1000

The slide tests may be completed within a few minutes and are therefore useful as a screening test in the office or clinic. The tube tests are a bit more sensitive but may require up to two hours to complete; however, they may be more useful for problem pregnancies, such as suspected ectopics or trophoblastic disease. Some psychotropic drugs, the presence of proteinuria and the postmenopausal state may result in false positive tests.

Immunochemical tests for the β-subunit of hCG do not cross-react with LH, are more sensitive and promise to replace all other tests for the presence of viable human trophoblast.

THE FIRST PRENATAL VISIT

As soon as pregnancy is suspected, a woman is advised to arrange for her first antepartum visit. This offers the physician an opportunity to detect any disease states or abnormalities, to institute therapy, and to educate the patient—preferably by written instructions—as to proper nutrition, the avoidance of teratogenic influences, the appropriate treatment of minor symptoms and potentially dangerous symptoms which should be reported. There are a variety of useful manuals or pamphlets in several languages and with varying degrees of sophistication available for distribution.

Similarly, a variety of printed prenatal record forms are available for the physician, such as that recommended by the American College of Obstetricians and Gynecologists. It is important to record data accurately on every item, not only for the intelligent management of the patient's pregnancy but also for the standardization of obstetrical data. Meticulous details about prior pregnancies are important, as well as the family history, past medical, surgical and psychiatric experiences, allergies or drug sensitivities, the marital history and a careful record of the menstrual cycles. A note should be added about the woman's reaction to her current pregnancy, because its confirmation by the physician is often the occasion for emotional responses indicating joy and anticipation, indifference, or downright fear or despair. A dietary history is useful in estimating the adequacy of her nutritional intake.

The initial visit is the proper occasion for a pelvic examination, cervical cytology, an accurate description of the size, shape and consistency of the uterus and a clinical evaluation of the bony pelvis, as described below. The minimal physical examination should include the heart and lungs, breasts, abdomen and extremities, weight and blood pressure. Minimal laboratory studies must include a serologic test for syphilis (ordinarily required by law), the hemoglobin concentration or hematocrit, blood typing for the Rh and ABO groups, screening for atypical antibodies, examination of the urine for protein and glucose, and the Pap smear. In most prenatal clinics, if not in all offices and clinics, an endocervical culture for gonorrhea should be done.

Although the need for routine nutritional supplements is still controversial, the majority of obstetricians favor the use of a prenatal supplement, as described later. A priceless supplement is reassurance and the alleviation of unfounded fears.

CLINICAL EVALUATION OF THE BONY PELVIS

The superior strait, or pelvic inlet, divides the false from the true pelvis and its average dimensions are 11 cm. antero-posterior (the true conjugate) by 13.5 cm., the maximum transverse diameter. The smallest cross-sectional area of the true pelvis, the midpelvis or plane of least dimensions, is bounded by the ischial spines, the tip of the sacrum and the lower border of the symphysis (Figures 13–1 and 13–2). The average A-P diameter is 11.5 cm. and the average distance between the ischial spines is 9.5 cm.

The true conjugate of the inlet cannot be measured directly except by radiologic techniques, but it can be estimated by measuring the distance

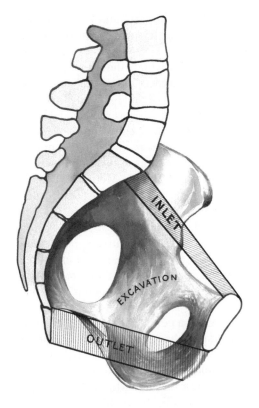

FIGURE 13–1. The pelvic inlet and outlet. (From Greenhill, Obstetrics, 13th Ed., W. B. Saunders Co., Philadelphia, 1965.)

between the inferior margin of the symphysis and the anterior surface of the first sacral vertebra, as shown in Figure 13–3. This is called the obstetric conjugate, and is 1.5 to 2.0 cm. longer than the true conjugate, depending upon the height and the inclination of the symphysis pubis. Normally, the sacral promontory can only be reached by depressing the perineal body with the index and middle fingers as illustrated, and even then, the average examiner can reach only 12.0 cm., whereas the normal obstetric conjugate should be about 12.5 cm. Inasmuch as this is the most important single measurement to be made clinically, it is necessary for each examiner to be familiar with his or her own "reach" and then to record "diagonal conjugate not reached at 11.5 cm." or "reached at 12.0 cm."

The second most important clinical measurement is the distance between the ischial tuberosities, which ideally should be 9 cm. or more. The Thoms pelvimeter or a series of metal bars shaped like tongue blades may be utilized, but with experience the four knuckles of one's hand serve equally well. If the distance should be 8 cm. or less (as would be true in a male pelvis), it is important to measure the posterior sagittal of the outlet, which is the distance between the midpoint of the intertuberous diameter and the tip of the sacrum (not the tip of the coccyx). Either the Thoms instrument (Figure 13–4) or some other outlet pelvimeter may be used. Thoms rule states that the sum of the intertuberous and posterior sagittal measurements should be at least 15 cm. for an adequate pelvis.

It is not that the pelvic outlet itself is a cause of cephalopelvic disproportion, but it is our best clinical estimate of the midplane diameters where obstruction is most likely to occur. There are other clinical measurements of value, such as the subpubic angle, the width of the sacrosciatic notch, the contour of the sacrum, and the height of the symphysis, which can best be learned by preceptorship in a prenatal clinic. The clinical estimate of pelvic capacity should be repeated near term, when one has the additional advantage of estimating fetal size and the degree of descent of the fetal head.

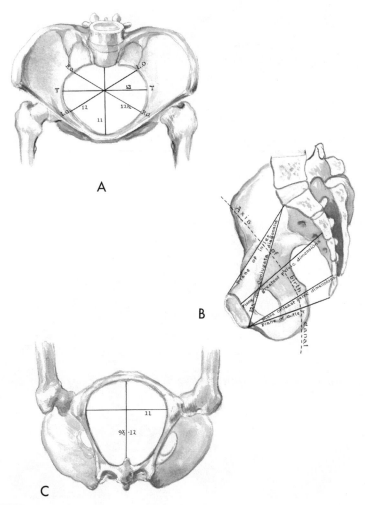

FIGURE 13–2. Average measurements for (A) the plane of the inlet, (B) the plane of greatest pelvic dimension and (C) the plane of the outlet. (From Greenhill, Obstetrics, 13th Ed., W. B. Saunders Co., 1965.)

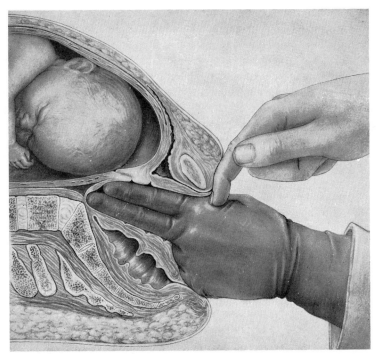

FIGURE 13–3. Measuring the diagonal conjugate. (From Eastman and Hellman, Williams Textbook of Obstetrics, 13th Ed., Appleton-Century-Crofts, New York, 1966.)

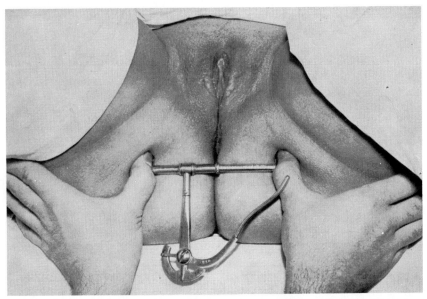

FIGURE 13–4. Measuring the pelvic outlet with Thoms pelvimeter. (From Eastman and Hellman, Williams Textbook of Obstetrics, 13th Ed., Appleton-Century-Crofts, New York, 1966.)

When the fetal head of a term fetus weighing 3400 gm. is well flexed, the two largest diameters presenting, the biparietal and the suboccipito-bregmatic, are each about 9.5 cm. (Figure 13–5). When these diameters are within the pelvic inlet (the superior strait), the head is said to be engaged. It so happens that the distance from these diameters to the vertex (the most dependent portion of the head), is 4 to 5 cm., and the distance from the superior strait to the ischial spines averages 5 cm. Therefore, when the fetal head descends to the level of the ischial spines, it is assumed to be engaged.

Variations of Pelvic Architecture

Many classifications of pelves have been described, but the one most commonly used is that devised by Caldwell and Moloy, which is based

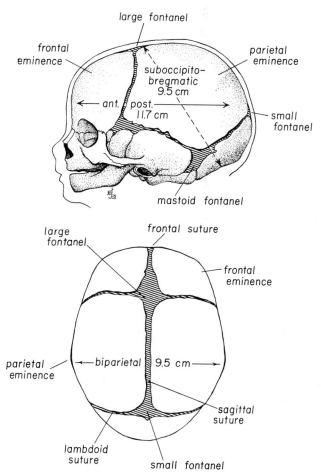

FIGURE 13–5. Average measurements of the full-term fetal head.

upon the shape of the inlet as determined by x-ray. The superior strait is divided into an anterior and a posterior segment by the widest transverse diameter, and it is the shape of the posterior segment which determines the parent type. The four major types are illustrated in Figure 13–6. Intermediate forms are more common than the "pure" types shown, and the anterior segment is used to designate the tendency (Figure 13–7).

GYNECOID

The gynecoid or true female pelvis is present in about half of the obstetrical population. The inlet is almost round, the posterior segment is roomy, the sacrosciatic notch is wide and the pubic rami form an angle approaching 90 degrees. This, of course, is the most favorable pelvis for normal delivery.

ANTHROPOID

The anthropoid (apelike) pelvis is probably the next most common type. The inlet is oval, with the A-P diameter exceeding the transverse one; the pelvis is deep because of a longer sacrum. Because of this pelvic shape, the fetal head often engages with the occiput directly anterior or posterior (or slightly oblique) and undergoes very little rotation during descent.

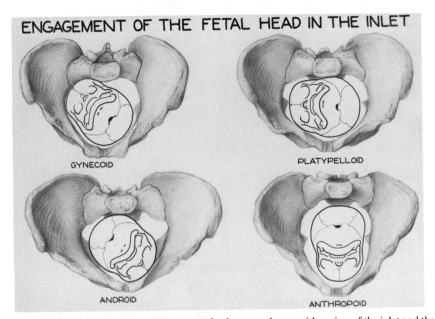

FIGURE 13–6.   The four parent types of pelves are shown with a view of the inlet and the usual mode of engagement of the fetal head.

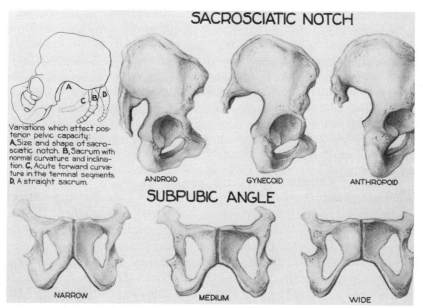

FIGURE 13–7. Variations of sacral inclination and curvature, sacrosciatic notch and subpubic angle. All of these affect the capacity of the mid-pelvis.

ANDROID

The android (male) pelvis has a heart-shaped inlet because the sacral promontory is pushed forward, so that the posterior segment is shallow. The pelvic side walls are usually convergent, rather than parallel, resulting in a "funnel pelvis" with contracted mid-plane measurements. This is the worst obstetrical pelvis because the cephalo-pelvic disproportion, if present, becomes increasingly more marked as the head descends deeply toward the perineum. Because progress in dilatation and descent still occurs during labor, the obstetrician is sometimes lulled into a false sense of security.

PLATYPELLOID

The platypelloid or flat pelvis is relatively rare. The true pelvic cavity is shallow and the fore-pelvis is wide, so that delivery is relatively easy, provided that the fetal head, which always engages transversely in this pelvic type, can pass through the inlet. Flat pelves are readily recognized clinically because of a short diagonal conjugate measurement. Disproportion, should it occur, is also recognized earlier in labor than with the android pelvis because of failure of the head to engage.

X-ray pelvimetry is requested much more rarely today than in former years because of an increasing awareness of the hazards of ionizing radiation to the maternal and fetal ovaries. Even when a moderate contracture is suspected clinically, a radiologic study of the pelvis is usually delayed until the time of labor, and then only when there is some apparent dystocia (difficult labor). This has the additional advantage of enabling one to study the size and precise position of the fetal head. Pelvimetry is indicated in primigravidous women when there is a persistent breech presentation or other malpresentation, and in multiparas when there is a history of previous dystocia, unexplained stillbirth or birth injury. Prior to labor, sonography has largely replaced x-ray pelvimetry for antenatal diagnosis.

THE HIGH-RISK OBSTETRICAL PATIENT

In dealing with large obstetrical populations, it is essential to identify those patients who may be expected to have an unfavorable outcome so that they may receive the most expert obstetrical care available. In many large hospitals serving underprivileged groups, special high-risk clinics have been established, so that the talents of obstetricians, nutritionists, nurses and social workers can be brought to bear upon the multiple problems. Nesbitt and Aubry utilized a semi-objective grading system for the derivation of a "maternal-child health care index." They estimated by their criteria that about 17 per cent of the obstetrical population of Syracuse, New York, should be classified as high-risk, and that this segment accounted for about 60 per cent of pathologic outcomes.

Among the many factors which determine high risk are these:

1. Age under 17 or over 35.

2. Parity of 0 or over 4.

3. Past obstetrical history of multiple abortions, premature or low birth weight infants, fetal or neonatal death, congenital anomaly or birth injury.

4. Maternal illnesses such as chronic hypertension, heart disease, diabetes, syphilis, severe anemia, malnutrition or marked obesity, chronic urinary tract infection or sensitization to the Rh factor.

5. Reproductive system disorders such as myomata uteri; the history of a prior cesarean section; contracted pelvis.

6. Psychiatric illness; a disrupted marital unit or single status; unfavorable environment.

Many of these factors, of course, are more prevalent in the non-white populations. Inasmuch as prematurity, mental retardation, neonatal mortality and morbidity are several times more common in the high-risk group than in the low-risk group, it is obvious that we must focus attention upon the care of these less fortunate mothers if we are to make any progress in the improvement of the obstetrical outcomes for our population as a whole.

ANTEPARTUM OBSERVATIONS

After the initial visit, the patient returns every four weeks until the seventh month, then every two weeks, and then every week during the last month. At each visit her weight and blood pressure are recorded, the morning urine sample is checked for glucose and protein, and specific questions are asked about any untoward symptoms. After the fourth month, an abdominal examination is made at each visit, the height of the fundus above the symphysis is recorded, fetal weight is estimated, and the fetal position and heart tones are noted. One or two weeks before the estimated time of labor, the pelvic examination is repeated for re-evaluation of the pelvic capacity and assessment of the cervix and the progress in the descent of the fetal head. Two months before term, the hemoglobin (or hematocrit) and the antibody screen are again determined.

MANAGEMENT OF COMMON PROBLEMS DURING PREGNANCY

In the previous chapter, a number of minor symptoms resulting from the normal physiological changes of pregnancy were described. Most of these do not require therapy, but they do require explanation and reassurance because the average patient cannot assess the seriousness of any particular symptom. There are a number of common problems, however, which can be alleviated by relatively simple measures.

When the *rate of weight gain* deviates significantly from the expected curve, every attempt should be made to determine the cause. The average total weight gain in pregnancy is 10.5 kg. (23 pounds); 1 kg. (2.2 lbs.) in the first trimester, 4.5 kg. (10 lbs.) in the second, and 4.7 kg. (11 lbs.) in the third, but the "normal" range (e.g. two thirds of women) may vary from 6 to 16 kg. (13 to 35 pounds) depending upon the size and body build of the individual. A rate of gain exceeding 0.5 kg. (1.1 lb.) per week is probably an indication of fluid retention, and such patients should sharply curtail their intake of sodium. The indiscriminate use of diuretics should be avoided. There is no valid evidence that the prophylactic use of thiazides prevents the onset of preeclampsia, and their potential abuse by patients may be harmful. A loss of weight during a four-week period may be the result of malnutrition or inept dieting, and a careful dietary history should be obtained.

Nausea with occasional vomiting affects 50 per cent of pregnant women during the first three months. It usually occurs when the patient arises in the morning with an empty stomach. The symptom should be viewed as normal; indeed, spontaneous abortion occurs more commonly in the women who do not experience nausea. Eating some toast or crackers before arising is a time-honored method of alleviating the symptom, just as frequent small feedings during the day and at bedtime may be of value. A variety of drugs have been used for the prevention of nausea, with variable success, but unless the symptom is severe, it is best to avoid the use of drugs completely during the period of embryogenesis.

Heartburn is a common complaint in the later months of pregnancy and results from the combination of the diminished gastric tone and pressure of the gravid uterus with regurgitation of the stomach contents into the esophagus. The aluminum antacids, either in tablet or liquid form, give safe symptomatic relief. The use of sodium bicarbonate should be prohibited.

Constipation is almost a universal symptom. Harsh cathartics, mineral oils and enemas should be avoided. The daily use of a "wetting agent" such as dioctyl sodium sulfosuccinate, with or without the active principle of cascara, is effective. A bulk-producing agent together with laxative foods such as prune juice may be sufficient.

Hemorrhoids may at times produce marked discomfort. If the pain is due to thrombosis within an external hemorrhoid, the clot may have to be evacuated in the office by a simple incision; if it is due to edema, an ice pack is effective. Anal ulcers or fissures may be treated with suppositories containing opium and belladonna. Hemorrhoidectomy is not recommended during pregnancy.

Vaginitis due to Candidiasis is not uncommon during pregnancy because of the high glycogen content of the mucosa. The use of topical agents as described in Chapter 17 should give rapid relief, but achieving negative cultures is difficult prior to delivery. *Trichomonas* vaginitis may, of course, occur during pregnancy. The discharge is thin, greenish-gray, and highly irritating, and the microscope shows it to be swarming with motile trichomonads. In the non-pregnant state, such infections are quite readily controlled by the oral use of metronidazole, but inasmuch as the manufacturer states, "At the present time, the use of Flagyl during pregnancy is contraindicated . . .," a physician cannot ethically utilize the agent even though no adverse effects upon the fetus have been reported. A variety of trichomonocides are available for topical use, however, and will yield symptomatic relief even though the recurrence rate is high.

Leg cramps occur commonly during the later months and may be quite painful. Almost invariably the spasm of the gastrocnemius muscle takes place when the woman awakens in the morning, lies on her back and stretches her legs with toes extended. The patient should be told to "point with the heels" whenever stretching the legs, and to avoid lying supine. If this does not suffice, she may take a soluble calcium preparation, such as calcium lactate, at bedtime to elevate the ionizable calcium levels in the plasma.

MATERNAL NUTRITION

From the standpoint of both maternal and fetal outcome, the importance of optimal nutrition during pregnancy can hardly be overstressed. The breeders of cattle and swine have perhaps paid more attention to prenatal diets, for economic reasons, than most obstetricians. This has been largely because of the continuing controversy about the relationship of maternal food intake to fetal growth and development in the human species.

The studies of Burke and her collaborators demonstrated that when accepted nutritional standards for the mother are not met, there are objective deficiencies in the weight, length and condition of the skeletal system at birth. Recent autopsy studies by Naeye on premature infants of comparable gestational age indicate that when the mother's economic status was below the poverty level established by welfare departments, the fetuses showed the organ weights and cellular changes of starvation. These and other studies should leave little doubt that despite the parasitic aspects of the human placenta, normal fetal growth and development are dependent in part upon an adequate maternal intake of proteins, minerals, vitamins and calories.

Pure undernutrition, such as that observed in periods of famine, leads to a high incidence of premature births as well as to intrauterine growth retardation. Faulty nutrition, a far more common condition in the United States than undernutrition, is strongly suspected of being associated with higher frequencies of preeclampsia and eclampsia, abruptio placentae, and other maternal complications, but the cause and effect relationship is difficult to establish with certainty. Faulty food habits are by no means limited to poverty groups, and are particularly prevalent among pregnant teenagers, who are prone to eat inadequate, even bizarre, diets.

Nutritional standards have been established by a number of scientific groups in various countries. The 1974 recommendations for the United States are given in Table 13–1. Requirements for vitamins not listed have not been firmly established. The requisite amounts of folic acid, iron and probably pyridoxine in pregnancy are not supplied by the average American diet and should therefore be added as supplements. Most obstetricians pre-

Table 13–1.   RECOMMENDED DIETARY ALLOWANCES FOR WOMEN
(FOOD AND NUTRITION BOARD, NATIONAL ACADEMY OF
SCIENCES–NATIONAL RESEARCH COUNCIL, 1974)

	Non-pregnant	Pregnant	Lactation
Calories	2000	2300	2500
Protein (gm.)	46	76	66
Vitamin A (I.U.)	4000	5000	6000
Vitamin D (I.U.)	400	400	400
Vitamin E (I.U.)	12	15	15
Ascorbic acid (mg.)	45	60	80
Folacin (mcg.)	400	800	600
Niacin (mg.)	13	15	17
Riboflavin (mg.)	1.3	1.6	1.8
Thiamin (mg.)	1.0	1.3	1.3
Vitamin B_6 (mg.)	2.0	2.5	2.5
Vitamin B_{12} (mcg.)	3	4	4
Calcium (mg.)	800	1200	1200
Phosphorus (mg.)	800	1200	1200
Iodine (mcg.)	100	125	150
Iron (mg.)	18	50–80	18

scribe a daily prenatal supplement containing the total minimal requirements of all vitamins and minerals, on the supposition that recommended diets may often be disregarded. While the professional nutritionist may regard this as unscientific, such an addition during pregnancy and lactation appears to be harmless and may well be beneficial for many women.

Recording a patient's dietary habits is an integral part of a medical history. The simplest approach is the 24-hour recall, in which an inventory is taken of everything eaten within the past 24 hours. A more sophisticated approach is to use the food record, a written diary of all calorie-containing foods and liquids consumed over a period of days. Its interpretation may involve the services of a nutritionist-dietitian, but the method is invaluable when dealing with women whose weight deviates markedly from the norm, or whose diet may be inadequate due to food-faddism, allergies or religious beliefs.

The *sodium intake* during pregnancy should probably be *ad lib* so long as there is no hypertension or generalized edema. The appearance of either symptom is an indication for the avoidance of excessive sodium intake. Diuretic drugs should be avoided if possible.

Assuming an adequate intake of vitamins and minerals through judicious supplementation, the major educational effort should be directed toward increasing the protein intake to as high as 85 gm. per day during the second and third trimesters. It must be remembered that the average storage of nitrogen during pregnancy is 2.5 gm. per day, and that the efficiency of fed nitrogen is only 25 to 30 per cent.

The period of pregnancy is not the proper time to treat obesity by sharp caloric restrictions, because the fetal outcome will be impaired. Diets containing less than 1500 calories are not compatible with a high-protein diet and sufficient carbohydrate to prevent ketosis. Pregnant adolescents, in particular, demand an intensive program of prenatal care, much of which is nutritional in nature.

EMOTIONAL ASPECTS OF PREGNANCY

The emotional response to the existence of pregnancy will depend initially, of course, upon whether the pregnancy was planned or unplanned, wanted or unwanted. The fact that a pregnancy was deliberately planned does not necessarily imply that it is wanted, because the motivations for becoming pregnant are multiple and complex.

The meanings which may motivate a woman to bear children include the social expectations of motherhood, identifications with her mother or with childhood memories of her family life, anticipated improvement in the marital relationship, holding a man or (if unmarried) "snaring" a man and even punishing her own parents. Even though the prospect of bearing a child may be terrifying, a woman may encourage conception because of prevalent childbearing among her peers, because it is expected of her, or because the husband simply wants to father a family.

Not all of these motivations are compatible with a tranquil acceptance of pregnancy, and the emotional response may result in hysteria or an acute anxiety neurosis on the one hand, or psychosomatic symptoms such as excessive vomiting on the other. Many physicians believe that a strong subconscious rejection of the pregnancy may actually produce repeated spontaneous abortions. Certainly unwanted pregnancies, such as the majority of those which occur out of wedlock, may result in acute situational depressions, sometimes leading to suicidal attempts. Pregnancy is a unique life situation because it requires adaptation rather than action, and the woman who is immature and who has frequently failed to cope adequately with other problems may fail to adapt in numerous ways. Even though an early pregnancy is an intangible abstraction, the concept may be so destructive or threatening that the woman may seek any means of terminating it, without regard to bodily harm or future fertility. In many instances, therefore, an unwanted pregnancy is truly a disease.

Late in pregnancy, early conflicts have ordinarily disappeared, and the woman's attitudes become positive and anticipatory, yet she is not without apprehension. There are fears of fetal deformity, of impending pain or mutilation, or feelings of inadequacy about motherhood. An important part of antenatal care is listening, explaining, comforting and reassuring. For primigravidous women, tours of the hospital and labor rooms, prenatal classes, motion pictures of normal childbirth and group discussions may be valuable adjuncts. The most capable obstetrician is one who tries to understand the workings of a woman's mind just as he tries to comprehend the functioning of her organs. If he obtains the trust and confidence of his patients, he becomes an important resource for the solution of patients' anxieties which might otherwise be destructive, or which might even culminate in acute puerperal psychotic episodes.

SUGGESTED SUPPLEMENTARY READING

Maternal Nutrition and the Course of Pregnancy. Report of the Committee on Maternal Nutrition, Food and Nutrition Board, National Research Council. National Academy of Sciences, Washington, D.C., 1970.
   The major report contains 200 pages of text, but a 23-page summary report is also available.
Pritchard, J., and Macdonald, P. Williams Obstetrics, 15th Ed. Appleton-Century-Crofts, New York, 1976.
   This classical text deals in more detail with diagnosis of pregnancy, the normal pelvis, fetal positions, antenatal care and the psychosomatic aspects of obstetrics.

Chapter 14

LABOR AND DELIVERY

THE DURATION OF HUMAN PREGNANCY

Based upon the thermal shift of basal body temperature at the time of conception, the mean duration of human pregnancy (excluding infants weighing less than 2500 gm.) is 266 days from conception (9 *true* lunar months), with a standard deviation of 12 days. In clinical practice, it is customary to refer to full term as 40 weeks from the first day of the last menstrual period, assuming that conception occurs two weeks after the onset of the menses. Thus, when we refer to a pregnancy of 16 weeks, the true duration is only 14 weeks from the time of ovulation. If we assume that 280 days from the onset of the last menses is term, then the distribution of births around this point is shown in Figure 14–1. A simple method (Naegele's rule) for calculating the estimated date of confinement (EDC) is to count back three months from the first day of the last menses, then add one year and seven days.

If a woman delivers before the thirty-seventh week, the infant is "premature by date" (as opposed to "premature by weight"). Post-maturity is a term used by some when the woman is two weeks past the EDC (which would include 12 per cent of mature infants), but others use it when she is three weeks past term (which would include 5 to 7 per cent).

THE CAUSE OF LABOR

The factors which trigger the initiation of labor vary from one species to another. In the rabbit, for example, the withdrawal of progesterone secondary to the decline of the corpus luteum of pregnancy triggers parturition. Such observations give rise to the progesterone block theory (Csapo), but in the human species, there is no fall of maternal plasma progesterone concentration at the onset of labor. Nevertheless, progesterone, which may inhibit the transmission of electrical impulses through the myometrium according to Csapo, must play a role in the maintenance of gestation in the human species.

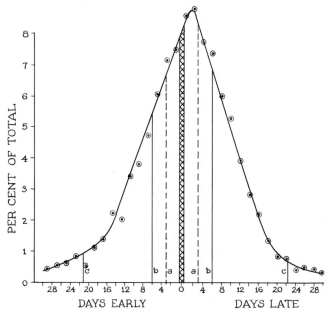

FIGURE 14–1. Distribution of births in two-day intervals before and after the 280th day after the first day of the last menstrual period. The vertical lines a, b and c include 25, 50 and 95 per cent of the 4121 observations. (By permission from Page, *in* Davis' Gynecology and Obstetrics, Volume 1, Chapter 4, Medical Dept. [Loose-leaf Section], Harper & Row Publishers, Inc., Hagerstown, Maryland, 1970.)

For many years it was assumed that the release of oxytocin from the maternal neurohypophysis was the most likely explanation of the initiation of labor, but the removal of this organ from a number of species did not prevent the onset of labor, although it did in many instances result in a prolongation of the labor process itself. Current opinion is that the intermittent release of oxytocin is of importance in determining the strength and duration of uterine contractions once labor is established, but is not crucial for the initiation of labor. Immunochemical assays of plasma oxytocin, however, have so far failed to detect any increased concentrations during labor.

Hippocrates taught that the fetus itself, when mature, initiated labor. In sheep, this theory has been borne out by the ingenious experiments of Liggins. Hypophysectomy of the lamb *in utero* prevented the onset of labor, but this effect could be countered by administering ACTH to the fetus. The removal of both adrenal glands from the lamb fetus also prevented the onset of labor. Premature labor could be produced by infusing ACTH into the intact fetus, or by the administration of cortisol to the fetus. Thus, it appears that the activity of the fetal zone of the lamb's adrenal glands is responsible in some way for triggering parturition.

In human reproduction it is almost certain that the fetus also initiates spontaneous labor and that the final biochemical event may be an increased rate of synthesis of prostaglandin $F_{2\alpha}$ in the maternal decidua. The following findings support this hypothesis: (1) The weight of the fetal adrenals is higher when spontaneous labor occurs than when cesarean section or induction of labor is employed. (2) The longest gestations on record have been associated with anencephalic infants in which the pituitary gland was absent or deficient and the fetal adrenal cortex was atrophied. (3) Increasing estrogen of feto-placental origin increases the synthesis and release of $PGF_{2\alpha}$ from maternal uterine tissues. (4) $PGF_{2\alpha}$ is formed from its immediate precursor, arachidonic acid, by the action of a multienzyme complex, prostaglandin synthetase. McDonald and his co-workers have shown that the intra-amniotic injection of arachidonate during the middle trimester induces abortion in all cases, without, incidentally, any gastrointestinal side effects. (5) Inhibitors of prostaglandin synthetase, such as aspirin or indomethacin, will prevent abortion following the instillation of arachidonate. (6) Aspirin and indomethacin prevent the onset of normal labor in monkeys, may delay labor in women and definitely prolong the interval between instillation of hypertonic saline and abortion in the middle trimester. (7) Both arachidonic acid and $PGF_{2\alpha}$ are found in increased concentrations in the amniotic fluid of women in early spontaneous labor.

Thus the initiation of labor in women appears to involve the fetal brain-pituitary-adrenal axis, the feto-placental steroid production and the intra-uterine synthesis of prostaglandin $F_{2\alpha}$. The precise sequence of events is yet to be elucidated, but its unfolding constitutes one of the most exciting chapters in obstetrical research.

BIOCHEMICAL AND ELECTRICAL CHANGES IN THE MYOMETRIUM

The biochemical reactions which provide energy for myometrial contraction are the same as those described for skeletal muscle, but the substrates for the reactions are present in different proportions and in lower concentrations. The quantities of creatine phosphate, ATP and ADP, and actomyosin all increase progressively in the myometrial cells as pregnancy approaches term.

A progressive increase in the ratio of RNA to DNA within the myometrium during pregnancy indicates that growth is due primarily to hypertrophy rather than to hyperplasia. This increase in RNA synthesis and total protein is brought about by the combined actions of estrogen and uterine distention.

In the absence of estrogen, the resting electrical potential of the myometrial cells is extremely low. Under the prolonged influence of estrogen, the potential rises to 50 mv. (inside negative). According to Kao, progesterone does not contribute to this high resting potential, and he disagrees with Csapo, who believes that progesterone blocks the transmission of impulses, thus rendering the myometrium incapable of contracting in a coordinated manner.

The pacemaker action that initiates uterine contraction is a series of action potential spikes which have a duration of 100 milliseconds and which become synchronous at the time of parturition. The frequency of the burst discharges, which appear to originate near the right or left utero-tubal junction, determine the frequency of uterine contractions, whereas the number of spikes in each burst controls the intensity of the contraction.

Oxytocin is able to initiate spike discharges in a quiescent myometrial strip and to increase the amplitude of the electrical discharge, possibly by increasing the number of sodium channels in the cell membrane. The reason the human uterus is relatively insensitive to oxytocin in the first six or seven months and becomes increasingly sensitive to it as term is approached has never been adequately explained.

THE PHYSIOLOGY OF UTERINE CONTRACTIONS

The following descriptions have been synthesized from the extensive writings of Caldeyro-Barcia and his Montevideo coworkers who for many years have studied the physiology of labor and the effects of labor patterns upon the fetus. Before discussing the course of normal labor and its many variants, it is important to understand the physiology of a single uterine contraction and its clinical correlates, as well as the various types of abnormal uterine contractions.

The normal contractile wave and pressure gradients of a single contraction during labor are illustrated in Figure 14–2. The large diagram of the uterus on the left shows the four points at which micro-balloons were inserted for the recording of intramyometrial pressures. The four corresponding pressure tracings are reproduced in time relationship to each other, to the spread of the contraction wave (as indicated by the shading on the small uterine diagrams above) and to the amniotic fluid pressure, which is recorded by an intra-amniotic catheter. In a normal contraction, the wave begins at one uterine cornu and spreads downward within the ensuing 15 seconds. Despite the fact that the contraction phase begins later in the lower portions of the uterus, the organ is normally so well coordinated that the peak of contraction is attained simultaneously in all portions. The intensity of the contraction also decreases from above downward and is essentially absent in the cervix, which is, after all, composed almost entirely of connective tissue.

These descending gradients of propagation, duration and intensity are essential for the efficiency of the contraction in dilating the uterine cervix. The synchronous relaxation of the uterus permits the amniotic fluid pressure to return to its normal tonus of about 10 mm. Hg pressure. *Hypertonus* refers to resting pressures from 12 to more than 30 mm. Hg (dangerous for the fetus), whereas *hyperactive labor* refers to contractions which exceed 50 mm. Hg at the peak or which occur more often than two minutes apart, or both. Hypotonus itself is of little significance. *Hypoactive labor,* defined as contractions with an intensity of less than 30 mm. Hg or more than five

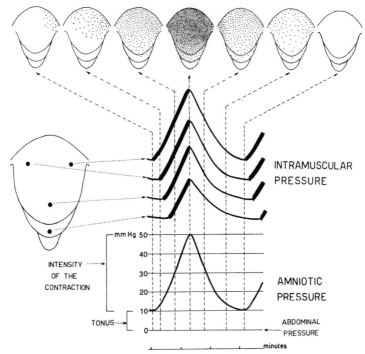

FIGURE 14–2. Schematic representation of the normal contractile wave of labor. Diagram of uterus on the left shows the four points at which the intramyometrial pressure is recorded with microballoons. (From Caldeyro-Barcia, Clin. Obstet. Gynec., *3*:394, 1960.)

minutes apart, or both, produces little progress and is a common cause of prolonged labor.

A clinician familiar with normal labor would immediately wonder how the contractile phase illustrated in Figure 14–2 could be called normal when its total duration is three minutes. He knows that a labor contraction lasting much longer than one minute is considered abnormal. This apparent discrepancy is explained in Figure 14–3, which shows schematically the clinical correlation between the contraction as recorded by physiological instruments and that which is palpated by the hand or perceived by the patient. The contraction of the fundus may be perceived by abdominal palpation when the intra-amniotic pressure reaches about 20 mm. Hg, so by this criterion the contraction lasts 70 seconds. The patient perceives pain when the pressure reaches about 25 mm. Hg (although this "pain threshold" will vary from one woman to another), so the duration of the "pain" is 50 to 60 seconds.

Uterine Work

Just as cardiac work may be defined in terms of force and frequency, uterine work may be defined as the product of the maximal pressures

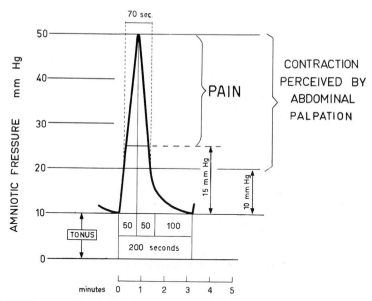

FIGURE 14–3. Correlation between abdominal palpation and intrauterine pressure tracing. (From Caldeyro-Barcia, Second World Congress F.I.G.O., Libraire Beauchemin, Ltd., Montreal, Canada.)

achieved during "systole" and the frequency of contractions. For study purposes, Caldeyro-Barcia and his coworkers defined a Montevideo unit as the number of contractions per 10 minutes times the intensity (or amplitude) in mm. Hg, as measured by the intra-amniotic pressure tracing. The uterine activity during the course of pregnancy, utilizing such units of work, is illustrated in Figure 14–4. The steadily increasing but painless contractions which occur during the month prior to active labor (known as Braxton-Hicks contractions) have an important function in causing the cervix to thin out or efface. The internal os of the cervix is retracted until there is no longer any cervical canal, and at this time the cervix is said to be 100 per cent effaced. In primigravidous patients, this process must occur before dilatation of the cervix takes place, whereas in multiparous women the cervix may remain as thick as 1 cm. or so until it is several centimeters dilated.

ABNORMAL UTERINE CONTRACTIONS

REVERSE GRADIENTS.   Abnormal waves may begin in the lower segment and spread upward, and under these circumstances the contractions may be stronger and persist longer in the lower uterine segment than in the fundus. Obviously, the cervix will not dilate under these circumstances; indeed, it will be noted to contract during the contraction. A reversed gradient may be suspected when the patient complains of lower abdominal and low back pain five or 10 seconds before the contraction is palpable over the

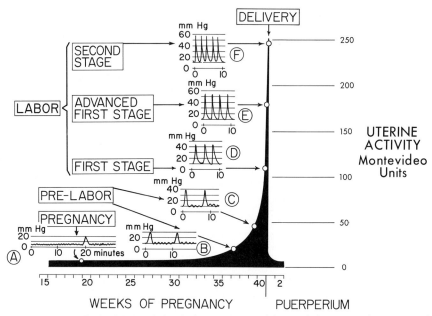

FIGURE 14–4. Uterine activity during pregnancy, labor and the puerperium (see text). (Redrawn from Caldeyro-Barcia.)

fundus. Resting the patient with a sedative and opiate may permit the institution of normal gradients later, or the use of paracervical anesthesia may block the contractions of the lower segment.

MILD INCOORDINATION.   This may result from the action of two asynchronous pacemakers, as illustrated in Figure 14–5. Clinically, one may observe small contractions alternating at irregular intervals with stronger ones. Such patterns are common in early labor and usually give way to more coordinated patterns as cervical dilatation advances, especially after rupture of the membranes.

SEVERE INCOORDINATION.   Severe incoordination (also called "colicky labor," "uterine fibrillation" or "hypertonic inertia") results from electrical waves beginning in various areas of the uterine wall at different times. The net result is a tetanic type of irregular hypertonus which may interfere with the utero-placental circulation for prolonged periods of time, leading to fetal distress. The use of oxytocin under these circumstances may exaggerate the pattern, and may therefore be harmful. The progress of labor is ordinarily arrested, and the patient may be in continuous discomfort and appear exhausted. The use of regional anesthesia (spinal, epidural or caudal) may improve the pattern, as will the use of sedation and analgesia; but if the situation is not corrected, and if there is evidence of fetal distress, cesarean section may be required.

Incoordination may be mimicked by the intravenous infusion of adrenal

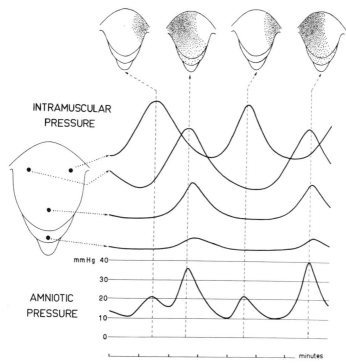

INTRAMUSCULAR
PRESSURE

mmHg 40
30
AMNIOTIC
PRESSURE 20
10
0

minutes

FIGURE 14–5. First-degree incoordination. The wave originating at the left cornu remains localized. Later, a contractile wave originates at the right cornu and spreads to most of the uterus. (From Caldeyro-Barcia, Alvarez and Poseiro, Triangle, the Sandoz Journal of Med. Sci., Vol. II, No. 2, 1955, Basel, Switzerland.)

medullary extracts containing 80 per cent epinephrine and 20 per cent norepinephrine. This has given support to the concept that excessive activity of the adrenal medulla, engendered by fear, pain or anxiety, may result in poor quality prolonged labors and might explain the beneficial effects of psychotherapy, sedation and analgesia.

HYPOACTIVE LABOR. True inertia or hypoactive labor is a persistence of the prelabor pattern in which the contractions are weak and infrequent, but are painful. If, after a variable number of hours, the pains cease, the episode is called "false labor." If the pattern ultimately merges into an active phase, it is called the "latent phase of labor." If, during the phase of dilatation, the contractions continue to be weak and far apart, there is a specific indication for the use of an infusion of highly diluted oxytocin solution intravenously. It must be remembered, of course, that ineffective uterine forces are only one of three major causes of *dystocia* (difficult labor); the other two are a faulty position of the fetus or an abnormal resistance due to some type of disproportion between the size of the passenger and the passage. Frequently there is a mixture of malposition, faulty pelvic architecture and inadequate uterine forces. The acme of obstetrical art is the prompt and proper diagnosis of the cause of a dystocia and the institution of a management appropriate for both mother and fetus.

THE STAGES OF LABOR

As noted above, the prelabor contractions ("insensible labor") are vital preparations for active labor, and the two forms merge so gradually that it is difficult to define the onset of true labor with precision. The *first stage* of labor begins when there is objective evidence of progressive cervical effacement or dilatation, usually with progressive descent of the presenting part. Inasmuch as the majority of women are first observed after the onset of true labor, we say (retrospectively) that the first stage began when her contractions became regular, less than 10 minutes apart and painful. The first stage ends with complete dilatation of the cervix. The average duration of the first stage is about 12 hours for primigravidas and about seven hours for multiparas, but there are wide variations. For example, when all conditions are favorable, that is, when the cervix is soft and effaced, the occiput is anterior and the head is engaged, and the contractions are strong and frequent, then the duration of the first stage may be as short as three hours for first pregnancies and two hours for multiparas.

The *second stage* of labor begins when the cervix is 10 cm. dilated (assuming that the baby is at term and of average size) and ends with the delivery of the infant. The average duration for primigravidas is 50 minutes and for multiparas, 20 minutes.

The *third stage* of labor ends with the delivery of the placenta, and the duration is ordinarily less than 10 minutes. The uterus contracts vigorously after it has expelled the fetus, aided by the customary parenteral administration of an oxytocic drug, and the sharp reduction in the area of placental attachment literally shears off the placenta, leaving a portion of the maternal decidua on the myometrium and a portion on the maternal surface of the placenta. The myometrial contractions constrict the uterine vessels which formerly supplied the intervillous space, preventing an immediate uterine hemorrhage.

THE CLINICAL COURSE OF LABOR

Normal Labor

When the cervical canal becomes obliterated and the external os dilates 1 or 2 cm. through the action of prelabor contractions, a blood-tinged mucus plug often escapes, and this is referred to by patients as the "show." This is frequently coincident with the onset of regular contractions. In about 15 per cent of patients, the membranes may rupture an hour or more before the onset of the first stage (premature rupture of membranes), and in another

15 per cent, the escape of amniotic fluid occurs about the time of onset of painful contractions. In the remainder of patients, the membranes break (or are ruptured artificially) later in the first stage or early second stage of labor. Patients should be instructed to enter the hospital if the membranes rupture or if it is believed that active labor has started.

The course of cervical dilatation with time follows an S-shaped curve, such as that described by Friedman (Figure 14–6). There is a latent phase which may last several hours, particularly in nulliparas. During this time, while the fundal contractions are becoming coordinated and more efficient, further effacement, thinning and softening of the cervix occur, but very little progress in dilatation is noted. The active phase then begins, and dilatation should proceed in a straight line when plotted against time, until only a rim of cervix remains (about 9 cm. dilatation). From this point on, the presenting part must descend through the cervix in order to obliterate the remaining rim, whereas up until this point the cervix has literally been pulled apart by the gradients of uterine forces. The deceleration phase, from a physiological standpoint, may therefore be considered a part of the second stage of labor, because if true cephalopelvic disproportion exists, the final rim of cervix may remain indefinitely.

Abnormal Labor

Major aberrations of the labor pattern are illustrated in Figure 14–7, and consist of a prolonged latent phase (Curve A), a primary dysfunctional labor (Curve B), or a secondary arrest of the first stage (Curve C).

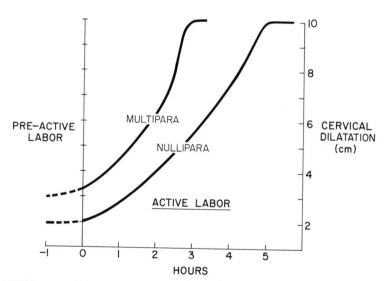

FIGURE 14–6. The relationship between cervical dilatation and hours of active labor. (After Friedman.)

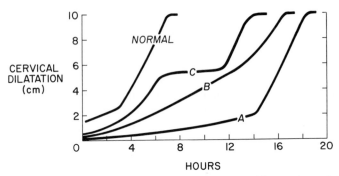

FIGURE 14–7.   Major aberrant labor patterns. *A,* Prolonged latent phase. *B,* Protracted active phase. *C,* Secondary arrest of dilatation. (Modified from Friedman.)

A *prolonged latent phase* is frequently due to the use of excessive analgesia too early in labor. Other than mild sedation, analgesic drugs should be withheld until the active phase of labor has begun. In other instances, the cervix has not "ripened," and a longer time must be permitted for softening and effacement. In only about one out of 10 cases of prolonged latent phase is the defect due to real myometrial dysfunction, so cesarean section is rarely indicated during the latent phase solely because of its prolongation. The use of a narcotic to permit a period of rest may result in a cessation of painful contractions altogether (leading to a diagnosis of false labor) or may permit the active phase to begin as soon as forceful contractions resume. Should neither occur, stimulation of labor with an infusion of oxytocin may be indicated.

Primary *dysfunctional labor,* with a protracted active phase, may result from fetal malposition, cephalopelvic disproportion, or inefficient myometrial activity, in that order of frequency. Under ordinary circumstances, only the last cause—that is, inefficient or hypoactive labor—indicates stimulation with oxytocin.

A *secondary arrest* of labor may occur at any time, and may be due to (1) excessive sedation or sometimes the use of conduction anesthesia, (2) fetal malposition or (3) cephalopelvic disproportion. The proper diagnosis of the cause, which may involve x-ray studies, continuous fetal monitoring, repeated vaginal examinations, and so forth, is of extreme importance, and will determine whether watchful waiting, rest by narcosis, stimulation by oxytocin infusion, forceps delivery or cesarean section is to be employed.

A *precipitate labor,* defined by Friedman as one in which the rate of cervical dilatation is greater than 5 cm. per hour, occurs most commonly in multiparas and is due to hyperactive labor. Inasmuch as this may be traumatic to mother or fetus, the use of general or conduction anesthesia is frequently indicated. A *precipitate delivery,* on the other hand, refers to an unusually short second stage. The infant is frequently delivered in an unsterile field. Under no circumstances should attempts be made to hold back the head until the obstetrician arrives, because the fetus may be injured. Unusually rapid labors or deliveries carry a high risk of lacerations to the birth canal, which must be carefully inspected.

Friedman has introduced a new graphic form which automatically diagnoses the various disorders of labor (Figure 14–8). The time of the onset of labor is inserted at point A and the clock time filled in to point B. The cervical dilatation and fetal station are plotted for each observation. When the dilatation line turns upward the next inclined line is darkened, because this represents the acceptable limit of the rate of active phase dilatation. Similarly, when the descent pattern turns downward, the next inclined line is darkened to indicate the acceptable limit for the rate of descent. If the dilatation line for nulliparas does not turn upward before reaching 20 hours, the diagnosis is *prolonged latent phase.* If the active phase dilatation crosses the darkened incline, the diagnosis is *protracted active phase;* or, if there is no progress for two hours, the diagnosis is *secondary arrest of dilatation.* If the descent curve crosses the darkened incline, the diagnosis is *protracted descent* if some progress is made, or *arrest of descent* if there is no progress for one hour. A similar chart differing only in the degrees of incline and the time limits is used for multiparas.

Following delivery of the infant, the woman enters the most dangerous phase of her entire pregnancy, because hemorrhage remains a leading cause of maternal death, and the hour after birth is the period when excessive blood loss commonly occurs. Careful measurements of total blood loss indicate that the average loss is about 350 ml., especially when incision of the perineum (episiotomy) is utilized. Postpartum hemorrhage is commonly

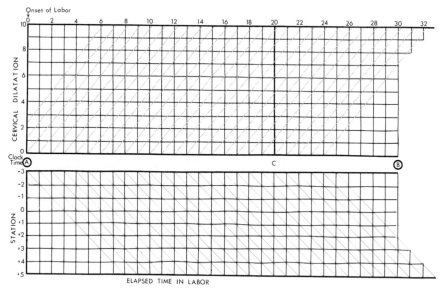

FIGURE 14–8. A modified Friedman labor graph for use in nulliparas (see text).

defined as a loss exceeding 500 ml., but would more properly be defined as a loss exceeding 1 per cent of the body weight (see Chapter 16).

Placental separation ordinarily occurs within three to six minutes after delivery, and may be recognized by the following signs: the uterine fundus becomes firm and globular, the umbilical cord advances 6 to 8 cm. (this can best be noted by applying continuous gentle traction on the cord) and there is usually a small gush of blood. When these signs appear, traction on the cord is increased, and the uterus is pulled upward and away from the separated placenta. This maneuver is accomplished by placing the fingers just above the symphysis, pressing downward on the lower uterine segment and raising the globular fundus toward the umbilicus (Figure 14–9). After the placenta is removed from the vagina, the lower uterine segment is then held firmly between the thumb and fingers until the entire uterus is firm and the bleeding is minimal. Oxytocic drugs may be administered intramuscularly as soon as the infant is delivered, or intravenously after delivery of the placenta to minimize blood loss from uterine atony. The placenta and membranes must be carefully inspected to be sure that they are intact. Should there be any suspicion that a portion is missing, an immediate manual exploration of the uterine cavity should be performed. The cervix is in-

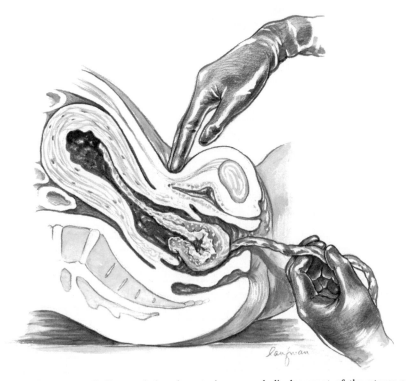

FIGURE 14–9. Delivery of the placenta by upward displacement of the uterus and moderate traction on the cord. (From Greenhill, Obstetrics, 13th Ed., W. B. Saunders Co., Philadelphia, 1965.)

spected for possible lacerations (especially after forceps deliveries or precipitate labors), and if any are found they are immediately repaired. The episiotomy or any perineal laceration is then carefully approximated with sutures. During the ensuing hour after delivery of the placenta, often referred to as the fourth stage of labor, the parturient is constantly observed, either in the delivery room or in a recovery room. The meticulous management of the third and fourth stages of labor will minimize blood loss and markedly reduce the incidence of postpartum hemorrhage.

In 2 or 3 per cent of deliveries, the placenta may be retained in the uterus for a period of 20 minutes or longer. The customary practice, after waiting 15 or 20 minutes, is to remove the placenta manually under light anesthesia. In most instances, the placenta will not be found adherent to the uterine wall, but rather completely separated and "trapped" by the contraction of the lower segment. The most common cause of the retained placenta is the failure to deliver it promptly after separation has occurred.

THE BIRTH PROCESS

Definitions

In describing the mechanism of labor, certain terms in common use must be clearly understood. The fetal *lie* refers only to the relation between the long axis of the fetus and that of the mother. Thus a fetal lie may be longitudinal, transverse or oblique. The *presentation* refers to that presenting part of the fetus which may be palpated on vaginal examination during the second stage of labor. The presenting part may be the vertex, brow, face, breech or shoulder. The *position* is the relationship between an arbitrary point of reference on the presenting part and the four quadrants of the maternal pelvis. The common points of reference are the occiput (O), the chin (M for mentum) and the sacrum (S). The abbreviation L.O.T., for example, indicates that we are dealing with a cephalic presentation in which the occiput is pointing left and midway between the anterior and posterior quadrants of the maternal pelvis, i.e., transversely.

The *station* of the fetus refers to the relationship between the most dependent or lowest portion of the presenting part and an imaginary line joining the two ischial spines. This is expressed in terms of centimeters above or below this line. A station +1 in a well flexed vertex presentation indicates that the occiput has descended 1 cm. below the ischial spines, also suggesting that the largest diameter of the fetal head has passed through the pelvic inlet, and that the head is therefore engaged (see Chapter 13).

Observations

An examination during the course of the first stage should include observations of the cervix, the position of the fetus and its station, the state of the membranes, the quality of labor, the fetal heart tones, the maternal blood pressure, and notes regarding the mother's general condition, her response to the labor, and records of all medications. Delivery room record forms ordinarily contain columns for all of these items, and should ideally provide a chart for the graphic recording of cervical dilatation with time, as illustrated in Figure 14–6. Sterile vaginal examinations have largely replaced rectal examinations in most medical centers.

Mechanisms of Labor

In vertex presentations, which constitute 95 per cent of all labors, the fetal head undergoes positional changes because of the relative shapes of the fetal head and the bony pelvis. The primary movements which constitute the mechanism of labor are classically described under the following headings: (1) descent, (2) flexion, (3) internal rotation, (4) extension, (5) external rotation, and (6) expulsion (i.e., of the body). Several of these positional changes, of course, are proceeding simultaneously. The mechanism of labor for a fetus in the R.O.A. position is illustrated in Figure 14–10.

Flexion occurs because the direction of force on the fetal head is transmitted through the cervical spine, which is eccentrically placed with respect to the occipito-frontal diameter: the "short arm of the lever," which is toward the occiput, is pushed downward. If, for some reason, the head is deflexed at the onset of labor, so that the brow presents, then the short side of the lever may extend from the occipital condyles to the chin, and the head extends rather than flexes, resulting in a face presentation. In the majority of nulliparous women, the head is engaged and fully flexed by the time the active phase of labor begins.

Internal rotation must occur as the head descends so that the long axis of the head, the suboccipito-bregmatic diameter, fits the long axis of the pelvic outlet, which is the antero-posterior diameter. Thus, when internal rotation is complete, and the vertex has descended to the perineal body and is ready for birth by extension, the occiput is either directly anterior or, less commonly, directly posterior.

Extension of the head is now the only movement by which the fetus in a vertex position can round the pelvic curve. The process is frequently aided by reducing the resistance to extension offered by the perineum, i.e., by episiotomy, and by upward pressure on the brow (the Ritgen maneuver). Should the fetus be in a face presentation with the mentum anterior, the head is born by flexion rather than by extension. Should the chin rotate posteriorly, delivery is ordinarily impossible unless the head can be rotated with the hand or with forceps to a mentum anterior position.

After the head is delivered, the long axis of the shoulders must now rotate to coincide with the long axis of the pelvic outlet, and as this takes place *external rotation* of the head occurs. At this point, gently depressing the fetal head permits the anterior shoulder to appear under the symphysis,

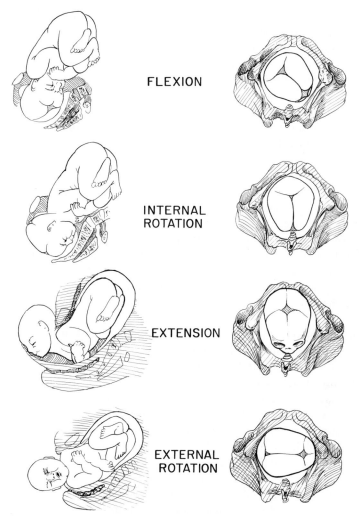

FLEXION

INTERNAL
ROTATION

EXTENSION

EXTERNAL
ROTATION

FIGURE 14–10. The mechanism of labor in a right occipitoanterior position.

then elevating the head assists in delivery of the posterior shoulder over the perineum. From this point on, expulsion of the body is rapid. Following expulsion, the infant should be held below the level of the vagina so that it may receive as much of the fetal blood contained in the placenta as possible. To receive such a "placental transfusion" (generally considered desirable) requires a delay of two or three minutes before clamping and cutting the umbilical cord.

Breech Presentations

These occur in about 4 per cent of all deliveries, but the rate depends upon the size of the fetus. With premature fetuses that weigh less than 1500 gm., the rate is 32 per cent; weight between 1500 and 2000 gm., 16 per cent; from 2000 to 2500 gm., 8 per cent; and over 2500 gm., 3 per cent. The high perinatal mortality of 10 per cent associated with breech deliveries is therefore largely due to the high frequency of premature infants. Nevertheless, the fetal mortality associated with breech deliveries of infants weighing over 2500 gm. is between 1 and 2 per cent, which is several times that of cephalic presentations.

The increased fetal risk is due largely to the fact that during the birth process, successively larger portions of the body must pass through the birth canal. In premature infants, this disproportion between the circumference of the head and that of the shoulders or hips is even greater, so that undue delays in the delivery of the head because of the incompletely dilated cervix are not uncommon.

There are three varieties of breech presentation, depending upon the position of the legs. In a frank breech, the feet rest alongside the head; in a complete breech, the legs are flexed in a sitting position; and in an incomplete breech, a foot or a knee is the presenting part. Foot presentations are more hazardous for the fetus because the irregular presenting part allows room for the cord to prolapse through the cervix when the membranes rupture.

The diagnosis of a breech presentation may usually be made by abdominal palpation, utilizing the *four maneuvers of Leopold,* which are routinely used for all presentations. (1) Using the fingers of both hands, the top of the fundus is gently palpated to determine whether it is occupied by the head or by the breech. The fetal head, of course, is round, hard and will bounce back against the fingers when pushed (ballottable). (2) The hands are now moved downward, pressing inward on either side of the uterus to locate the fetal back on one side or the small parts on the other. (3) The examiner now grasps the lower uterine segment between thumb and fingers to determine whether it is occupied by the head or the breech. (4) Finally, the four fingers of each hand are placed alongside the lower segment, pointing toward the symphysis, and the presenting part is palpated. In cephalic presentations, the first resistance to be encountered should be the forehead, indicating that the head is well flexed. Should such a cephalic prominence be on the same side as the fetal back, the examiner would suspect a face presentation.

There are times, of course, when the diagnosis of a breech is not made until the presenting part is actually palpated by sterile vaginal examination performed during labor. Under these circumstances, a decision must be made whether to permit labor to continue or whether a cesarean section is indicated. This may require the use of x-ray pelvimetry, particularly in nulliparas. The frequency of cesarean section varies between 10 and 35 per cent in breech presentations. The operation is particularly indicated when the fetus is judged to be in excess of 8 lbs. in weight, when the patient is an elderly primigravida, when the pelvic measurements are borderline, or in the presence of a dysfunctional labor.

Other Abnormal Presentations of the Fetus

PERSISTENT OCCIPUT POSTERIOR

When the fetal head engages in the R.O.P. or L.O.P. position, the occiput will undergo an internal rotation of 135° to the direct anterior position nine times out of 10. Should there be difficulty in rotation during the second stage, the process may be assisted by placing one or two fingers against the lambdoidal suture and turning the posterior fontanelle toward the symphysis during the height of a contraction.

When the occiput rotates 45° to the direct posterior position, complete flexion of the head is often delayed, adding somewhat to the length of the second stage of labor, especially in primiparas. Under these circumstances, the head may sometimes be rotated manually through a 180° arc; should this prove difficult, the head may be permitted to be born spontaneously or with outlet forceps in the O.P. position. Rotation of the head with forceps utilizing a double application (the Scanzoni maneuver) is least desirable because of the marked increase in vaginal lacerations. When the root of the nose, rather than the brow, becomes the pivotal point under the symphysis, then the occipito-frontal diameter of 11.5 cm must pass through the outlet, and this requires a more extensive episiotomy incision.

FACE PRESENTATION

Extension rather than flexion of the head during descent may occur two or three times in each 1000 deliveries, resulting in a face presentation. Fortunately, in two thirds of these the chin will rotate to a direct mentum anterior position (Fig. 14–11). Under these circumstances, the undersurface of the mandible becomes the pivotal point and the head is born by flexion. Inasmuch as no greater diameters are presented than in a vertex presentation, few difficulties are encountered. Should the chin rotate to the hollow of the sacrum, however, the delivery of a term-sized infant is essentially impossible. Should efforts to rotate the mentum posterior to the anterior position fail, then a cesarean section must be performed in the interests of both the mother and the fetus.

BROW PRESENTATION

On rare occasions, the brow may present, but fortunately at least half of these will spontaneously convert to either a vertex or a face presentation. The remainder pose a problem of absolute cephalopelvic disproportion, because the presenting diameter is the mento-occipital, which measures 13.5 cm. in the average term fetus, and no such pelvic diameter is available. A persistent brow presentation which cannot be converted to a vertex or to a mentum anterior position is therefore an indication for cesarean section.

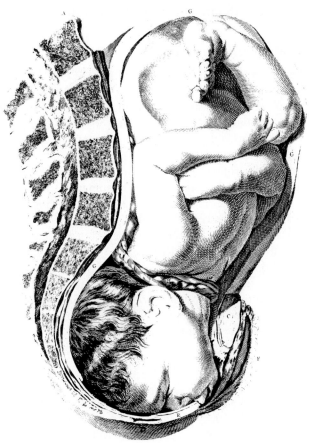

FIGURE 14–11. Face presentation with mentum anterior. (From Smellie, A Set of Anatomical Tables with Explanations and an Abridgement of the Practice of Midwifery with a View to Illustrate a Treatise on that Subject and Collection of Cases, London, 1761.)

TRANSVERSE LIE OR SHOULDER PRESENTATION

Presentation of the shoulder or the arm may occur once in 300 to 400 deliveries and poses a serious problem. Among the predisposing causes are placenta previa, grand multiparity with a pendulous abdomen, the presence of a twin, a truly contracted pelvis or an anomalous uterus (e.g., an arcuate or partially septate fundus). When this lie is recognized before or during early labor, an external version to a cephalic or breech presentation may be successful. If it is recognized for the first time during the second stage of labor, and the membranes are intact, an internal podalic version under full anesthesia may be accomplished. If, on the other hand, the membranes should rupture and the transverse lie persists, then an immediate cesarean section is ordinarily done.

The Use of Forceps or Vacuum Extraction

The obstetrical forceps were invented by Peter Chamberlen the Elder and were kept a secret throughout the entire 17th century, being handed down through three generations of the Chamberlen family. The instrument came into general use early in the 18th century, and since then literally hundreds of varieties of forceps have been described. Today, over a third of all infants born in U.S. hospitals are delivered by forceps, usually as an elective procedure. The obstetrical forceps are used primarily for traction and for extension of the fetal head, but they may occasionally be used for internal rotation.

There are certain conditions which must be fulfilled before the forceps are applied. These prerequisites are as follows: (1) the cervix must be fully dilated; (2) the head must be engaged; (3) the pelvic midplane or outlet must not be contracted; (4) the presentation and position of the fetal head must be accurately known; (5) the membranes must be ruptured; (6) some form of local, regional or general anesthesia should be utilized; (7) the bladder or rectum should not be full (desirable but not mandatory); and (8) an episiotomy should be a part of the procedure.

Forceps operations are classified according to the degree of descent and the position of the fetal head. The classification adopted by the American College of Obstetricians and Gynecologists is as follows:

1. *Outlet forceps.* The application of forceps when the scalp is or has been visible at the introitus without separating the labia, the skull has reached the pelvic floor (perineal body) and the sagittal suture is in the anteroposterior diameter of the pelvis.

2. *Midforceps.* The application of forceps when the head is engaged but the conditions for outlet forceps have not been met. In the context of this term, any forceps delivery requiring artificial rotation, regardless of the station from which extraction is begun, shall be designated a "midforceps" delivery. The term "low midforceps" is disapproved.

3. *High forceps.* The application of forceps at any time prior to full engagement of the head. High forceps delivery is almost never justifiable.

A purely elective or "prophylactic" forceps delivery should always be an outlet forceps as defined. There should be some fetal or maternal indication to justify a midforceps delivery. When a midforceps operation is done, a record should be made of the position and the station of the head when the delivery is begun. In addition, a description of the various maneuvers and of any difficulties encountered in the forceps application or in the extraction of the infant should be recorded, as well as the indication.

Outlet forceps are frequently used purely as an elective procedure in order to shorten the second stage of labor or to permit the use of regional or general anesthesia. The instrument is used only for extension of the head (Fig. 14–12). Forceps are also used electively for the delivery of the after-coming head in breech presentations in order to prevent undue traction upon

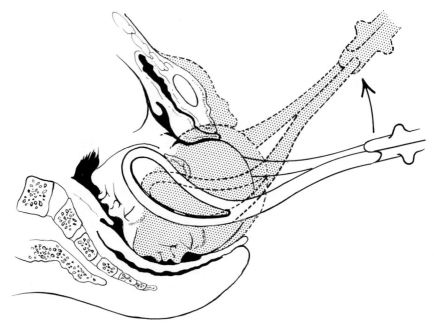

FIGURE 14–12. Extension of the head in an outlet forceps delivery. (From Benson, Handbook of Obstetrics and Gynecology, Lange Publications, Los Altos, Calif., 1971.)

the neck and to permit a controlled and deliberately slow delivery of the fetal skull. A special instrument such as the Piper forceps is frequently used for this purpose.

All obstetrical forceps have a cephalic curve to fit the fetal head, and most varieties also have a pelvic curve to match the axis of the birth canal. The classic forceps are those designed by Simpson or Elliott or the numerous minor modifications of these types. Special forceps have been designed for instrumental rotation of the head (e.g., Kielland or Barton) or for increased traction (e.g., Tarnier or Dewey forceps, or Bill's axis traction handle).

The usual maximal traction force employed during a midforceps delivery is between 25 and 40 pounds, and if the force does not exceed 40 pounds the risk of fetal injury is negligible. By comparison, it should be noted that the downward force of a uterine contraction during the second stage is about 9 pounds, and this may be doubled as the result of a voluntary bearing-down effort. Traction on the obstetrical forceps is used during the contraction, with or without the cooperative bearing down efforts of the mother, so that the forces of second stage labor are augmented rather than replaced. Midforceps deliveries rise in frequency with the increased use of caudal, epidural or spinal analgesia.

VACUUM EXTRACTION

The Malmström vacuum extractor is essentially a metal suction cup which is affixed to the fetal scalp with a negative pressure of 0.7 to 0.8 kg.

As a substitute for the obstetrical forceps to apply traction, it is more popular in Europe than in the United States. Special indications for its use include scalp traction in cases of marginal placenta previa, the arrest of labor while the head is high in the midpelvis, transverse arrest of internal rotation with the head at midpelvis, or as a substitute for midforceps when there is secondary inertia in the second stage.

Among the advantages of vacuum extraction are: (1) use of the instrument may obviate a hazardous midforceps operation; (2) compression of the fetal head is avoided; (3) anesthesia is not required, so that patient cooperation is retained; (4) the instrument requires less technical skill than the application of forceps; and (5) the instrument may be applied at any time after the cervix is half dilated and at any station of the head. Among the disadvantages are these: (1) the instrument is limited to vertex presentations; (2) ecchymosis of the scalp occurs in all cases and true cephalhematomas are common; (3) retinal hemorrhages are alleged to be more common; (4) it cannot be used for axis traction or rotation; and (5) the application requires more time than the application of forceps, so that vacuum extraction is not as useful in the event of fetal distress.

Despite the controversy in the United States regarding the usefulness of the vacuum extractor, it has established its place in obstetrics and should be available for use when indicated. There are times, for example, when women who are imbued with the concepts of "natural childbirth" reject all forms of anesthesia, and even forbid the application of forceps, but will accept the assistance of the vacuum extractor when help is needed.

ANALGESIA DURING LABOR AND DELIVERY

The use of drugs for the relief of pain during labor is complicated by the fact that all sedatives and analgesics affect the fetus to some degree and may therefore interfere with the establishment of normal respiration. Drugs may also affect uterine motility and thus alter the pattern of labor. Conduction anesthesia carries some risk of maternal hypotension, which is hazardous for the fetus. For these reasons, no completely satisfactory protocol of obstetrical analgesia has evolved. The abolition of apprehension and anxiety through constant reassurance is an essential ingredient of any program adopted. Similarly, the psychoprophylactic preparation of the patient for natural childbirth is valuable even when the use of analgesia is planned.

It is wise to avoid the use of drugs until the active phase of labor is reached, preferably until a cervical dilation of 3 or 4 cm. has been attained. At this time, 75 to 100 mg. of meperidine (Demerol) may be administered intramuscularly, or a combination of meperidine, 50 mg., and a tranquilizer such as Phenergan (25 mg.) or Valium (5 to 10 mg.). If labor is proceeding very rapidly, 50 mg. of meperidine may be slowly administered intravenously. This drug is recommended over morphine and related opiates because of lesser neonatal depression.

A barbiturate such as sodium pentobarbital or secobarbital, in an oral dose of 0.1 to 0.2 gm., is useful in the early stages of labor for allaying apprehension and for a sedative-hypnotic effect. If it is desired to produce amnesia, then 0.3 to 0.4 mg. of scopolamine is given hypodermically, and repeat injections of 0.1 to 0.2 mg. are used one or two hours later if needed. The popularity of scopolamine-induced amnesia has diminished over the last 20 years, partly because a certain number of patients became quite unmanageable under its influence, and partly because the popularity of self-participation in the labor and delivery process has reduced the number of patients' demands for an obliteration of memory.

In evaluating the effectiveness of these various agents, a sedative-hypnotic action may be judged by the degree of drowsiness that the patient manifests between contractions; the degree of amnesia is inversely proportional to the lucidity of her conversation; and the degree of analgesia is estimated by her physical response to the labor pain.

Late in the first stage or for the delivery itself, some form of conduction analgesia may be chosen. Whether this be a low spinal or "saddle block" epidural injection, continuous caudal analgesia, a paracervical block, a bilateral pudendal nerve block or simply an infiltration of the perineum will depend largely upon the ability and training of the obstetrician or anesthetist. Anesthesia for vaginal or abdominal delivery is the proper province of the obstetrical anesthesiologist and will not be discussed here.

The Effect of Drugs Upon the Uterus

OXYTOCIC DRUGS

Oxytocic drugs are those which stimulate uterine contractions. The only one which reliably produces physiological contractions simulating normal labor is *oxytocin* itself. This is a nonapeptide, now produced synthetically (Pitocin, Syntocinon), which is so potent that one molecule per muscle fiber (according to our calculations) is enough to stimulate a highly responsive gravid uterus at term. The original method of administration was intramuscular, but this is now obsolete. The intranasal or transbuccal routes are still used in some institutions, but the rate of absorption cannot be predicted and the method is therefore unreliable. The intravenous infusion of a highly diluted preparation was introduced in 1942, and is still the safest and more reliable technique. Its use is described later under "Induction of labor." The increasing responsiveness of the uterus during the second half of pregnancy is illustrated in Figure 14–13. The use of concentrated oxytocin intravenously during the third stage is ill-advised because maternal hypotension may develop and because the duration of action is less than five minutes.

Some of the prostaglandins are highly oxytocic. Indeed, clinicians have speculated that the absorption of these compounds from seminal fluid in the vagina may be related to the etiology of premature labor.

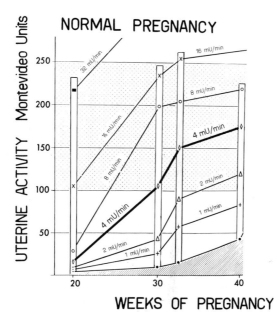

FIGURE 14–13. Uterine activity plotted against duration of pregnancy (in weeks). Shaded area indicates spontaneous activity prior to labor. Lines show the uterine response to intravenous infusions of oxytocin, expressed in milliunits per minute. Note the increased uterine sensitivity as pregnancy progresses. (From Caldeyro-Barcia and Sereno, Oxytocin, Pergamon Press, Oxford, 1961.)

The greatest oxytocic activities have been exhibited by PGE_2 and $PGF_{2\alpha}$. These substances have been used by intravenous infusion to induce labor at term in well over 500 cases, but at the present time they do not appear to have any advantages over oxytocin. The major difference is that the PG compounds act on the gravid uterus at all stages of gestation, whereas oxytocin is clinically effective only in the last several weeks. The prostaglandins may therefore be utilized for therapeutic abortion in the first two trimesters (see Chapter 10).

Ergot derivatives, such as ergonovine (Ergotrate) and methylergonovine (Methergine), all share the same property: a complete disruption of the normal pattern of uterine motility, with a sustained hypertonic contraction of long duration. Such drugs should never be used before delivery, but are useful in the third stage of labor, during the puerperium and in selected cases of incomplete abortion.

Sparteine sulfate achieved brief popularity as an agent for the induction or stimulation of labor, but in recent years it has been shown to exert unpredictable responses, sometimes with a delayed overstimulation.

Norepinephrine causes an increased frequency and intensity of labor contractions, with elevated tonus and varying degrees of uterine incoordination. *Acetylcholine* appears to be a good oxytocic agent when used as an infusion, but has had little clinical usage.

RELAXANTS

There are a number of drugs which relax the uterus. *Epinephrine* is a potent relaxant but its cardiovascular effects render it impractical for clinical use. *Isoxsuprine* is a synthetic compound similar to epinephrine in its uterine action, but the cardiovascular effects are less, and consist of a fall in the blood pressure with tachycardia. Prolonged infusions have been used to stop premature labor, with variable success.

Adrenergic β-receptor stimulants are important drugs for inhibiting uterine contractions. The receptors in the uterus are referred to as β_2, in contrast to β_1 receptors in the cardiovascular system. The ideal drug for stopping premature labor would be a specific β_2 receptor stimulant. Among the promising agents under investigation are ritodrine, fenoterol, salbutamol, berotec and terbutaline. Even the contractions of advanced labor may be inhibited by terbutaline with minimal maternal cardiac effects. In all likelihood the year 1976 will see one or more of these agents introduced into general obstetrical practice.

Ethyl alcohol inhibits uterine motility, allegedly by "turning off" the secretion of endogenous oxytocin, although a direct effect upon the myometrium may be the explanation. When given by intravenous infusion in quantities sufficient to bring the blood level to about 0.12 per cent, it may abolish early labor contractions. Such infusions have proven useful in stopping some cases of premature labor, but should be used only when the membranes are intact.

Progesterone has not been found useful for the inhibition of uterine contractions, even in huge doses. *Relaxin* has also been found to be clinically useless.

Morphine, despite clinical impressions to the contrary, does not diminish the motility of the uterus even though it may abolish the pain of uterine contractions. *Meperidine* has a slight oxytocic effect when administered intravenously. Intravenously administered *magnesium sulfate* exerts a transient relaxing effect upon the uterus, as well as a hypotensive action when blood levels are high. It is employed primarily for its anticonvulsant activity in cases of eclampsia. *Ether* and *halothane* are excellent relaxants of the uterus when full anesthesia is induced, rendering these agents useful for the rapid treatment of uterine tetany in labor or as anesthetic agents when intrauterine manipulations are required.

The Induction of Labor

The artificial induction of labor may be done electively under specified circumstances, or may be indicated because of maternal complications, such as preeclampsia or diabetes mellitus. Elective inductions purely for convenience are generally frowned upon, because in large series, about one fetal death in 200 inductions has been directly attributable to the procedure.

When the fetal head is fixed in the pelvis and the cervix is soft, at least half effaced, and readily admits one finger, the simplest procedure for the induction of labor is artificial rupture of the membranes. If active labor should not begin within the next six hours, it is then customary to begin an intravenous infusion of oxytocin.

Synthetic oxytocin is commercially available in 1 ml. ampules containing 10 international units. The initial rate of intravenous infusion should not exceed 3 milliunits per minute, because for about one in 10 women at term, this will be sufficient to induce strong contractions and any greater quantities would constitute an overdosage. To achieve this rate, 5 units of oxytocin (0.5 ml.) may be added to a liter of 5 per cent dextrose in water, so that each ml. now contains 5 milliunits. Inasmuch as 15 drops is approximately 1 ml., adjusting the intravenous "drip" for a rate of 10 drops per minute delivers 3 milliunits each minute. If available, a constant infusion pump gives the most reliable results. The rate may be safely doubled each 20 minutes until an adequate response is obtained. Every time an oxytocin infusion is administered, it is an experiment in pharmacologic titration, and requires a competent observer who must remain at the bedside. An overdosage may cause uterine tetany, or even uterine rupture. The same principles hold true when the infusion is used for the stimulation of a hypoactive labor.

The Diagnosis of Acute Fetal Distress

The term *acute fetal distress* is commonly used to denote variable degrees of fetal hypoxia, with or without acidosis. Fetal hypoxia with resulting acidosis may be manifested by (1) a steadily rising basal heart rate, (2) episodes of bradycardia which persist after the uterine contraction is over, (3) meconium staining of the amniotic fluid or (4) an abnormally low pH of the fetal blood and base deficit as determined by a sample obtained from the scalp during labor.

The most accurate way of monitoring the fetal heart rate, of course, is with continuous tracings utilizing phonocardiography, Doppler ultrasonic detectors or electrocardiography. By listening continuously with a head stethoscope and following the fetal heart rate through two consecutive contractions, the clinician may train his ear to recognize the two forms of transient bradycardia which may occur during labor. The first is a brief V-shaped slowing (or Type I dip) which coincides with the peak of the contraction and disappears before the contraction is over. This is a vagal reflex, believed to be due to compression of the fetal head, and is therefore not a sign of hypoxia. The second is a U-shaped bradycardia (or Type II dip) which may begin near the end of the contraction and persists at a rate below 120 for 20 to 60 seconds or longer. This is brought about by fetal hypoxia resulting from a prolonged diminution of the uteroplacental circulation (Fig. 14–14.) In severe cases, especially when the labor contractions

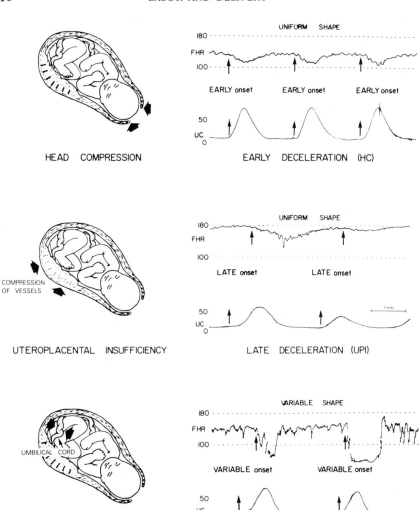

FIGURE 14–14. Patterns of fetal bradycardia. *A*, Early deceleration due to head compression (HC). UC is the tracing of uterine contraction. This is also known as a Type I dip. *B*, Late deceleration due to uteroplacental insufficiency (UPI). This is also known as a Type II dip. *C*, Variable deceleration due to cord compression (CC). (From Hon, An Atlas of Fetal Heart Rate Patterns, Harty Press, Inc., New Haven, Conn., 1968.)

are long and close together, the Type II dips may merge without recovery to the base line tachycardia, so that continuous bradycardia is observed. This is a serious sign which may precede intrauterine death. Decelerations which are variable both in time of onset and in shape result from umbilical cord compresssion, according to Hon.

In 1962, Saling developed a technique for sampling capillary blood from the scalp of the fetus with a long glass capillary tube. By this method, 0.1 to 0.2 ml. of blood may be obtained, which is sufficient for measurement of the pH and for calculation of the base deficit. In general, a pH greater than 7.27 is associated with normal Apgar scores (see Chapter 15), whereas a persistent pH of less than 7.22 is associated with Apgar scores of less than 6.

It is obvious that fetal scalp blood sampling cannot be done routinely, but the method has been of value in determining whether or not an immediate cesarean section should be done during the first stage of labor, or an immediate forceps delivery during the second stage, when other signs of fetal distress are present. The method is also useful when one is dealing with a high risk of fetal asphyxia, as in patients with preeclampsia, diabetes mellitus or Rh incompatibility, in elderly primigravidas or in truly postmature pregnancies. The normal basal fetal heart rate is between 130 and 150 beats per minute. Persistent rates above 160, episodes of Type II bradycardia below 120, and the passage of meconium-stained fluid in a cephalic presentation are danger signals which demand constant monitoring of the fetal status.

A summary of the effects which a uterine contraction may have upon the fetus is shown in Figure 14–15. The diagnosis of *chronic fetal distress* is considered in the next chapter.

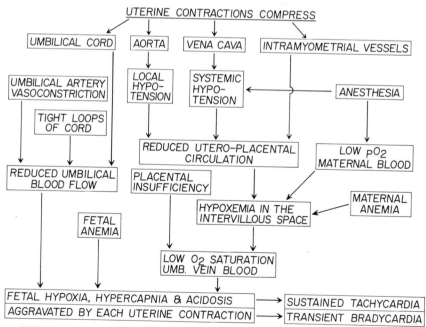

FIGURE 14–15. Schema illustrating the potential causes of acute fetal distress resulting from uterine contractions in labor. (After Caldeyro-Barcia.)

THE PUERPERIUM

The puerperium lasts from the delivery until six weeks postpartum, at which time the involutionary changes of the uterus are complete. It is for this reason that the customary postpartum examination is scheduled six weeks after delivery. An additional reason for consultation with the patient at this time is the fact that ovulation almost never occurs in less than six weeks, even in non-lactating mothers, and this becomes an appropriate time for discussion of contraceptive methods.

Involution of the uterus occurs rapidly under normal circumstances, the weight of the organ decreasing from 1000 gm. to its normal 50 to 60 gm. at the end of the six-week period. During the first two days, the fundus remains at a level about halfway between the symphysis and the umbilicus, a firm, rounded, and normally hard mass which is undergoing vigorous contractions, sometimes painful and referred to as "after pains." The blood vessels and sinuses of the placental site become rapidly thrombosed. The placental site itself is a shaggy area containing remnants of the decidua, which gradually become necrotic and are shed with the lochia. The area becomes progressively smaller and is replaced by new endometrium so that no scar remains.

The morphological changes in the *urinary tract* gradually recede during the puerperium. Immediately after delivery, however, the bladder is often bruised and edematous, sometimes with a disturbed neural control; so that overdistention must be carefully watched for. Infections of the urinary tract are not uncommon during the first two weeks; indeed, pyelonephritis is one of the triad of febrile complications, the other two being endometritis (with febrile illness most commonly beginning on the third day), and mastitis, which occurs at some later time in nursing mothers.

The current custom is to begin ambulation on the day of delivery, a practice which has definitely reduced the frequency of thromboembolic complications. Thirty years ago, women were kept in the hospital for 10 to 14 days after delivery. The custom of discharging patients three to five days postpartum arose during World War II, and has persisted because few harmful results were observed. To be sure, nursing has not been well established by the time of early discharge, but with the decreasing numbers of women who wish to nurse and the high cost of hospitalization, early discharge home has remained popular.

Except for the day of delivery, when the body temperature may be mildly elevated, there should be no fever exceeding 38° C. (100.4° F.) during the puerperium. The so-called "milk fever" occurring at the time of breast engorgement three days after delivery is almost always due to puerperal infection.

From the second through the sixth day postpartum, there is a pronounced diuresis, and most of the additional extracellular water accumulated during late pregnancy is lost. This is accompanied by a loss of 3 to 6 pounds of body weight, and the total loss—which should be observed and recorded

—is an indication of the degree of generalized edema which existed prior to delivery.

LACTATION

Anatomy of the Mammary Gland

The mammary glands of different species of mammals show wide variations in number, size, shape and location. Within the mammary glands are two types of tissue, the true glandular tissue, or parenchyma, and the supporting tissue, or stroma. The parenchyma is composed of minute saclike structures termed alveoli, the walls of which consist of a single layer of epithelial cells, which secrete the milk. The alveoli occur in clusters and each alveolus opens into a small duct; these join to form larger ducts which eventually open to the exterior at the tip of the nipple. In the human breast the ducts eventually join to form a common duct, or galactophore, which leads directly through the nipple to the exterior. Some 12 to 20 galactophores pass through the nipple and each, at the base of the nipple, expands to form a sinus.

In the fully developed and functional mammary gland the alveolar walls are formed by a single layer of epithelial cells, the shape of which varies with the amount of secretion being stored in the lumen. When the lumen is empty, the cells are tall; when the alveolus is full of secretion, the cells are low and stretched. At the base of the epithelial cells is a network of star-shaped myoepithelial cells which envelop the alveolus like a basket. Adjacent to the myoepithelium is a network of capillaries supplying the alveolar cells with the necessary precursors for the synthesis of milk.

The alveolar cells have numerous microvilli on their free surface, visible by electron microscopy. Each cell is joined to its neighbor by junctional complexes just below the luminal surface. The bases of the alveolar cells abut on the myoepithelial cells or on the basement membrane, and are indented into a system of clefts; these, by increasing the surface area of the cell, probably facilitate the absorption from the bloodstream of the precursors for the synthesis of milk. In the cytoplasm an abundant endoplasmic reticulum is arranged as arrays of flattened sacs; these are arranged parallel to one another and their outer surfaces are covered with ribosomes. This endoplasmic reticulum is situated mostly in the basal two-thirds of the cell. The mitochondria are large with conspicuous cristae, and the cells have a large Golgi apparatus located in the apical part of the cell adjacent to the nucleus. The myoepithelial cells underlying the epithelial cells have a contractile function, being instrumental in the expulsion of milk from the lumen of the alveolus into the duct system.

Development of the Breast

During fetal development two linear thickenings appear as ridges in the embryonic ectoderm on either side of the midline. The ridges become interrupted into a series of nodules of ectodermal cells, the number and position depending on the species. These nodules, the first rudiments of the mammary gland, sink into the dermis to become mammary buds. The bud grows and elongates into a cordlike structure, the primary mammary cord. The base of the cord remains attached to the epidermis while its distal end penetrates deep into the dermis and branches into a number of secondary buds that elongate and become hollow, forming the primary mammary ducts.

At birth the mammary apparatus is a rudimentary duct system leading to a small nipple. The duct system grows very little until the onset of reproductive cycles in the female. As thelarche approaches, the mammary glands, stimulated by the rising titer of estrogen, begin to grow actively and the duct system branches until the future lobules are indicated by collections of fine ductules and alveoli surrounded by stromal tissue. In man and most mammals full growth of the mammary gland is not achieved until the end of the first pregnancy or early in lactation. Beginning about midpregnancy there is rapid formation of lobules of the alveoli, which take over some of the space formerly occupied by stromal tissue.

Endocrine Regulation of Mammary Function

The growth and development of the mammary gland after puberty and the production and secretion of milk after parturition are controlled by a complex sequence of endocrine events. The initial development of the breast and the proliferation of glandular elements, stimulated by estradiol, require the presence of insulin and growth hormone. Progesterone provides further stimulation of the growth of the alveolar system of the glands at the time of puberty. During pregnancy the further development of the mammary gland is stimulated by estradiol and progesterone produced primarily in the placenta. The alveoli and the ducts continue to develop, becoming secretory. The premature production of milk is prevented by an inhibition of the process by progesterone. After parturition the sudden decrease in progesterone due to the removal of the placenta permits milk production to begin. If the breast is regularly emptied of milk the pituitary continues to secrete prolactin; however, if the milk is not removed the pituitary ceases to secrete prolactin. Nerve impulses from the nipple to the hypothalamus stimulate the secretion of prolactin releasing factor, which passes to the pituitary and increases the production of prolactin.

Prolactin, a secretion of lactotropes in the adenohypophysis, is essential for lactation. Immunochemical studies have proven that human prolactin is a separate protein hormone distinct from human growth hormone. The hypothalamus is responsible for a tonic inhibition of prolactin release through a dopamine-stimulated synthesis of a prolactin inhibiting factor (PIF). Thyrotropin releasing hormone (TRH), on the other hand, specifically stimulates the release of prolactin.

Serum prolactin concentrations steadily rise during pregnancy from 30 ng./ml. to about 200 ng./ml. of serum. During the puerperium, suckling or mechanical stimulation of the breast induces episodic rises in the serum concentration of prolactin, a response which is essentially lost about two months after delivery.

In addition to pregnancy and the puerperium, prolactin levels are high in various syndromes characterized by amenorrhea and galactorrhea. High concentrations of prolactin are believed to render the ovaries refractory to gonadotropins, hence the association of these two symptoms. The administration of L-dopa, which stimulates PIF, or an ergot alkaloid (2-Br α-ergocryptine), which probably acts on the lactotropes, will suppress prolactin, with a resulting return of ovulation and a cessation of the lactation.

The secretion of milk appears to be controlled solely by hormones. The mammary gland of the goat can be transplanted into the neck without decreasing its ability to secrete milk, suggesting that nerves play no role in controlling milk secretion. Anterior pituitary hormones, such as ACTH, TSH and growth hormone, in addition to prolactin, may play a role in controlling both the growth of the mammary gland and its secretion of milk. In the pregnant rat the lactogenic activity of pituitary prolactin and of placental lactogen is blocked by direct action of progesterone on the alveolar epithelium. With a fall in the amount of circulating progesterone after parturition, the lactogenic effects of prolactin come into play and milk secretion is stimulated.

The Milk Ejection Reflex

The transport of milk from the alveolus to the nipple, where it can be removed by the suckling infant, is triggered by the milk ejection reflex, a neurohormonal reflex. Shortly after the infant is put to breast the mammary gland suddenly seems to fill up with milk, which comes under pressure and may spurt from the nipple. Milk flow can occur in anticipation of the suckling stimulus. The movement of milk results from the contraction of the myoepithelial cells that squeeze the alveoli and expel their contents. The act of suckling triggers nerve impulses from receptors in the nipple; these pass up the spinal cord to the hypothalamus, where they cause the release of oxytocin from the posterior lobe of the pituitary. On reaching the mammary gland oxytocin causes the myoepithelial cells to contract, increasing intramammary pressure and resulting in the ejection of milk. The recording of intramammary pressure by cannulation of a mammary duct has been used to assay circulating oxytocin in human subjects. Thus the milk ejection response is a reflex, but the afferent arc is nervous and the efferent arc is hormonal. Like other reflexes, it can be conditioned.

Attempts to augment lactation in women by administering bovine prolactin have met with little or no success. In the past, estrogens, either alone or with progesterone, have been used to inhibit lactation in women who do not want to suckle their children. The effectiveness of this treatment has been questioned, and some doubts have been raised about its safety in view of the possible association of estrogens with thromboembolism. Oral contraceptives do not impair lactation in women who are breast feeding. Apparently once lactation is established the dose levels of estrogen and progestin in the oral contraceptives now in use are too low to have any deleterious effect on milk production.

The differentiation of alveolar cells in culture is regulated by insulin, cortisol and prolactin, which cause the epithelial cells to become arranged in an orderly fashion around the enlarged alveolar lumen. The hormones lead to an increased synthesis of the milk protein casein and of lactose synthetase. The enzyme lactose synthetase has two components, a galactosyl transferase and another milk protein, α-lactalbumin, which specifies that glucose will be the acceptor of the galactosyl moiety. In the absence of α-lactalbumin the galactosyl transferase will transfer glucose to other acceptors, such as N-acetyl glucosamine. In the presence of α-lactalbumin, however, glucose is specified as the acceptor and galactose is transferred from UDP galactose to glucose to form lactose. There is an increase in galactosyl transferase activity all through pregnancy in response to hormonal stimulation, but only at the end of pregnancy, when the concentration of progesterone is decreased and its inhibition of α-lactalbumin is removed, does the α-lactalbumin concentration increase rapidly and form lactose.

Lactating mothers must be taught to wash the nipples with soap and water both before and after nursing. Small fissures frequently develop and may provide access to bacteria from the baby's mouth, notably *Staphylococcus aureus,* which is the chief cause of puerperal mastitis. Cleansing the alveolar area after nursing and the prompt treatment of fissures may prevent breast infections.

PREMATURE LABOR

A premature labor may be arbitrarily defined as occurring before the thirty-seventh week of gestation (again assuming that 40 weeks represents term). A "premature infant" has been arbitrarily defined as one weighing less than 2500 gm. (5.5 lbs.) at birth, but it must be recognized that from 34 to 40 per cent of infants weighing less than 2500 gm. are more than 37 weeks of gestational age. (A more detailed discussion of low birth weight infants is contained in the next chapter.) Nevertheless, the use of 37 weeks as a dividing line, together with birth weight classes of 500 to 1500 gm., 1500 to 2500 gm. and over 2500 gm. forms a simple classification, as shown by Yerushalmy (Figure 14–16). In the five groups illustrated, the neonatal death rate varies from 0.5 per cent for the mature infant weighing over 2500 gm. to 71 per cent for infants weighing between 500 and 1500 gm.

The causes of premature labor are multiple, and in at least one out of four instances there appears to be no associated factor which might be held responsible. If we go one step further and admit that premature rupture of

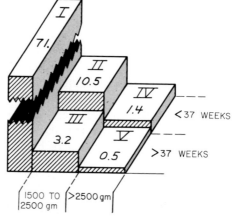

FIGURE 14–16. Neonatal death rate in per cent by duration of pregnancy and fetal weight. (From Page, based on data of Yerushalmy. Clin. Obstet. Gynec., *13*:80, 1970.)

the membranes is also of unknown etiology, then one-half of all premature labors are unexplained. Factors which are associated with birth weights of less than 2500 gm., the incidence of the factor in the general obstetrical population and the percentage of all low birth weight infants associated with each factor are shown in Table 14–1.

Table 14–1. SOME FACTORS ASSOCIATED WITH LOW BIRTH WEIGHT INFANTS*

Factors	Incidence, Total Sample	% of all LBW Newborns
Iatrogenic		
Elective induction	7.1	3.1
Elective (previous) cesarean	3.2	3.9
Placenta and membranes		
Abruptio placentae	0.75	6.1
Placenta previa	0.3	1.8
Other antepartum bleeding	0.72	2.3
Premature rupture membranes	10.4	24.0
"Placental insufficiency"	0.2	1.6
Uterine accommodation		
Twins	1.2	12.0
Uterine anomaly	2.3	3.6
Fetal factors		
Rh hemolytic disease	0.7	1.5
Other hemolytic disease	0.5	1.1
CNS and CV anomalies	0.6	2.5
Hydramnios	0.4	1.5
Maternal diseases		
Toxemia	4.5	10.0
Epilepsy	0.35	0.6
Psychiatric disease (severe)	0.8	1.9

*Adapted from Page, Clin. Obstet. Gynecol., *13*:79, 1970. Data are derived from Obstetrical Statistical Cooperative for the years 1961 and 1962.

In addition to the selected factors shown in Table 14–1, it must be recognized that some mothers habitually deliver small babies. Twenty-five per cent of all babies weighing less than 2500 gm. are born to the 10 per cent of mothers who give a history of a prior premature infant. The teenage mother and the patient who weighs less than 100 pounds at the start of pregnancy also have an increased chance of delivering a small infant.

The prevention of prematurity is one of the greatest challenges in obstetrics, because herein lies the key to any further material reduction in the perinatal mortality rate. The abolition of poverty and maternal malnutrition and the universal provision of high quality obstetrical care would be the major steps toward a reduction of prematurity, but these are functions of government and beyond the reach of the individual practitioner. The obstetrician can, however, bring the frequency of low birth weight infants to a minimum among his own patients by following 12 simple rules:

1. Make certain that the patient's diet is adequate in protein, vitamins and iron, and avoid the temptation to undernourish her for cosmetic reasons, particularly during the second half of pregnancy.

2. Avoid cesarean section prior to labor or the elective induction of labor if there is any doubt about fetal maturity.

3. Administer anti-D antibody to all Rh-negative unsensitized mothers at risk after each abortion or delivery (see Chapter 15).

4. Urge effective birth control measures for high-risk patients who do not desire further pregnancies.

5. Utilize legal abortion, where available, for women with serious psychiatric disturbances, or when anomalies of the fetus are known to exist *in utero*.

6. Adopt a conservative management of placenta previa prior to the thirty-seventh week of pregnancy (see Chapter 16).

7. Advise pregnant women to avoid or at least to sharply reduce cigarette smoking, which is known to interfere to some extent with fetal growth.

8. Diagnose multiple pregnancy early and institute a program of enforced rest during the third trimester.

9. Discover and treat bacteriuria during pregnancy in order to prevent the occurrence of pyelonephritis.

10. Detect preeclampsia early and hospitalize promptly.

11. Immunize susceptible (i.e., non-immune) women with rubella vaccine at least three months prior to conception (see Chapter 10).

12. Identify high-risk patients for early evaluation and, if indicated, refer them to an obstetrician who is associated with a large, well-organized and fully equipped modern obstetrical unit.

If such measures could be universally applied, the frequency of low birth weight infants (i.e., those under 2500 gm.) could probably be reduced from our national average of 10 per cent to something well below 5 per cent, with a corresponding reduction in the perinatal morbidity and mortality.

SUGGESTED SUPPLEMENTARY READING

Caldeyro-Barcia, R., and Poseiro, J. J. Physiology of the uterine contraction. Clin. Obstet. Gynecol., *3*:386, 1960.

Friedman, E. A., Hunt, H. B., and Sachtleben, M. R. Disordered labor: simplified method for recognition. Ob/Gyn Digest, *17*:12, 1975.

The use of both the conventional and simplified labor graphs is described in detail.

Greenhill, J. P., and Friedman, E. A. Biological Principles and Modern Practice of Obstetrics. W. B. Saunders Co., Philadelphia, 1974.

Section two of this textbook is noteworthy for its descriptions of labor and delivery.

Oxorn, H., and Foote, W. R. Human Labor and Birth, 2nd Ed. Appleton-Century-Crofts, New York, 1968.

The outline form and good line drawings make this a useful handbook for anyone learning the clinical art of obstetrics.

Page, E. W. Pathogenesis and prophylaxis of low birth weights. Clin. Obstet. Gynecol., *13*:79, 1970.

A discussion of the causes of prematurity and what a physician may do to reduce its incidence.

Shnider, S. M. (Ed.). Obstetrical Anesthesia: Current Concepts and Practice. Williams & Wilkins Co., Baltimore, 1970.

This includes sections on fetal distress and neonatal problems as well as the methods of analgesia and anesthesia.

Yuen, B. H., Keye, W. R., Jr., and Jaffe, R. B. Human prolactin: secretion, regulation and pathophysiology. Obs. Gyn. Survey, *28*:527, 1973.

Chapter 15

THE FETUS AND NEWBORN: PERINATAL MEDICINE

The recent realization that the fetus is a discrete patient within the reach of diagnostic and therapeutic procedures has led to a substantial growth of information on human development prior to birth. This in turn has led to the emergence of a group of physicians and investigators whose interest lies in the study of the human fetus and newborn. The subspecialty combines certain aspects of obstetrics and of pediatrics and has been called perinatal medicine. The astonishing growth of this area of medicine suggests that its consideration should occupy an important place in any core curriculum of obstetrics and pediatrics.

Each system of the fetus must proceed by orderly differentiation and growth to a stage of maturity that will permit extrauterine life. We know some of the influences that lead to a cell's becoming differentiated, such as the effect of adjacent cells, but the marvelous interdigitation of the controls of embryonic development remains a mystery. Usually, differentiated cells acquire new functions; however, the more specialized a cell is, the less likely it is to divide. Thus, the human embryo undergoes cellular division and differentiation and starts on the long road to eventual death because of the loss of capacity to divide which differentiation frequently entails. Some of the factors involved in the nutrition and growth of the human fetus were discussed in Chapter 9. In this chapter we shall discuss the assessment of the maturity of the fetus and newborn, as well as the development and function of various key systems involved in the transition from an intrauterine fetus to an extrauterine newborn.

THE ASSESSMENT OF FETAL MATURITY

There are many occasions when it is necessary to estimate the gestational age of the fetus as closely as possible. This is true when an elective cesarean section or induction of labor is contemplated, or when there is an indication for the premature termination of the pregnancy, such as maternal diabetes, hemolytic disease of the fetus, hypertensive complications in the mother or intrauterine growth retardation with chronic fetal distress. No single method is highly reliable; therefore a combination of methods must be used. The following eight methods are available:

1. *Menstrual history.* In past years, the date of the last menstrual period has been overly maligned as an indicator of the length of gestation because it was not realized how frequently low birth weight infants may be born at term. When, in the patient's opinion, the last menses was normal and the date is known, the probability of error is less than 5 per cent. An exception would be those women who conceive shortly after discontinuing oral contraceptives, because the interval between stopping the hormones and ovulation is highly variable.

2. *Serial measurements of uterine growth.* Inasmuch as intrauterine growth retardation or hydramnios does not ordinarily begin before 28 weeks, the estimates of uterine size in terms of gestational weeks during the first two trimesters are reasonably reliable, unless one is dealing with a multiple pregnancy.

3. *Audible fetal heart tones.* When fetal heart tones can be counted with the stethoscope, it may be assumed that the patient is 20 or more weeks pregnant. Obesity may interfere, however, and the relative infrequency of prenatal visits in the middle trimester restricts the value of this sign.

4. *Quickening.* When the patient first detects fetal movements, the mean time to date of delivery is 147 days, but with a standard deviation of 15 days. The variability is too great to inspire confidence, but should quickening first be recorded at 14 or 22 weeks instead of at 18, it may be the first suggestion that the date of conception does not coincide with the menstrual history.

5. *Size of fetus.* During the third trimester, the examiner records his estimate of the fetal weight together with the height of the uterine fundus at each visit. Depending upon his experience and the number of examiners involved, the information is of value as the first arousal of suspicion that the fetal growth rate or the amniotic fluid volume is deviating grossly from the average curve. A radiologic measurement of the length of the lumbar spines yields a close estimate of the crown-to-rump length of the fetus. An ultrasonic scan of the fetal head will yield a highly accurate measurement of the biparietal diameter, and if this is found to be more than 8.5 cm., the probability is high that the fetus is mature. The difficulty is that these are all measurements of fetal size, which is not necessarily correlated with gestational age. It must be remembered that a fetus may weigh over 2500 gm., but if the gestational age is less than 37 weeks, the risk of neonatal mortality is increased.

6. *Epiphyseal development.* The distal femoral epiphysis may appear as early as 32 weeks and is nearly always present at term. The proximal tibial epiphysis is rarely seen prior to 36 weeks and is present in from half

to three-fourths of fetuses at term. Absence of the first makes maturity doubtful and presence of the latter makes maturity very likely.

7. *Estimates of placental metabolism.* Three maternal serum enzymes have been used as a measure of placental function: oxytocinase, diamine oxidase and the heat-stable alkaline phosphatase. During the first 20 weeks of pregnancy, the measurement of oxytocinase by a bioassay method proved to be an accurate method of dating the gestation, but the method is tedious. In the last trimester, the normal scatter of values, as with hPL hormone, is so great that gestational age cannot be predicted. Unusually low values of any of these proteins may be more valuable in the diagnosis of placental insufficiency than of fetal maturity.

8. *Amniotic fluid studies.* As the fetus approaches maturity, there is a rise in the creatinine concentration to 2 mg. or more per 100 ml., a disappearance of the bilirubin content to an optical density different from the baseline at 450 nm. of 0.01 units or less, and an increased number of lipid-containing fetal cells, as demonstrated by orange staining with Nile blue. Each of these parameters is of differing significance, depending upon the week of gestation. The use of all three methods on a single sample of amniotic fluid is highly accurate for the estimation of fetal maturity. Wiser and Thiede have devised a scoring method which, in their initial study of 65 patients, produced no falsely positive results; that is, if the total score was greater than 0, no fetus proved to be less than 37 weeks of gestational age. The scoring system is as follows:

Score for each:	−2	−1	0	+1	+2	+3
Lipid-staining cells (per cent of total)	<20	20	25	30	40	>50
Creatinine (mg. %)	−	−	<2.0	2.0	>2.0	−
Bilirubin (△O.D. @ 450 nm.)	>0.01	0.01	0.00	−	−	−

Such studies of amniotic fluid are helpful in assessing fetal maturity, provided the volume of fluid falls within the usual range of 500 to 1500 ml. Oligohydramnios and hydramnios may give false high or low values, respectively.

Fetal maturity may also be measured by the concentration of "surfactant" (p. 332) in amniotic fluid. The tests actually measure lecithin (phosphatidyl choline), formed by the fetal lung in rapidly increasing amounts at about 35 to 36 weeks' gestation. The concentration of lecithin in amniotic fluid, believed to reflect the lecithin in the fetal lung, can be compared with the concentration of sphingomyelin, a lipid which changes very little during gestation and does not arise in the lung. The ratio of lecithin to sphingomyelin in amniotic fluid, the L/S ratio, is often valuable in assessing the maturity of the fetal lungs. Gluck found that premature infants with an L/S ratio greater than 2.0 have little or no tendency to develop respiratory distress syndrome (p. 339). Infants with an L/S ratio less than 2.0 are at risk.

The surface active properties of the lecithin in amniotic fluid can also be assessed by the simpler "shake test." Successive dilutions of amniotic fluid are prepared, then shaken in a tube containing alcohol. The presence of bubbles is noted 15 minutes later. In general, an infant is mature enough to sustain extrauterine existence without developing respiratory distress syndrome if the shake test is positive at 1:4 dilution of amniotic fluid. A positive bubble stability test indicates a very low probability of the development of neonatal respiratory distress syndrome.

Another reflection of fetal age is the concentration of estriol (p. 241) in amniotic fluid. The precursors of estriol arise chiefly in the fetal liver and adrenal; the amount of these precursors increases as these organs increase in size and activity. The placenta converts these precursors to estriol, which appears in increasing amounts in maternal blood and urine as well as in amniotic fluid. The concentration of cortisol in amniotic fluid also increases late in gestation. Fencl and Tulchinsky found that prior to 34 weeks the mean amniotic cortisol values were less than 40 ng./ml., whereas after 34 weeks the mean values were 2.4 times greater. In pregnancies over 40 weeks in duration the concentration of cortisol in amniotic fluid increased further to more than 120 ng./ml. The concentration of cortisol and the L/S ratios in amniotic fluid showed parallel rises with increasing gestational age. No incidence of respiratory distress syndrome occurred in babies whose amniotic fluid had contained over 60 ng. cortisol per ml.

The several amniotic fluid tests for fetal maturity were considered by Doran and co-workers (Figure 15–1). From 18 biochemical parameters and one cytologic test they arrived at the following three best tests of fetal maturity: (1) per cent lipid positive cells; (2) L/S ratio; and (3) creatinine. From a graph of the mean values of these three key parameters it is possible to

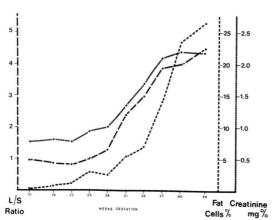

FIGURE 15–1.   Tests for fetal maturity based on analyses of amniotic fluid: percentage of lipid positive cells, creatinine content, and L/S ratio. (Reproduced, with permission, from Doran, T. A., Benzil, R. J., Harkins, J. L., Owen, V. M. J., Porter, C. J., Thompson, D. W. and Liedgren, S. I. Amniotic fluid tests for fetal maturity. Am. J. Obstet. Gynecol., *119*:833–837, 1974.)

estimate the period of gestation with 95 per cent accuracy for any individual sample. It seems clear that no single test of fetal maturity is as accurate as the consensus from a combination of tests.

DIAGNOSIS OF CHRONIC FETAL DISTRESS

The diagnosis of *acute* fetal distress during labor was discussed in the previous chapter. *Chronic* fetal distress results from an impairment of oxygenation, or nutrition, or both, and the several factors which may cause interference with normal fetal growth and development were outlined in Chapter 9. At the present time, methods of assessing jeopardy to the fetus prior to labor depend upon serial estimates of fetal growth rate (see method 5), measurements in maternal serum of constituents produced in part or wholly by the placenta, estimates of the rate of estrogen metabolism within the feto-placental unit, as described in Chapter 11, or the effects of "stress tests" upon the fetal heart rate pattern.

Most of these methods require two or more serial observations a week or so apart, so that the diagnosis of intrauterine growth retardation or uteroplacental insufficiency can rarely be substantiated by studies conducted on any given day. If an amniocentesis should yield fluid stained with meconium, or should direct observation of the fluid through the cervix with an amnioscope reveal meconium, fetal distress may be suspected. Measurements of gas tensions or of the pH of the amniotic fluid have not correlated well with fetal condition.

OXYTOCIN CHALLENGE TEST

This test, conducted in the hospital, is designed to assess the fetus at risk during the third trimester in a high-risk pregnancy situation. Uterine activity is recorded with an external tocograph, and the fetal heart rate is monitored continuously with an external monitor. An intravenous infusion is started, and after a baseline period of 30 minutes a dilute solution of oxytocin is infused with an electrical pump, slowly increasing the rate until uterine contractions occur every 3 to 4 minutes over a 30-minute period. Should a late deceleration pattern develop (p. 310), the fetus is considered in jeopardy, and labor should be avoided. A negative oxytocin challenge test should be repeated weekly until term.

Pomerance *et al.* have suggested a simplified maternal exercise test for uteroplacental insufficiency, utilizing a stationary exercise bicycle. The fetal heart rate was determined before and after the exercise; if the rate changed by more than 16 beats per minute in either direction, 4 out of 5 infants demonstrated fetal distress at birth. It was suggested that this might be a method for screening supposedly low-risk obstetrical populations 3 to 5 weeks before term, but more clinical experience with the method is required.

Obstetricians have accepted the challenge of determining the state of health of a tiny patient concealed from direct examination and are constantly exploring every possible means of improving the methodology, but many further advances are needed before jeopardy to the fetus can be diagnosed with accuracy.

ASSESSMENT OF THE NEWBORN

Most infants breathe spontaneously at birth and gentle aspiration of the mouth and nose to clear the airway is all that is required. The attending physician, however, must be prepared to deal with any respiratory problems that may exist in the critical minutes or hours following delivery.

APGAR SCORE

It is helpful in encountering the newborn for the first time to have some means of evaluating his physiological status quantitatively. The Apgar scoring system is used for this evaluation in many hospitals. Table 15–1 lists the physical signs to be noted.

A score of 10 (two points for each of the five signs) indicates the best possible condition. Obviously, several of the signs are more important than others, and this fact has led many to criticize the usefulness of the scoring. Infants with a high initial score may subsequently do poorly and others with low scores may do well. Thus, repeated scorings are important. At present, scorings after one minute, five minutes, and ten minutes are recommended as a guide to resuscitation.

The newborn with an Apgar score of 7 to 10 usually requires only supportive measures. He should be placed in a slightly head-down position and the mouth and oropharynx should be cleared of fluid by gentle suction. He should be observed and scored for the first 10 minutes at least.

Resuscitation of the Newborn

The moderately depressed infant (score 3 to 6) with a heart rate of over 100 but with weak respirations may be treated with 100 per cent oxygen

Table 15–1.   APGAR EVALUATION METHOD

Sign	Score: 0	1	2
Heart rate	Absent	<100/minute	>100/minute
Respirations	Absent	Weak, irregular	Strong, regular
Muscle tone	Limp	Some flexion of extremities	Active movement, well flexed extremities
Reflex response to stimulation of feet	No response	Some motion	Cry
Color	Blue; pale	Body pink, extremities blue	Completely pink

blown over his face. If the respirations become more feeble, or if the heart rate decreases, intermittent positive pressure breathing with oxygen is indicated. Frequent checks for fluid in the respiratory passages are mandatory. It must be remembered that the fragile lung of the newborn may readily be ruptured by over-vigorous resuscitation. Pressures should not exceed 35 cm. of water, in short bursts of less than one-half second's duration.

An infant who is limp, with a heart rate less than 100, and shows little or no respiratory effort (score 0 to 3) requires immediate endotracheal intubation. The baby is placed on his back on a flat surface and his head is slightly extended (Figure 15–2). The blade of the laryngoscope is passed along the dorsum of the tongue to the epiglottis. The laryngoscope blade is inserted into the glottis, which is cleared by gentle suction. The endotracheal tube is then passed through the glottis for about 2 cm. The laryngoscope is withdrawn carefully, holding the endotracheal tube in position. The lungs are inflated by careful mouth-to-tube insufflation. A second tube delivering oxygen to the resuscitator's mouth may be used. In those patients with cardiac arrest, mouth-to-tube resuscitation should be coupled with external cardiac massage. Cardiac massage is accomplished by two-finger compression of the middle third of the sternum and should be discontinued with each lung inflation. Intracardiac injection of 0.05 to 0.1 mg. of epinephrine (1:10,000) may also be helpful in restoring cardiac activity. Nalorphine (0.25 to 0.5 mg.) may be injected into the umbilical vein in babies who are severely depressed by morphine. Simultaneous efforts must be taken to

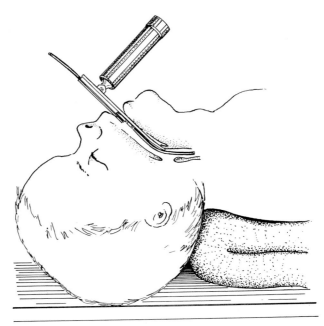

FIGURE 15–2.

correct any existing acidosis by the intravenous administration of an appropriate dose (usually 5 to 10 ml.) of 5 per cent sodium bicarbonate.

GENERAL EXAMINATION OF THE NEWBORN

Once respiration is well established and the pulse and heart rate are normal, the newborn has made his immediate emergency adjustments to extrauterine life. Other systems and conditions can then be examined. The routine examination of the newborn is summarized in Table 15–2. This examination should be performed as soon as possible after birth and certainly within the first 24 hours.

Table 15–2.   ROUTINE EXAMINATION OF THE NEWBORN

General	1. Sex, skin color, posture, type of cry, weight, height, head circumference
Head	1. Note size and tension of fontanelles and width of sutures 2. Examine eyes: red reflex, macula, cloudy membrane of retrolental fibroplasia, cataracts 3. Examine ears: normal position or low set? 4. Mouth: color of lips, cleft lip or palate? 5. Throat: tracheoesophageal fistula?
Chest	1. Degree and nature of expansion and air exchange 2. Ausculate both lung fields (crepitation often heard just after birth) 3. Heart: check rate and heart sound, check for murmurs 4. Feel for femoral pulses
Abdomen	1. Outline border of liver and spleen 2. Feel for intra-abdominal masses, check kidney size 3. Check cord for number of vessels and umbilical hernia
Spine	1. Check for abnormalities: dimple at lower end of spine or actual pilonidal sinus
Extremities	1. Count toes and fingers; check for talipes (clubfoot) 2. Check for congenital hip: hips should be completely rotatable to frog position and skin folds on posterior upper legs should be symmetrical 3. Check upper extremity for Erb's palsy (arm held in adduction and internal rotation with forearm pronated) 4. Walking movements when supported 5. Prehensile grasp, peripheral reflexes
Genitalia and rectum	1. Look for abnormalities: hypospadias, undescended testicles 2. Check patency of rectum by inserting clinical thermometer 3 cm. 3. Have meconium and urine been passed?
Skin	1. Soft, fine, smooth; check for birth marks: capillary nevus, hemangiomata
Neurological	1. Moro and tonic neck reflexes

The abrupt entrance of the newborn into extrauterine life requires re-markable alterations in his metabolic and physiologic functions. Some sys-tems must make greater adjustments than others, but all are subject to aber-rations of one sort or another, the results of which may compromise neonatal existence.

RESPIRATORY SYSTEM

Immediately at delivery, the respiratory system must change from a dormant, liquid-filled state to a system capable of breathing air. Prior to about 28 weeks' gestation, fetal lungs are not capable of breathing air. The alveolar surface area is insufficient and the vascular portion of the lung is underdeveloped. The development of the fetal lung must ensure a large enough alveolar surface area and a sufficient number of capillaries in contact with the alveolar surfaces.

DEVELOPMENT OF LUNGS

The lungs begin to develop relatively early in embryonic life. Around the twenty-fourth day of gestation (Table 15-3), an interaction between the endodermal epithelium and the surrounding mesenchyme of the pharynx occurs that results in a ventral budding. By the process of centrifugal branching from this bud, the bronchial tree eventually takes form. At four weeks, the right and left primary bronchi form. By five weeks, the right bronchus has divided into three main branches and the left bronchus into two main branches, thus foreshadowing the presence of three lobes on the right side and two on the left. The main bronchi divide repeatedly, and by the end of the sixth month approximately 17 generations of subdivisions have been formed, the last ones known as respiratory bronchioli. By birth there are about 27 generations of subdivisions.

Terminal air sacs or alveoli appear as outpouchings of the bronchioles and are lined with a cuboidal epithelium. These epithelial cells lining the potential air spaces of the fetal lung begin to change at around 16 to 20 weeks' gestation. They flatten, and coincidentally capillaries grow in to es-tablish contact with this epithelium. A continuous squamous epithelium is present, and between 20 and 24 weeks two types of cells differentiate. One type is the flat epithelial alveolar cell and the other is polygonal in shape and tends to lie between capillary loops. Lamellar osmiophilic inclusion bodies are present in the cytoplasm of the latter cell and are thought to represent the production of a surface-active lipid (surfactant).

The large alveolar cells contain enzymes active in glycolysis, the pentose pathway and oxidative phosphorylation. Macrophages are also present in the lungs, which are particularly rich in various hydrolases. Thus, in extrauterine life, they can function to protect the lungs from foreign particles by digesting material, which they engulf. These macrophages are unique in their very high oxygen consumption and their preference for oxygen as a gas phase. Active synthesis of protein, fatty acids and phospholipids has been demonstrated in alveolar cells. Early in gestation (less than 16 weeks), the fetal lung is rich in glycogen, but after this time glycogen gradually disappears.

At 26 to 28 weeks, capillary networks proliferate close to the developing alveoli. The alveoli increase in number to form multiple pouches from a common chamber. Considerable quantities of fluid are produced by the fetal lung during the last half of gestation. This fluid has an osmolality like plasma, but a lower pH and protein content. Because of a high pulmonary vascular resistance and the presence of a wide ductus arteriosus, pulmonary blood flow is low during fetal life.

The fact that most infants survive the critical transition from placental to pulmonary respiration is a tribute to the prenatal preparation, which provides not only an adequate cardio-respiratory system but also the chemical and neural controls which will facilitate gas exchange. The carotid and aorticopulmonary bodies are identifiable in the 50- to 60-day human embryo. Peripheral chemoreceptors are apparently present also and are capable of discharge.

ONSET OF BREATHING

The work of Dawes and his colleagues has shown that normally the fetal sheep breathes *in utero* about 40 per cent of the time; the breathing is related to REM sleep. Human fetal chest wall movements *in utero* have been recorded by ultrasonic scans. These fetal movements are often transmitted through the maternal abdominal wall. Broddy and Mantell have shown that fetal breathing in the human is episodic, occurs at a frequency of 30–70 breaths per minute, and is normally present for about 65 per cent of the time. It is observed more often in multiparous women. These investigators believe that fetal breathing movements *in utero* may provide an indication of fetal health both before and during labor. However, failure to observe maternal abdominal signs of such movements does not necessarily mean that the fetus is not breathing.

Prior to birth, the lungs are filled with fluid; however, with delivery of the head, pressure exerted on the thorax by the birth canal forces some of the fluid out of the lungs. Elastic recoil of the thoracic cage occurs with delivery of the chest, admitting air to replace the fluid displaced.

Varying degrees of hypoxia are present at birth. Oxygen levels in the umbilical arterial blood at birth range from zero to 70 per cent saturation. The average carbon dioxide tension was found to be 58 mm. Hg, and average pH, 7.26. In general, lower oxygen concentrations are associated with a lower pH and higher carbon dioxide tension. James has proposed that the final stages of labor and delivery are associated with decreased exchange of oxygen and carbon dioxide across the placenta, producing

Table 15–3.   DEVELOPMENT OF ORGAN SYSTEMS IN THE HUMAN FETUS

Time (wks)	Respiratory Tract	Erythropoietic System	Lymphoreticular System	Kidney	Nervous System	Gastrointestinal Tract
2		Erythropoiesis begins. Primitive erythroid cell lines				
3	Lung buds present	"Blood islands" in yolk sac				Liver anlage seen
4	Right and left primary bronchi				Fore-, mid- and hind-brain present	Hepatic buds and ducts
5	Five main bronchi (3 right, 2 left)		Granulocytes appear	Metanephros begins to develop	Brain now has five regions. First reflex arc	Hepatic lobes recognizable
6		Definitive erythropoiesis in yolk sac and liver. Erythrocytes have Hb	IgG levels = 10% of adult values			
7	Dichotomous branching continues. Glands form and airway is lined with ciliated columnar epithelium (continues through twelfth week)		Lymphocytes seen in thymus	Nephrogenesis begins	Cerebral hemispheres develop	Gastric pits present
8		End of synthesis of primitive erythroid cells			Cerebellar rudiments begin. First recordable electrical activity	
10		Liver = major site of erythropoiesis, but gradually regresses	IgM synthesis detectable	Pelvis and calyces form. A.F. = 30 ml.	Medulla exerts some control over spinal cord	Glycogen and bilirubin synthesis begins in liver. Intestine capable of peristalsis and of absorbing glucose
12		Erythropoiesis in spleen begins	Fetal liver and G.I. tract can synthesize IgG	Glomeruli and prox. convol. tubules. Urine formed	Cerebral hemispheres cover diencephalon	Parietal cells and later chief cells
14		Begin gradual transition from Hb_F to Hb_A		Loop of Henle functional but short. 20% of nephrons mature	Some mesencephalic control	Bilirubin found in A.F. Liver can synthesize fatty acids

Table 15-3. DEVELOPMENT OF ORGAN SYSTEMS IN THE HUMAN FETUS (Continued)

Time (wks)	Respiratory Tract	Erythropoietic System	Lymphoreticular System	Kidney	Nervous System	Gastrointestinal Tract
16	Alveolar epithelial cells flatten			30% of nephrons mature		Mucus formed by epithelial cells of stomach
18			IgG synthesized by spleen			
20	Capillaries around potential air sacs proliferate	Erythropoiesis in bone marrow begins	Lymphoid tissue appears in spleen	350,000 nephrons. A.F. vol. = 350 ml.	Myelinization in brainstem	
24	Alveolar cells start to make surfactant		IgG levels rise rapidly, reaching maternal values			
26					Beginning of alertness and neurovegetative behavior. Hypotonia. Rooting and incomplete Moro reflex	
28	Lungs capable of breathing air	Erythropoiesis in spleen ends; bone marrow becomes major site				
30					More muscle tone in legs and trunk. Pupillary light reflex. Moro reflex	
32		Erythropoietin formed				
35					Flexed limbs; firm grasp. Spont. orientation to light. Crossed ext. reflex incomplete	
40			Cord blood: IgM = 10% of adult, IgA = little or none. IgG = maternal origin primarily	A.F. vol = ca. 1000 ml. 830,000 nephrons	Myelinization reaches level of hemispheres. Crossed ext. reflex complete. Hypertonic upper and lower limbs	

various degrees of asphyxia at birth. It is known that a fall in arterial oxygen tension and pH plus a rise in carbon dioxide tension can induce gasping *in utero* as well as after birth. Thus, it may be that the onset of breathing is triggered by chemoreceptors. Thermal and tactile stimuli may also play a role in the first inspiration.

A majority of healthy infants start to breathe within a few seconds of being born, even before separation from the mother. Catheters have been placed in the esophagus to measure the negative intrathoracic pressure during the first breath. Values of from −20 to −70 cm. of water have been found and are accompanied by an inflow of 20 to 80 ml. of air. Much of this air remains in the lung as residual volume. The first inspiration is usually followed by a cry as the infant expires against a partially closed glottis. After the first few breaths, the lungs are almost completely expanded. Residual fluid in the lungs is presumably absorbed as pulmonary blood flow increases during the first few minutes after birth.

SURFACE TENSION

With the initial entrance of air into the trachea, an air-fluid interface is created. Whereas molecules in the interior of a homogeneous liquid are free to move and are attracted by surrounding molecules equally in all directions, molecules at the surface of the liquid, where it touches the air, are attracted downward and sideways more than upward. They are therefore subjected to unequal stress. The force with which the molecules are bound is called the surface tension of the liquid. The relationship between the pressure across a curved surface (P), the surface tension at the air-liquid interface (γ) and the radius of curvature (r) is given by the La Place equation:

$$P = \frac{2\gamma}{r}$$

Thus, the "collapsing forces" at the gas-liquid interface in a system of differently sized alveoli will be much higher if the alveoli are small than if they are big. Because of inequalities of alveolar size at any given moment, and because of changes in size during respiration, the distribution of forces at the air-liquid interface varies and produces an inherent alveolar instability. Smaller alveoli tend to collapse, and each new breath would require increased effort to reopen closed alveoli.

SURFACTANT

At about 24 weeks' gestation, alveolar cells in the human fetus start to produce a unique surface-active agent, called surfactant, that is capable of lowering the surface tension at the air-fluid interface and thus maintaining alveolar patency. John Clements and his colleagues found that the largest (41%) component of the surface active material is saturated lecithin, and that most of this is dipalmitoyl lecithin. Some monoenoic lecithins, as well as cholesterol and small quantities of other phospholipids, are present in surfactant. Nine per cent by weight of the isolated material turned out to be protein. Clements believes that dipalmitoyl lecithin functions as the principal molecular component of the surface active material in the pulmonary alveoli. Furthermore, he maintains that it bears the brunt of setting up tightly-packed stable surface films which can withstand high compression and in this way give low values for surface tension in the alveoli. The lecithin is complexed with other material, such as protein, unsaturated phospholipids and neutral lipids, which enables it to form a film rapidly. The protein component is unique to the lung.

The lung has the ability to synthesize saturated phosphatidylcholine compounds like dipalmitoyl lecithin. Such saturated compounds have greater surface activity than their unsaturated counterparts. Other tissues form phosphatidyl cholines with a saturated fatty acid in the α' position and an unsaturated fatty acid in the β position. The pathways of synthesis are shown in Figure 15–3. According to Gluck, the two pathways involved in the synthesis of phosphatidyl choline develop at different times during gestation. The methylation of phosphatidyl ethanolamine involves the transfer of three methyl groups from three molecules of S-adenosyl methionine. This pathway develops in human fetal lung as early as 22 to 24 weeks. The pathway utilizing diglyceride and CDP-choline, though detectable by 18 weeks' gestation, is inadequate to produce sufficient phosphatidyl choline to stabilize alveoli before 35 weeks. At 35 or 36 weeks, the amount of phosphatidyl choline in amniotic fluid rises sharply, reflecting the increased activity of the diglyceride pathway in the fetal lung.

Of the three enzymes in the choline incorporation pathway (Figure 15–3) choline phosphotransferase is the rate-limiting catalyst. Glucocorticoids have been shown by Farrell to induce the synthesis of this enzyme in rabbit fetal lung. Thus the maturation of this pathway near term may depend upon adrenal steroids.

HYPOXIA AND ACIDOSIS OF THE NEWBORN

The newborn incurs an oxygen debt during the birth process. His initial hypoxia alters the pattern of metabolism from that of oxidative phosphorylation to that of glycolysis with an ensuing accumulation of lactate, producing a metabolic acidosis. At the same time, the accumulation of carbon dioxide causes a respiratory acidosis. The pH falls. The normal newborn is capable of repaying his oxygen debt by his increased respiratory efforts during the first 10 to 20 minutes of life.

TEMPERATURE REGULATION IN THE NEWBORN

Because the newborn infant has a larger surface area to body weight ratio than the adult, he has a greater physical problem in maintaining his

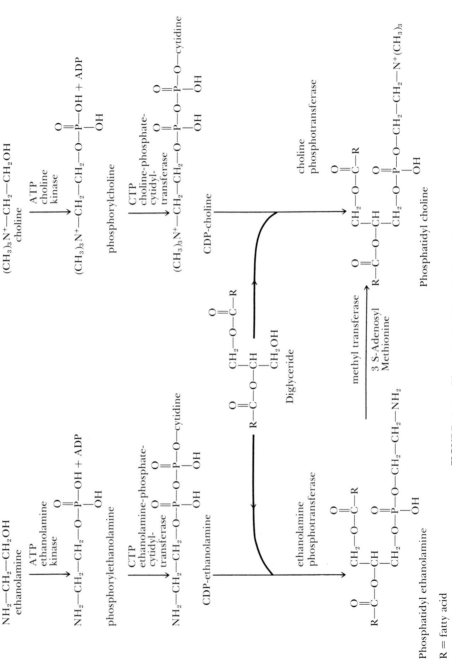

FIGURE 15–3.   Biosynthesis of phosphoglycerides.

body temperature. The fall in body temperature of the newborn kept at 20 to 23°C. is initially extremely rapid. This does not imply that the infant is poikilothermic. As in other homeotherms, the newborn will increase his metabolic rate upon exposure to cold. However, heat production is insufficient to make up for heat loss, and the body temperature falls. The newborn responds by increasing his ventilatory effort.

The newborn is aided in his efforts to maintain body temperature by deposits of brown adipose tissue between the shoulder blades, around the neck, behind the sternum and around the kidney and adrenal. This kind of adipose tissue has a high content of mitochondria, giving it a distinctly brown hue. Adult humans have little of this tissue but it is prominent in the newborn. Brown adipose tissue acts as a means of producing heat; the function of ordinary adipose tissue is to serve as a reservoir of potential energy. The metabolism of this tissue is deliberately inefficient, so that the rate of oxidation (and therefore of heat production) is not geared to the formation of high-energy phosphate. The mitochondria of brown adipose tissue lack the usual mechanism for coupling oxidation and phosphorylation in the electron transport chain. The constant utilization of ATP for the acylation of free fatty acids during the cyclic synthesis and hydrolysis of triglycerides (Figure 15–4) causes further oxidation of fatty acid (not shown) with the release of a large amount of heat and a small amount of ATP.

The glycerol phosphate used in the synthesis of triglyceride is obtained from the metabolism of glucose. The glycerol released in the breakdown of the triglycerides cannot be phosphorylated in adipose tissue and is therefore released to the bloodstream. A rise in blood glycerol suggests that hydrolysis of triglycerides is occurring in adipose tissue. Dawkins and Scopes have observed that cold exposure caused the plasma glycerol concentration in babies to double, although there was no significant rise in plasma-free fatty acids. Thus, the newborn probably oxidizes fat in brown adipose tissue rather than releasing free fatty acids to be oxidized elsewhere. At least 11 different hormonal factors are known to be lipolytic; however, in the newborn it is likely that the catecholamines (epinephrine and norepinephrine) play the major role in mediating the thermogenic response to cold.

A more prolonged increase in heat production depends upon thyroid hormone. Fisher and Odell have shown that during delivery and the early postpartum period, the concentration of TSH in the infant's blood rises rapidly, reaching a maximum of 86 μU./ml. at 30 minutes of age. Between 30 minutes and 4 hours, serum TSH decreases and thereafter gradually levels off at a value of 13 μU./ml. by 48 hours. The half-time of disappearance of TSH in the newborn is similar to that of administered radioactive TSH in the adult. Thus the rapid rise and fall in serum TSH concentration in the newborn may represent release of stored pituitary TSH.

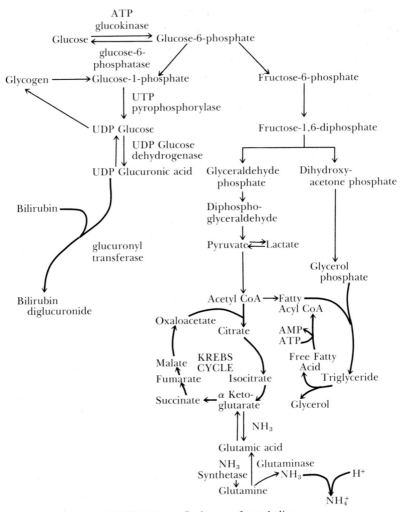

FIGURE 15–4. Pathways of metabolism.

MONITORING PULMONARY FUNCTION IN THE NEONATAL PERIOD

The normality of the respiratory effort in the newborn can be monitored by a variety of means. A useful measure of ventilatory adequacy is the partial pressure of carbon dioxide in the arterial blood (pCO_2). Capillary blood from a warmed extremity or ear lobe has the same gas tensions as arterial blood and is used extensively for this measurement. For a complete picture of the acid-base balance of the infant, either the pH or the carbon dioxide content of the blood should be measured also. A more complete discussion of acid-base balance will be found in a later section.

Hypoxia may be measured directly by determining the arterial oxygen tension. By the use of appropriate equations, the extent of right to left shunt can be calculated. Measurements in normal newborn babies show that about 15 to 20% of the cardiac output is shunted.

FETAL CIRCULATION AND ITS CHANGES AT BIRTH

Adequate respiration in the neonatal period is highly dependent upon the circulatory changes which occur at birth. In order to comprehend such disorders as the respiratory distress syndrome, in which pulmonary hypo-perfusion plays such a prominent role, it is necessary to understand the physiology of the fetal circulation and the remarkable alterations which develop within the first hours of life.

The Fetal Circulation

Much of the quantitative data relative to the distribution of blood flows in the fetus has been derived from studies in the fetal lamb at differing periods of gestation, but available human data suggest that the patterns are very similar. Quantitative data mentioned here are derived from Rudolph's studies on pregnant ewes, in which he measured the distribution of radio-active microspheres from each of the major fetal veins.

A schematic representation of the fetal circulation is shown in Figure 15–5. Blood returning from the umbilical vein is well oxygenated, having a pO_2 averaging 30 mm. Hg. About half of this blood passes through the liver capillaries, and the rest goes through the ductus venosus into the inferior vena cava. Blood from the latter is largely deflected through the foramen ovale to the left atrium and thence into the left ventricle. Essentially all of the superior vena caval blood passes through the right atrium into the right ventricle. From here, only 10 to 15 per cent reaches the pulmonary circula-tion, the rest being diverted away from the lungs by the ductus arteriosus to the descending aorta.

The inferior vena cava receives all of the blood coming from the pla-centa, and most of this is shunted to the ascending aorta; this ensures well oxygenated blood (pO_2, 25 to 28 mm. Hg) for the coronary and cerebral circulations. From the descending aorta, 40 to 50 per cent of the blood returns to the placenta (pO_2, 19 to 22 mm. Hg), and the remainder circu-lates through the lower half of the body.

It will be noted that only 5 to 10 per cent of the total cardiac output goes through the lungs. This is due to a very high pulmonary vascular resistance, which is associated with the low pO_2 of the blood. In response to this high pulmonary arterial pressure, the pulmonary arteries develop an increasing thickness of their muscular walls as gestation continues.

FETAL CIRCULATION

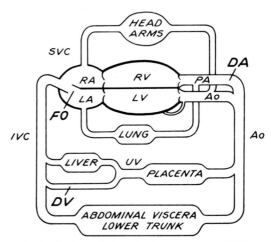

FIGURE 15–5. In the fetal circulation, the ductus venosus (DV) serves as a bypass for umbilical venous blood to enter the inferior vena cava directly. The foramen ovale (FO) carries well-oxygenated blood from the inferior vena cava into the left atrium and left ventricle. The ductus arteriosus (DA) carries the major portion of blood ejected from the right ventricle to the descending aorta and mainly to the placenta. *SVC* = superior vena cava; *IVC* = inferior vena cava; *RA* = right atrium; *LA* = left atrium; *RV* = right ventricle; *LV* = left ventricle; *PA* = pulmonary artery; *Ao* = aorta. (From Rudolph, Circulation, *41*:343, 1970.)

Changes after Birth

Ventilation of the lungs is associated with a dramatic fall in pulmonary vascular resistance, a marked increase in pulmonary blood flow and a progressive thinning of the walls of the pulmonary arteries. The fall in resistance is the result of increased oxygenation, but since most arteries in the body respond to higher oxygen by constriction, it is believed that the effect is mediated by some substance released from the parenchyma of the lung.

The general rule about arterial constriction with increased pO_2 applies to the ductus arteriosus, which becomes functionally closed within 10 to 15 hours after birth, although in premature infants or in cases of persistent hypoxia it may remain open for much longer periods of time.

Elimination of the placental circulation a few minutes after birth causes a considerable increase in overall systemic vascular resistance, resulting in a reorientation of circulatory flow patterns. Atrial pressures are reversed, resulting in a functional closure of the foramen ovale by apposition of the right and left atrial flaps near the opening. In a few infants, some left to right atrial shunt can be detected for several months. Cessation of umbilical flow terminates flow through the ductus venosus (which becomes the ligamentum venosum) and the intra-abdominal portion of the umbilical vein (which becomes the ligamentum teres).

The change from a fetal to an adult circulatory pattern is not a sudden event, but occurs over a period of hours and days, and may take weeks under abnormal circumstances. During the transitional period, there may be a continued flow, right to left, through the foramen ovale, a continued patency of the ductus arteriosus with shunts in either direction and an unusual sensitivity of the pulmonary arterioles, which constrict vigorously in response to hypoxia or to acidemia. The occurrence of asphyxia or metabolic acidosis in the perinatal period thus sets the stage for development of the respiratory distress syndrome.

RESPIRATORY DISTRESS SYNDROME (HYALINE MEMBRANE DISEASE)

Hyaline membrane disease is a leading cause of death in liveborn premature infants. In the typical case, respiratory distress develops at or shortly after birth and is not due to any presently known condition (infection, central nervous system depression, aspiration, and so on). The diagnosis of hyaline membrane disease can be made only at autopsy; however, respiratory distress syndrome is a more general diagnosis that can be made clinically. This respiratory distress syndrome characteristically progresses for the first one to three days, ending in death (10 to 40 per cent) or spontaneous, but not always complete, regression.

Incidence

Infants born prematurely are most prone to develop respiratory distress syndrome. In general, birth weights of 1 to 1.5 kg. are the most common. Also, infants of diabetic mothers are predisposed to respiratory distress syndrome. The second-born of twins is at greater risk than the first born, perhaps because of the greater chance of asphyxia. In general, the risk of recurrence of respiratory distress in subsequent premature infants is nearly 90 per cent, whereas the risk of occurrence after one normal infant is less than 5 per cent.

Clinical Diagnosis

The infant with respiratory distress syndrome fails to maintain adequate oxygenation or carbon dioxide removal. Within the first few hours of life, the infant develops rapid and labored breathing, with retractions of the soft tissues during inspiration, flaring of the nares and an expiratory grunt. In the severely affected infant, hypoxia progresses and respiratory and metabolic acidosis develops. The magnitude of right to left shunt in respiratory distress syndrome may reach two-thirds of the cardiac output; in general, the less the ventilation the greater the shunt.

The most useful single diagnostic aid is a roentgenogram. The classic films show a diffuse reticulogranular pattern of the lungs, with the air-filled trachea and bronchi standing out against the relatively opaque perihilar areas. Cardiac enlargement is frequently found. Indeed, changes in the roentgenogram may precede prominent clinical disease. The roentgenogram is essential in the distressed newborn in order to rule out treatable conditions (e.g., pneumothorax) as well as to assist in the diagnosis of respiratory distress syndrome.

Systemic hypotension has been well documented. The infants may look pale and dusky, even with hematocrit values of 60 per cent or greater. The dependent areas may have a darker color because of pooling of blood from poor peripheral circulation. Peripheral edema is usually present. Cyanosis is a constant finding in severely affected infants, and is one of the most grave prognostic signs.

Death from uncomplicated hyaline membrane disease almost always occurs before 72 hours of age. The lungs of such infants are strikingly different from those dying of non-pulmonary causes. They do not expand as fully as the non-diseased lung. Atelectasis and capillary engorgement are prominent. Usually, but not always, fibrin (hyaline) membranes line the alveolar ducts.

Etiology

Though a specific deficiency in surfactant has been described, it is not known whether this abnormality is cause or effect. The recent work of Gluck has shown that the concentration of lecithin (phosphatidyl choline) in amniotic fluid rises abruptly at about 35 to 36 weeks' gestation. In contrast, the concentration of another lipid substance, sphingomyelin, changes very little. Gluck finds that prematures with a lecithin/sphingomyelin ratio of over two have little or no tendency to develop respiratory distress syndrome. Infants with ratios less than two are at risk. These studies with amniotic fluid may very well reflect the surfactant activity of fetal lungs. Undoubtedly the deficient pulmonary surfactant in hyaline membrane disease accounts for the severely deranged mechanical properties of the lung, which have been demonstrated both *in vivo* and at autopsy.

Infants with respiratory distress syndrome were shown to have only 5 per cent of the carbonic anhydrase activity in their bloods that an adult has. Such a deficiency might result in CO_2 retention, which could produce a respiratory acidosis. This in turn could cause increased pulmonary vascular resistance, shunting of blood away from the lungs, and therefore decreased pulmonary blood flow. Less pulmonary blood flow would compromise the synthesis of surfactant, which would lead to atelectasis and further CO_2 retention. It is clear that any one of the elements of this vicious circle, or indeed any etiologic agent bearing on such an element, could be the responsible factor in respiratory distress syndrome.

The fetal lung is extremely vulnerable to asphyxia. Pulmonary blood flow may practically cease as a consequence of pulmonary vasoconstriction. Asphyxial damage to alveolar cells could very well result, since impairment of pulmonary surface properties follows ligation of the pulmonary artery in experimental animals.

Recently, Chu et al. showed that marked pulmonary vasoconstriction was present in infants with hyaline membrane disease. This vasoconstriction caused a reduction of pulmonary blood flow to one-third of normal. They postulate that loss of surfactant may be caused by continuing pulmonary ischemia from birth.

Asphyxia *per se* is probably not enough to explain the disease, since many severely asphyxiated infants, once resuscitated, have no evidence of respiratory distress syndrome. Prolonged intrauterine asphyxia may, however, set up irreversible changes in alveolar cell function, so that the cell is no longer capable of forming sufficient surfactant, carbonic anhydrase, or any number of other biochemical products essential for normal respiratory function.

Though aspiration of amniotic material was once considered the cause of the hyaline membrane that is typical of the respiratory distress syndrome, the failure to see this membrane early in the disease and the high fibrin content of the material suggest endogenous origin. Indeed, this high fibrin content has suggested a deficiency of fibrinolysins to many investigators. There is no definitive evidence as yet that any one of the aforementioned conditions is the primary cause of respiratory distress syndrome, and we must conclude that as yet the etiology of this disorder is unknown. The clinical syndrome may reflect an immaturity of the mechanism for synthesis of surfactant, or a normal mechanism damaged by prenatal or intrapartum asphyxia, or a combination of both.

Therapy

In view of our poor understanding of the basic underlying cause of respiratory distress syndrome, therapy is largely supportive. The infant should receive intensive care, preferably in a unit specifically designed for such. Closed assisted ventilation has been used in many hospitals and appears to be remarkably helpful. Indeed, Gregory et al. in San Francisco have reported 36 infants with severe respiratory distress syndrome treated with continuous positive airway pressure, with survival of 32. Oxygen should be given in sufficient quantities to maintain arterial tensions at around 60 to 70 mm. Hg. Any metabolic acidosis should be corrected with sodium bicarbonate, and 10 per cent glucose may be given to spare catabolism. Transfusions for infants in shock and maintenance of proper thermal environment are all part of a supportive regimen that may save some lives.

Howie and Liggins have conducted clinical trials to determine the efficacy of antepartum cortisol treatment in the prevention of respiratory distress syndrome. They concluded that glucocorticoid treatment of the mother (24 to 48 hours before delivery) reduced the incidence of respiratory distress syndrome in infants born before the thirty-second week of pregnancy. Glucocorticoid receptors have been demonstrated in human fetal lung, and it would be anticipated from animal studies that glucocorticoids could stimulate the synthesis of surfactant in the lung. Obviously the use of glucocorticoids in the prevention of respiratory distress syndrome is still at a preliminary stage. Until we know more about the effects of cortisol on all fetal tissues this therapy should be used with great caution.

A recent study at the Boston Hospital for Women indicated that the infant of the diabetic mother is particularly prone to respiratory distress syndrome. The hyperinsulin state of these infants may block cortisol induction of surfactant. Such an insulin-cortisol antagonism has been described by Weber for the induction of liver enzymes in perinatal rats. One would anticipate a lower L/S ratio in the infant of the diabetic mother; this has been shown by some investigators but not by others. At any rate, the use of prenatal glucocorticoids in these children may be beneficial in preventing respiratory distress syndrome.

Prematurity and Respiration

The immaturity of the respiratory system in the preterm baby may be reflected in a variety of ways. The incidence of irregular breathing was 75 per cent in one series of premature infants. One of the most common respiratory patterns in the premature is called periodic breathing. Brief periods of apnea occur at fairly regular intervals, lasting usually 5 to 10 seconds. The ventilatory periods, alternating with the apneic periods, last 10 to 15 seconds. Periodic breathing is rarely seen in the first 24 hours of life, and is more common and lasts longer in smaller infants. It seems probable that neural pathways involved in integrating respiration are not well developed in these infants.

Apneic spells (in contrast to periodic breathing) are alarming and call for immediate attention. The duration of apnea is of more than 20 seconds and the spells are usually associated with cyanosis and a decrease in heart rate. These episodes may indicate the presence of pulmonary disease, intracranial hemorrhage, infection, hypoglycemia, hypocalcemia, bilirubin encephalopathy, congenital malformation or drug intoxication.

HEMOGLOBIN

Just as the development of the respiratory system must prepare for aeration of the newborn lungs, and the developing circulatory system provide for extensive exposure of the pulmonary circulation to inspired air, so the ultimate in cellular respiration must be ensured by the development of a circulating cell whose sole function is the transport of the respiratory pigment, hemoglobin. Angioblast cells can be found as early as the blastocyst stage, but erythropoiesis is considered to begin at about day 14. In the primitive streak stage, primitive erythroid cells can be found in the body stalk and wall of the yolk sac, forming so-called blood islands. Thus, these first erythroid cell precursors appear in extraembryonic membranous structures. By day 25, the circulation is established, and at this time there is a shift from extravascular to predominantly intravascular erythropoiesis. In the human embryo, primitive erythroid cell lines develop first and continue through the ninth week of gestation. This cell type is larger and has a small ratio of nuclear volume to cytoplasmic volume, and the chromatin is relatively homogeneous. The primitive erythrocyte retains a pycnotic nucleus. Although primitive erythrocytes are no longer formed after nine weeks' gestation, they are found in the circulation until the end of the third month of gestation.

Definitive erythropoiesis begins in the yolk sac and liver at about six weeks' gestation. These erythrocytes are first seen to contain hemoglobin at about seven weeks. By 10 weeks, the liver is the major site of erythropoiesis, and thereafter it slowly regresses. The spleen is an erythropoietic organ between the third and seventh months of gestation. At five months, the bone marrow begins erythropoietic activity, and by the seventh month it is the major site. At birth, over 90 per cent of hematopoietic activity occurs in the bone marrow, and normally continues to do so throughout life.

During the first week after birth, erythropoiesis diminishes rapidly. Coincidentally, the production of the hormone, erythropoietin, decreases rapidly after birth. Analyses of cord blood from both premature and term infants show that erythropoietin production in the human fetus is well established by the thirty-second week of gestation. In the first weeks of life, the normal infant has little or no erythropoietin activity in its plasma; however, high levels are found in cord blood from anemic erythroblastotic infants and in cases of severe dysmaturity with high hemoglobin concentrations. Thus, the human fetus can respond to anemia and chronic hypoxia by elevating the production of erythropoietin, which in turn increases hemoglobin synthesis in the erythrocyte series.

The stem cell in bone marrow, termed the hemocytoblast, gives rise to progeny which divide for a time, undergoing sequential changes from hemocytoblast to proerythroblast to normoblast. This last cell extrudes its nucleus to become a circulating reticulocyte. Further maturation produces the definitive erythrocyte. Hemoglobin concentration is first seen to increase in the proerythroblast. Its synthesis continues in the normoblast and reticulocyte. No hemoglobin synthesis occurs in the adult cell.

It has been known for a long time that human fetal blood has a higher affinity for oxygen than does adult blood. The difference is usually ascribed to the hemoglobin F synthesized by the fetus. Normal adult hemoglobin molecules are made up of four subunits. Each subunit (or polypeptide chain) has its own heme group. It is this heme group (ferrous protoporphyrin complex) which is responsible for the reversible combination of hemoglobin with oxygen and carbon monoxide. The adult hemoglobin A is composed of two α chains (each containing 141 amino acid residues) and two β chains (146 amino acids each). Fetal hemoglobin (HbF) also has two α chains, but in place of the β chains there are two polypeptides (γ chains) of 146 residues each. The γ chains differ from the β chains in 39 of the 146 residues. This difference in amino acid composition residing in the γ chains of HbF alters the physicochemical properties of the molecule, including alkali resistance and shifts in the isoelectric point and wavelength for spectrophotometric analysis. Despite the observed differences in oxygen affinity of fetal blood and adult blood, HbA and HbF have similar oxygen affinities if the hemoglobin solutions are dialyzed beforehand. Thus, the chemical milieu of the red cell is important in maintaining the higher oxygen affinity of fetal blood.

One clue to the problem of oxygen affinity of Hb resides in the recent observations of Tyuma and Shimizu. They showed that if HbA and HbF were freed of all phosphate and if the pH were maintained between 7.0 and 7.8, HbA had a consistently higher oxygen affinity than HbF. The addition of 2, 3-diphosphoglycerate (DPG) and ATP dramatically decreased the oxygen affinity of HbA. Similar effects of phosphates were observed with HbF, but to a lesser extent. Thus, in the presence of excess DPG or ATP, the oxygen affinity of HbA is consistently lower than that of HbF. This would favor the uptake of oxygen by fetal erythrocytes over that of maternal erythrocytes.

It is known that α chains do not bind DPG, whereas β chains do. Apparently DPG acts as a cross linking agent between the two halves of the Hb molecule, thus stabilizing it. The five ionic groups of DPG form salt bridges with cationic groups of the β chain at the N-terminal valine groups and with the histidine (position 143) and lysine (position 82) residues. Only reduced Hb binds DPG, and this reversible combination shifts the oxygen dissociation curve to the right, decreasing oxygen affinity. Both thyroid hormone and androgens cause an increase in formation of DPG in the red cell, thus facilitating its ability to release oxygen. The pH of the cell is another important factor in regulating the amount of DPG formed. An increase in pH activates phosphofructokinase and favors the formation of 1,3-DPG with subsequent conversion to 2,3-DPG. Thus in various states of hypoxia, which commonly produce a respiratory alkalosis, DPG levels in the red cell would rise and bring about decreased oxygen affinity; acidosis would result in the reverse effect.

The transition from fetal to adult hemoglobin *in utero* begins in the second trimester. Prior to that time, very nearly 100 per cent of the Hb is alkali resistant (HbF). The percentage of fetal hemoglobin in fetal blood remains at values over 90 per cent until the last four to five weeks of gestation, when there is a precipitous fall as well as a spread of values. At birth (20 per cent HbF), there is a continuing steep decline in fetal hemoglobin, leveling off at about 12 weeks after birth. Small amounts (up to 25 per cent) of HbF may persist in the first year of life; however, HbF cannot ordinarily be detected after age $2\frac{1}{2}$ years.

Oxygenation

The role of hemoglobin in binding and transporting oxygen is well known. Oxygen is transferred across the placenta (see Chapter 9) by simple diffusion. Its uptake by red cells involves diffusion through plasma, across the red cell membrane, and through the interior of the red cells to final combination with hemoglobin. In the tissue capillaries, the hemoglobin becomes less saturated as it loses oxygen to the tissues. There is no good evidence that normal fetal tissues *in utero* are hypoxic.

Tissues utilizing oxygen are also generating carbon dioxide. Of the carbon dioxide released from tissue cells, a small fraction dissolves physically in plasma water and another small fraction combines with hemoglobin and is transported to the lungs in the form $HbCO_2$, but the bulk of the carbon dioxide combines with water, forming carbonic acid and causing a fall in blood pH (see the section on acid-base balance). This drop in pH is counteracted by the uptake of H^+ on Hb. The H^+ displaces oxygen from the protein (Bohr effect), causing an increase in the release of oxygen from oxyhemoglobin beyond that which would have occurred at constant pH and at a given oxygen tension. The greater the output of carbon dioxide (and therefore the greater amount of H^+), the greater the release of oxygen. Thus oxygen release is adjusted according to its demand, since a greater carbon dioxide output results from an increased utilization of oxygen. The Bohr effect, therefore, permits blood to take up CO_2 without a change in pH. This sequence of events is reversed in the lungs. The exposure of Hb to the higher oxygen tension causes oxygen to be taken up with the displacement of H^+. The H^+ is utilized to convert bicarbonate to carbonic acid, and the resultant CO_2 diffuses into the gas space of the alveoli.

A variety of factors may alter the ability of the fetus to oxygenate his blood. Obviously, any reduction in blood flow through the cord will lower arterial oxygenation, since the blood in fetal veins will be passing directly into the arterial circulation with little or no admixture with oxygenated blood flowing from the placenta in the umbilical vein. Large differences in oxygen saturation may be found between umbilical artery and vein blood. Reduction of the effective area of diffusion (e.g., placental infarction or partial separation), reduction in uterine blood flow (e.g., compression of mother's inferior vena cava) or reduction of maternal gas exchange may lower the oxygen saturation of fetal blood.

It is of interest that during the first ten weeks of gestation a unique hemoglobin chain (ϵ chain) is formed in a separate line of nucleated erythropoietic cells of yolk sac origin. Some of the embryonic Hb is made up of four ϵ chains, whereas other embryonic Hb consists of two α chains and two ϵ chains. The role of this unique ϵ chain, as well as that of embryonic Hb, is unknown. The cells forming this kind of Hb disappear at about ten weeks' gestation.

The primary role of respiration and oxygenation in the survival of the newborn has been discussed at some length. The blood, however, is not composed of red cells alone. The role of the white cells in the defense of the organism brings us to the immunological system of the fetus and newborn.

FETAL AND NEONATAL IMMUNOLOGY

Immediately upon exposure to extrauterine life, the newborn must prepare to confront a host of foreign proteins, including microorganisms. During intrauterine life, maternal immunoglobulins belonging to the IgG class cross the placenta and afford the fetus passive immunity. The abrupt change at birth from an environment virtually free of microbial challenge to one of potentially overwhelming invasion makes the immunologic competence of the newborn a subject of great importance.

The first line of defense against invasion of the body by bacteria is the circulating white blood cells, particularly the neutrophils. These cells have the ability to phagocytize foreign particles, and their number rises at the onset of infection. Reticulo-endothelial cells are closely allied to white blood cells, but they are primitive, sessile cells. They line the sinusoids of the liver, spleen, lymph nodes and bone marrow, and filter out bacterial and other foreign particles absorbed into the bloodstream or lymph.

The second line of defense of the body against invasion is called immunity. Antibodies may be formed against specific antigens and either remain inside tissue cells (tissue immunity) or circulate in the blood (humoral immunity). The primary immune response to an antigen consists of the sequential appearance in the serum of antibodies of three main classes. All proteins with antibody activity are designated immunoglobulins (Ig), and the main classes by capital letters (A, G, M) according to their molecular size and carbohydrate content. In the adult, IgM antibodies appear first in the serum, within three to six days after immunization. After an interval, IgG antibody appears, both classes persisting for a time, with the eventual earlier decline of IgM. IgA appears after the other two. It appears that in the human fetus and newborn, most antigens elicit only IgM antibody. The transition to the more adult response occurs in the latter part of the first year. IgA immunoglobulins first appear in the serum in the second week of extrauterine life.

Development of the Lymphoid System

The development of the mammalian lymphoreticular system begins early in gestation. During the early yolk sac phase of hematopoiesis, a few histiocytes may be seen in the blood islands. Granulocytes appear in the blood about the second month of gestation. Hematopoietic stem cells migrate via the bloodstream to the thymus, where they acquire surface alloantigens. The cells undergo a process of maturation, becoming resistant to corticosteroids and acquiring immunocompetence. These are the "thymus-dependent" or T lymphocytes. They leave the thymus to populate the mid- and deep cortical areas of lymph nodes and the periarteriolar regions of the spleen. In the human fetal thymus lymphocytes are first seen at nine weeks' gestation; they appear in fetal blood at about the same time. The thymus grows rapidly, reaching its maximum size shortly before or after birth. Thereafter involution occurs, presumably as the function of the thymus is taken over by other organs. Postnatal encounters with foreign substances have little or no effect on the rate of lymphopoiesis in this organ.

Lymphoid tissue first appears in the spleen at about 20 weeks; however, it is not until after birth that a rapid increase in primary follicles occurs. Peripheral lymphoid tissue, though present by the end of the first trimester, is poorly differentiated until after birth.

Evidence of functioning bone marrow has been found at 11 to 12 weeks' gestation. From this time until term, the lymphoid compartment of the bone marrow forms about 25 per cent of the total nucleated cells. Marrow hematopoiesis attains its maximal activity at 30 weeks' gestation. Lymphocytes from bone marrow have surface markers and can produce immunoglobulins.

Upon exposure to foreign antigens, the lymphocyte is transformed into the plasma cell, a highly differentiated cell with well-developed endoplasmic reticulum and Golgi apparatus. Each plasma cell is differentiated to produce large amounts of a single immunoglobulin molecule. The immunoglobulin molecules consist of paired polypeptide chains—two "light" and two "heavy" chains. A portion of each chain has a variable amino acid sequence which is probably responsible for forming the antibody-combining site. The IgG immunoglobulins are of the 7S sedimentation class, and bivalent in antigen-combining ability. They are the most prevalent of all immunoglobulins and have the longest half-life (about 23 days). They are the most efficient in neutralizing toxins and viruses but less efficient in killing bacteria.

IgA is the chief immunoglobulin of the gastrointestinal and respiratory tracts and of breast secretions. It has the property of neutralizing viruses and probably kills bacteria. The IgA in the circulation is a 7S globulin, whereas the IgA which is secreted onto the mucosal surfaces is 11S. Both the 7S and the 11S, as well as a third part, a β-globulin, must be intact and functioning to protect the mucosa against infection.

IgM is of particular interest since this class of antibody is dominant in the response of the fetus and newborn to antigen. Typical antigens for this class of antibodies include ABO and Rh agglutinins. IgM immunoglobulin is far more effective than IgG in mediating complement-dependent bacteriolysis or hemolysis. This class is also most efficient in mediating chemotaxis, phagocytosis and intracellular destruction after phagocytosis. The IgM antibodies are made in the spleen.

The development of lymphoid tissue depends in large measure on the degree of activation of these centers by exposure to foreign protein *in utero.* Maternal infections (rubella, cytomegalovirus, toxoplasma, syphilis) may produce chronic intrauterine infection, which is difficult to detect. One hope lies in the detection of immunoglobulins characteristic of the fetus (IgM) in the amniotic fluid or fetal blood. A concentration of IgM in cord or neonatal serum in excess of 20 mg. per 100 ml. is considered presumptive evidence of intrauterine infection.

Concentrations of Antibody in Serum

We know that there is a selective transfer of maternal immunoglobulins across the placenta. Of the three major classes, IgG immunoglobulins are the only ones that are transferred to the fetus in considerable amounts. By the middle of the first trimester, IgG levels in the fetus equal about 10 per cent that in the adult. At about 20 to 24 weeks, there is a sudden rise in IgG in the fetal blood, reaching concentrations that are essentially the same as those in the maternal blood. This rise is felt to be due to increased placental transfer of globulins.

Not all IgG in fetal blood is of maternal origin. Gitlin has shown that by 12 weeks' gestation the fetal liver and the fetal gastrointestinal tract are capable of synthesizing IgG in culture. At 17 to 18 weeks, IgG can be synthesized by fetal spleen. Concentrations of IgG in amniotic fluid over the period from $9\frac{1}{2}$ to 38 weeks' gestation rise from 1.7 to 44.0 mg./100 ml.

Synthesis of IgM in the fetus can be detected as early as $10\frac{1}{2}$ weeks. This is of interest, since lymphocytes first become distinguishable in human lymphoid tissue at eight to nine weeks. IgM is first detected in fetal serum at about 20 weeks and averages 8 mg./100 ml. It does not vary much as gestation proceeds.

Response of Newborn to Antigens

Since, for all practical purposes, only IgG antibodies cross the placenta, we can assume that the fetus and newborn are protected from antigens eliciting an IgG response. However, those antigens eliciting an IgM or IgA response, particularly early in gestation, before fetal immunocompetency, could endanger the developing organism.

At birth, we assume that nearly all the IgG in cord blood is of maternal origin and the IgM (10 per cent of that in adult blood) of fetal origin. Very little (5 per cent) or no IgA is normally present in cord blood. After birth, IgG levels fall and IgM immunoglobulins rise very rapidly. When young animals are kept in a germ-free state, neither the IgM immunoglobulins nor IgM antibodies rise. This IgM response probably represents the interaction of the newly born with bacterial flora of the intestinal tract. The very low levels of antibody to enteric bacteria present in the newborn are of type IgG from the mother. The newborn infant is therefore without immunologic experience with enteric flora and has little maternal protection. Thus, the newborn is singularly vulnerable to generalized infection with enteric organisms.

The newborn response to antigens may be altered by feedback inhibition by maternal antibodies. These antibodies may have been transferred across the placenta before birth or may have been imbibed by the newborn in colostrum.

In summary, although the fetus undoubtedly possesses the biological apparatus for making phagocytizing and antibody-producing cells, the maternal protection during gestation frequently renders the use of this system unnecessary *in utero*. However, the fetus can respond to intrauterine infection appropriately and present investigations of IgM levels in amniotic fluid suggest that such states may be more common than previously was imagined. The antigenic challenges of birth call forth activity in the immunological systems of the newborn. A period of grace is allowed, during which the surviving maternal antibodies render the newborn passively immune. In addition, maternal antibodies from breast milk may afford some protection, particularly from coliform intestinal infections. Apparently delayed-type hypersensitivity is not transmitted from mother to fetus (although maternal cells do cross the placenta). The newborn can respond to BCG antigen with a delayed-type immunity, but it is not known how early in gestation delayed hypersensitivity can be induced.

Antigens of Maternal Origin

Maternal infection can provide exogenous antigens to the fetus. The response of the fetus will depend upon his immunologic competence at the time of exposure. Maternal rubella in the first trimester stimulates no antibody response in the fetus and because this is the time of organogenesis there is a high probability of defective organ development (see Chapter 10). *Treponema pallidum, Toxoplasma gondia* and cytomegalovirus infecting human fetuses evoke a response in the lymphoreticular organs. Germinal centers develop, plasma cells are formed and antibody titers rise.

Hemolytic Disease of the Fetus

For centuries this disease of the fetus and newborn, also referred to as erythroblastosis fetalis, was a mysterious ailment which caused recurring tragedies in seemingly random families and accounted for some 10 per cent of all cerebral palsy victims. Today we are on the verge of eliminating the disease altogether. The story can best be told by recounting the events of the last 40 years.

In 1932, we thought we were dealing with four separate diseases: the fatal hydrops fetalis, erythroblastosis, icterus gravis neonatorum and congenital anemia of the newborn. In that year, Diamond et al. showed that all four were simply different stages of the same disease process, although of unknown etiology. In 1938, Darrow wrote a remarkable paper, in which she concluded that the disease was due to the formation of a maternal antibody against some component of fetal red blood cells.

In 1940, Landsteiner and Wiener discovered the Rh antigen of red blood cells by injecting Rhesus monkey red cells into rabbits and producing an antiserum which was found to agglutinate the red blood cells of most human beings. Those whose red blood cells were not affected were called Rh negative. In the following year, Levine et al. demonstrated that 93 per cent of cases of erythroblastosis fetalis resulted from isoimmunization to the Rh antigen.

The etiology of hemolytic disease of the fetus now became clear. The isoimmunization occurred either when Rh negative women were transfused with Rh positive blood or when there was a transplacental passage of Rh positive fetal red blood cells into their circulation. The resulting maternal antibodies then crossed the placenta and hemolyzed the red blood cells of the fetus, resulting in a variety of pathological changes.

The Rh factor was later found to consist of three pairs of allelic genes, and in 1944 Fisher and Race proposed the designations CDE/cde to designate the antigens. The most potent antigen is D (called Rh_0 in Wiener's terminology), and for practical purposes women who are dd are called Rh negative. When the father is homozygous, or DD, all of the offspring will be Dd or Rh positive. When the father is Dd, the chances are 50 per cent that any one offspring will be D positive. Population studies revealed that 15 per cent of a mixed white population in the United States are Rh negative. In some groups, such as the Amish of southeastern Pennsylvania, the incidence is as high as 30 per cent, whereas among Chinese, Japanese, American Indians and the black races of Africa the incidence is as low as 1 per cent.

Not all pregnant Rh negative women carrying a D positive fetus become immunized, and it was noted that when the mother and her fetus were ABO incompatible, the frequency of hemolytic disease was markedly reduced, presumably because of antigenic competition. The ABO incompatibility itself will cause hemolytic disease of the fetus quite frequently, but in most cases the disease is mild and may go unnoticed. Other red cell antigens such as Duffy, Kell, Kidd and so forth, may cause serious hemolytic disease, but their occurrence is so rare that the problem is of little magnitude.

During the 1940's, it became customary to perform Rh typing on all prospective donors and recipients of blood transfusions, and this essentially eliminated the sensitization of women from that source. It also became mandatory to screen all obstetrical patients for the Rh factor, to screen all husbands of the Rh negative group, and then to study the sera of those pregnant women at risk periodically for the development of Rh antibodies in order to be forewarned. The only treatment, however, was the planned early induction of labor in women whose antibodies appeared for the first time during the last 12 weeks of pregnancy.

In 1946, Wallerstein used the first exchange transfusion for treatment, and over the next few years Diamond's group popularized this method for infants born with hemolytic disease. This reduced the neonatal mortality from about 50 per cent to less than 5 per cent, and vastly reduced the frequency of kernicterus, which was the cause of the major neurological defects.

Treatment of the fetus *in utero* by any means had to await some method of accurately predicting both the existence and the severity of the hemolytic process. Predictions based upon the behavior of the antibody titers in the maternal serum were fraught with uncertainties. In 1952, Bevis proposed analyzing the amniotic fluid for its bilirubin content as a means of predicting the seriousness of the disease, and the method was refined by a number of workers during the next several years. The technique most commonly used is based on a spectrophotometric tracing between 350 and 550 nm. of the centrifuged fluid. In normal fluids, a fairly straight line is observed, but with affected infants there is a hump which is maximal at 450 nm., and the height of this deviation from the baseline (called the ΔO.D. 450) parallels the severity of the hemolytic process (Figure 15–6). By correlating these values with the fetal outcome in a large number of cases, zones could be established which distinguished with reasonable accuracy the degree of anemia which existed in the fetus *in utero* (Figure 15–7). In 1963, Liley performed intrauterine, intraperitoneal transfusions in a desperate attempt to save those infants in whom the amniotic fluid studies indicated a severe process early in the third trimester—such infants would ordinarily be doomed. Extensive experience with the method suggests that about 40 per cent of such infants may be salvaged, but the procedure itself carries a substantial risk of causing fetal demise.

In the meantime, methods were developed for detecting the number of fetal cells which escaped into the maternal bloodstream. This began with the introduction of the acid elution method by Kleihauer et al. in 1957. Subsequent studies proved that the time of entry of the major number of fetal cells was during parturition, or to a lesser degree during spontaneous, induced or therapeutic abortion, and that the risk of isoimmunization was proportional to the number of fetal cells found. The stage was now set for attempts to block the process of isoimmunization. In 1960, Freda and his co-workers in New York and Finn et al. in England independently began to use human anti-D gammaglobulin injections during the immediate postpartal period to block the formation of maternal antibodies. The two groups worked on different premises, but both concluded that the method was successful. As little as 300 μg of the antibody administered within 72 hours

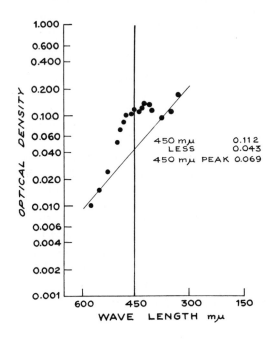

Rh SENSITIZED
34 WEEKS

450 mμ   0.112
LESS   0.043
450 mμ PEAK 0.069

FIGURE 15–6. Spectrophotometric analysis of amniotic fluid illustrating the method of deriving the
450 nm. peak. (From Westberg and
Margolis, Am. J. Obstet. Gynecol.,
92:583, 1965.)

after delivery (or abortion) will prevent isoimmunization in all but rare cases
of large feto-maternal "bleeds." The product is commercially available as
RhoGAM.

Injections of RhoGAM are useless when the mother is already sensitized and are contraindicated during pregnancy because the passively administered antibodies are capable of harming the fetus. Its use should be
mandatory when the mother is D negative and has no anti-D antibodies in
her serum and when the infant is D positive and has a negative direct
Coombs test, indicating an absence of hemolytic disease. The use of Rho-
GAM after each abortion in a D negative woman is also advisable, even
though the risk of isoimmunization is far less.

Thus, in the brief span of one physician's professional life, a disease
which has threatened the lives and the health of newborns since recorded
medical history is at the threshold of becoming eradicated in all populations
receiving adequate obstetrical care.

KIDNEY

Development

Because the placenta adequately rids the fetal blood of metabolic
wastes, and maintains (via maternal lungs and kidneys) electrolyte, water,

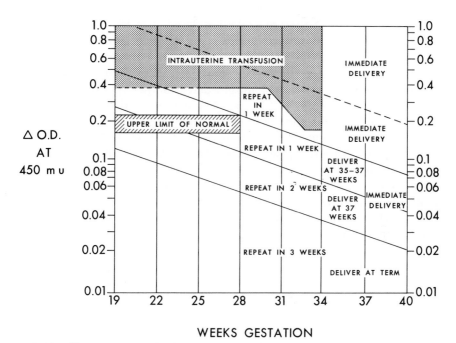

WEEKS GESTATION

FIGURE 15–7. A modified Liley graph which correlates the 450 nm. peak at each week of gestation with the recommended method of management. The phrase "repeat in 1 week" refers to a repeat spectrophotometric analysis of the amniotic fluid. (Courtesy of Dr. E. R. Jennings, from #4 Supplement of Family Health Bulletin, Winter 1970–71, State of California Department of Public Health.)

and acid-base balance, there is no real need for renal function in the fetus, and indeed a fetus can develop without any kidneys. However, the kidneys must be capable of assuming their excretory and regulatory roles promptly at birth. The permanent (or metanephric) kidney begins to develop in the fifth week of prenatal life. Ureteral buds (derived from the mesonephros) migrate cephalad to join the caudal end of the nephrogenic ridge (metanephric blastema). Early branching of the ureteral buds is responsible for the formation of the pelvis and major and minor calyces of the definitive kidney. This process is complete at about 10 weeks' gestation. Further branchings of the ureteral buds are surrounded by diffuse masses of nephrogenic cells. The contact of the ampullary ends of these branches with nephrogenic cells induces the formation of nephrons. Nephrogenesis begins at seven or eight weeks and continues until 32 to 34 weeks. The glomerular tuft of the nephron is probably formed from extensive proliferation of an incoming branch of the renal artery.

By the third month of gestation, well-differentiated glomeruli and proximal convoluted tubules are present. Glomerular filtration, tubular reabsorption of sodium ion and of chloride ion, and urine formation are all evident. The fetal urine is hypotonic with respect to fetal plasma, probably because of the low concentration of electrolytes, primarily sodium ion and chloride ion.

By the fourteenth week of gestation, the loop of Henle appears to be functional; however, its overall length is short in the fetus and infant and increases with advancing age. Not all nephrons are mature and functional. It is estimated that at 11 to 13 weeks' gestation, some 20 per cent of the nephrons are morphologically mature, whereas 30 per cent are in a similar stage of maturation at 16 to 20 weeks. The number of nephrons more than doubles from 20 weeks' (350,000) to 40 weeks' (820,000) gestation.

After birth, although no more tubules are formed, the kidney continues to grow. Glomeruli increase in diameter. The tubules increase in length and tortuosity. At birth, the ratio of glomerular to tubular size is greater than it is later in life. The rates of glomerular filtration in young infants are well below those observed in adults. This is probably due to the relatively incomplete morphologic and functional development of the filtration apparatus.

Water Balance

A mechanism exists for changing the degree of concentration of the urine by increasing or decreasing the amount of water reabsorbed from the collecting tubule. This mechanism is the medullary countercurrent multiplier system and involves the rapid osmosis of water from the collecting tubule as a result of the highly concentrated interstitial fluid of the medulla. The pores of the collecting tubules are sensitive to ADH, and in the presence of this hormone they permit free passage of water, providing there is an osmotic gradient. In the presence of excess water, osmoreceptors are not stimulated and ADH is not released from the neurohypophysis

Infants in the first three days of life do not respond to a water load with diuresis. They are relatively ineffective in concentrating their urine after water deprivation. ADH has been detected in the plasma following the injection of hypertonic saline, but it is not certain whether the collecting tubules in the very young infant are responsive to ADH. It has been suggested that the limited concentrating capacity of the newborn kidney may be due to the low rate of urea excretion in the newborn. Other possibilities are the shortness of the loop of Henle in the newborn and unresponsiveness of the collecting tubules to ADH.

Electrolyte Balance

Most of the studies of electrolyte reabsorption in fetal kidneys have been performed with sheep. Nevertheless, the basic observations are probably true in the human fetus as well. Both sodium ion and chloride ion reabsorption increase as gestation progresses. Because electrolyte reabsorption exceeds water reabsorption, the fetus puts out a hypotonic urine. Muscle at birth contains relatively more water, sodium and chloride, but relatively less potassium, than does adult muscle. There is a greater total sodium content in the fetus than in the newborn. After birth sodium is released by muscle in exchange for potassium. Thus in the first day of life there is a flux of sodium into the extracellular space, and a relatively high fractional sodium excretion is observed in urine.

The newborn kidney is limited in its ability to excrete a solute load. Therefore, when a high sodium load is presented to the kidney of the neonate, some of it is not cleared and the concentration of sodium in the plasma rises. Cellular dehydration occurs owing to movement of water from the intracellular space to the extracellular space. This low renal excretory capacity is probably the result of a combination of factors, including (1) low glomerular filtration rate, (2) low hydrostatic pressure and (3) glomerular-tubular imbalance with relatively smaller tubular mass.

Acid-Base Balance

The acid-base balance of the fetus *in utero* is regulated by the maternal lungs and kidneys, via the placenta. There appears to be a correlation between maternal and fetal pCO_2, with a carbon dioxide gradient of about 5 mm. Hg (see Chapter 9). Immediately after birth there is a rise in carbon dioxide tension, resulting in acidosis. A return to normal pH values in the immediate hours after delivery is accomplished primarily by increased pulmonary ventilation with the elimination of carbon dioxide. Renal regulatory mechanisms come into play later.

After the first day of life, the infant shows low (below normal adult values) carbon dioxide tension and plasma bicarbonate concentration. The blood pH may be normal or slightly low. These values tend to persist throughout the first month of life. This physiological acidosis of the newborn is further complicated by the ingestion of food, for in the course of metabolism of food, more non-volatile anions are produced than cations. These excess metabolic acids must be eliminated from the body to maintain a constant pH. As an immediate and first line of defense, the body buffers combine with any excess acid or base. For example, the bicarbonate buffer system (a mixture of carbonic acid and bicarbonate ion) can combine with any strong metabolic acids. The hydrogen ion of the acid combines with the bicarbonate ion (HCO_3^-) to form carbonic acid (H_2CO_3). The latter can be eliminated as carbon dioxide and water. The anion of the acid remains, however, and must be eliminated by the kidneys.

Renal excretion of acid is accomplished by the tubular excretion of hydrogen ion in exchange for sodium ion. The hydrogen ion, now in the tubular fluid, reacts with bicarbonate ion (filtered through the glomerulus) as follows:

$$H^+ + HCO_3^- \longrightarrow CO_2 + H_2O$$

Thus, the tubular reabsorption of bicarbonate ion probably occurs in the form of carbon dioxide. As the extracellular fluid becomes more acidic (excess metabolic organic acid), the quantity of bicarbonate ions in the glomerular filtrate is decreased (remember, bicarbonate was used to combine with the hydrogen ion of the acid). Now there is more hydrogen ion than bicarbonate ion, and the excess hydrogen ion combines with excess anions either from the metabolic acids or from the phosphate buffer system, and is lost from the body in the urine. The price is a fall in pH in the urine comparable to the amount of acid excreted.

During the first few days of life, infants do not excrete strongly acidic urines. After the first week or two, their capacity to acidify the urine is comparable to that observed in older children and adults. The response of an infant to an acid load depends in large measure on his intake of protein and phosphate. Phosphate is an important buffer in the body, and can combine with hydrogen ion in the tubular fluid to provide a means for ridding the body of excess hydrogen ion. At urine pH 6.0, almost all the hydrogen ion excreted is accounted for by phosphate (titratable acidity). Cow's milk contains more phosphate than does human milk. Babies receiving cow's milk can excrete more titratable acid than those on human milk; however, it seems quite likely that the newborn, and in particular the premature infant, is less able to excrete hydrogen ion because of a decreased ability to excrete ammonium and phosphate into the tubular lumen. Urinary ammonia is formed via the conversion of glutamine to glutamate in the kidney cell (see Figure 15–3). The ammonium ion that is excreted carries with it a hydrogen ion that would not have been excreted if the amino group of the glutamine had been metabolized by the urea pathway. This is still another means by which the kidney rids the body of the excess acid. Foods rich in ammonia precursors (glutamine) enhance the infant's ability to excrete ammonia.

During the first year of life, infants have a significantly lower bicarbonate threshold than adults. As maturation of the kidney proceeds, this threshold rises to adult values, and the serum bicarbonate ion reaches normal adult values also. A combination of low renal threshold for bicarbonate ion plus relatively poor excretion of ammonia and phosphate probably is sufficient to account for the physiological acidosis of the newborn.

Amniotic Fluid

The fetal urine enters the amniotic sac; thus, the kidney is concerned with the regulation of the volume of amniotic fluid. In renal agenesis, the volume of the amniotic fluid is small; with functional kidneys but esophageal atresia the volume is pathologically large. It is known that the fetus swallows amniotic fluid, which is subsequently absorbed by the gastrointestinal tract, and it is estimated that fetal intake of amniotic fluid amounts to 400 ml. per day near term.

The total volume of amniotic fluid is variable. At about 10 weeks' gestation, the average volume is about 30 ml., at 20 weeks, 350 ml., and by 38 weeks it is close to 1000 ml. The pH is close to 7.0; pCO_2 is slightly higher and bicarbonate slightly lower than in maternal or fetal blood. Carbohydrates (principally glucose) are present in concentrations considerably below those found in maternal serum. Amino acids are present in amounts comparable to maternal serum, but other non-protein nitrogenous compounds, such as urea, uric acid and creatinine, are present in steadily increasing amounts from the end of the first trimester to term. Very small amounts of fatty acids, cholesterol, phospholipids and steroids are also found.

Sampling of the amniotic fluid during pregnancy has been used increasingly as a means of detecting abnormalities of fetal and placental function. In anoxic conditions, the fetus passes meconium into the amniotic fluid, and the characteristic green stain of the fluid may be a valuable indication of such distress. In Rh-sensitized pregnancies, the products of fetal hemolysis may be detected in the amniotic fluid. Attempts to measure carbon dioxide and oxygen tensions in amniotic fluid have yielded such variable results that it seems unlikely that such measurements could serve as a guide to the adequacy of placental gas exchange. A variety of hormones and hormone precursors have been detected in amniotic fluid.

It is generally assumed that the cells found in human amniotic fluid originate from both fetal ectodermal structures and amniotic epithelium. Such cells have been used for enzyme studies and for detection of certain chromosomal abnormalities. Prenatal sex prediction has been found to be accurate in as high as 87 per cent of cell cultures derived from amniotic specimens.

NERVOUS SYSTEM

Development

The brain develops rapidly in the embryo—much more rapidly than almost any other organ. In the early stages the future brain is merely an expanded part of the neural tube (see Chapter 11). With continued increase in size the anterior end of the brain bends downward (cephalic flexure) separating the prosencephalon (forebrain) from the remainder of the neural tube. Somewhat later a second flexure in a reverse direction occurs posteriorly, separating the mesencephalon (midbrain) from the rhombencephalon (hindbrain) (see Figure 11-3). These three primary brain vesicles are present by four weeks' gestation.

Meanwhile, cells lining the wall of the neural tube (ependymal layer) divide repeatedly. Some of these cells migrate into an adjacent layer of cells called the mantle, and there they differentiate into neuroblasts or spongioblasts. Further growth and differentiation of the neuroblast into a mature neuron occurs in the mantle layer. The spongioblast is the forerunner of the glial cells (astrocytes). Whereas neurons undergo no further division after differentiation, glial cells continue to proliferate throughout life. The outermost layer of the neural tube, the marginal layer, is made up of axons (from the neurons in the mantle) and glial cells. This layer becomes the white matter of the maturing nervous system, whereas the mantle is the forerunner of the gray matter.

Further development transforms the tripartite brain into one of five regions. By five weeks' gestation, both the forebrain and hindbrain have subdivided, the former into telencephalon (including the cerebral hemispheres) and diencephalon (the thalamus and hypothalamus) and the latter into metencephalon (cerebellum, pons) and myelencephalon (medulla oblongata). It is at about this time that the first reflex arc becomes functional in response to tactile stimulation around the mouth, where sensory receptors are present.

At an early stage (seven weeks), the telencephalon produces two anterolateral bulges which become the cerebral hemispheres. In each bulge is a cavity, the lateral ventricle of the brain. Outward migration of neurons occurs, with the migrating cells accumulating near the surface to form the cortex of the cerebral hemispheres. The two eye vesicles develop as lateral outgrowths of the diencephalon. The cavity within the diencephalon is known as the third ventricle. A depression in the floor of the diencephalon forms the infundibulum, a portion of which fuses with an outgrowth of the stomodeum (Rathke's pouch) to form the pituitary gland. The walls of the diencephalon differentiate into the hypothalamus.

The development of the cerebellum starts rather late. In the human embryo, the growth of the cerebellar rudiments begins during the second month of pregnancy. At the end of two months, the swellings are very distinct. Neurons also migrate, as in the cerebrum, from their place of origin toward the surface, producing the cerebellar cortex.

At the beginning of the third month of pregnancy, the cerebral hemispheres constitute by far the largest part of the brain. They almost completely cover the diencephalon. By four months, the cerebral hemispheres have grown so large that they extend back to touch the developing cerebellar hemispheres. Apart from the lateral fissure, the surface of the cerebral hemispheres is smooth. During the second half of pregnancy, the surface becomes wrinkled and folded.

ELECTRICAL ACTIVITY

The brain of the human fetus exhibits recordable electrical activity as early as the eighth week. By the seventh or eighth month of prenatal life, clear signs of maturation appear in the EEG tracing, the most striking of which is the shift from an "active sleep" pattern to a "quiet sleep" pattern. From birth until 13 years of age (when the mature EEG is established), spontaneous electrical activity continues to increase in voltage and in the ratio of quiet/active sleep patterns.

MYELINIZATION

Myelinization of the central nervous system falls behind cellular maturation. Myelogenesis really begins only in the second half of intrauterine life, but thereafter it continues essentially for life. It begins in the brainstem and reaches the level of the hemispheres by term. The afferent bundles are myelinated first, as are most of the cranial nerve roots and the extrapyramidal system. The motor pathways do not become heavily myelinated until after birth, when myelogenesis accelerates rapidly.

REFLEXES

Neurological maturity is programmed and will proceed whether the infant is born prematurely or continues to develop *in utero*. The gestational age of an infant can thus be determined by neurologic study alone. There is a gradual perfection of the primary reflexes in a cephalocaudad direction and an increase in muscle tone in a caudocephalic direction. For example, at a gestational age of 26 weeks, which is the limit of possible survival, the infant shows the beginnings of alertness and certain sensory (neurovegetative) patterns of behavior. General behavior is still dominated by a state of profound sleep, but this sleep is less intense and alternates with phases of torpor or even wakefulness. The rooting and Moro reflexes are present, although abduction and crying are absent from the latter. There is marked hypotonia in the 26-week-old infant. There is little resistance to movement and excessive mobility of all joints. At the distal ends of the lower limbs, the beginnings of muscle tone may be noted at this age. There is definite resistance to flexion of the foot, and the leg muscles have palpably greater tone than do those in younger fetuses.

By 30 weeks' gestation, the gradual increase in muscle tone has reached the thighs and trunk. There are vigorous movements of the whole body, especially of the trunk. The oldest primary reflexes — rooting, grasp, Moro and walking — are complete. The pupillary light reflex (not present at 26 weeks) is readily observed. Now the infant has definite periods of wakefulness.

By 35 weeks, the premature infant presents a neurologic picture quite close to that of the newborn infant at term. All four limbs are in an attitude of flexion. Grasping is steady and firm. There is spontaneous orientation to light. The flexors and extensors of the upper portion of the body are still weaker than those of the lower portion of the body. The crossed extension reflex is almost complete now, with flexion and extension components but no adduction of the contralateral limb and fanning of the toes.

The newborn of 39 weeks now has hypertonia of the upper limbs as well as the lower. The crossed extension response is completed now with its third component, adduction. The premature infant born at 26 weeks and maintained in an incubator until 39 weeks has a similar but slightly different neurologic picture. His muscle tone is good, but not hypertonic, as is the full-term baby. On the whole, though, it appears that muscular and nervous maturity is a function of age alone, and is relatively independent of environmental stimuli and experience.

Development of Neurologic Control

In the early stages of development the human fetus is essentially a spinal animal. By seven to 10 weeks, the medulla probably exerts some functional control over the more mature spinal cord. Sectioning of the medulla just above the cord at this stage abolishes certain reflexes, whereas removal of the cerebral cortex has little effect.

Later, at 13 to 14 weeks' gestation, transection at the midbrain weakens certain reflexes, indicating some mesencephalic control at this stage. Stimulation of the pons at 20 to 21 weeks produces a response of the muscles innervated by the facial nerve.

The cerebral cortex probably plays little if any role in the control of nervous activity during intrauterine development. Stimulation of the cerebral cortex up to the seventh fetal month elicits no response. Most reactions of the newborn can be observed in an anencephalic infant, which possesses only a brainstem and certain components of the basal ganglia. These neonatal reflexes disappear as voluntary control of movement and posture develop.

Growth of CNS

The embryonic and fetal periods are the most critical growth period of the central nervous system. By birth, the brain constitutes about 12 per cent of the body weight, and it doubles its birth weight in the first year of life. By the time the child is 5 or 6, the brain has tripled its weight, reaching 90 per cent of the adult level. After this, growth slows. In the adult, the brain is only 2 per cent of the body weight. Thus, the growth pattern of the brain (less than five-fold increase in weight after birth) is distinctly different from other organs, which increase in weight from birth to maturity by much greater amounts, such as muscle (30- to 40-fold), lungs (20- to 25-fold) and heart (15- to 20-fold). Unfavorable environmental factors during gestation and neonatal life may impair development of the nervous system.

METABOLISM OF CNS

Analysis of the blood entering and leaving the brain has shown that the main source of energy under normal conditions is glucose. In the embryo, glucose is utilized mainly by glycolysis, but in the adult it is almost entirely oxidized to carbon dioxide. It is estimated that the oxygen consumption of the brain accounts for over half the total oxygen consumption in the child (age 3 to 10) at rest. It is interesting that under conditions of hypoxia there seems to be no change in the rate of oxygen utilization by the brain. The cerebral blood vessels quickly dilate in response to a decrease in oxygen tension or an increase in carbon dioxide, and the brain continues to extract the normal amount of oxygen it needs from the blood.

In comparison with other metabolically active organs, the brain is poorly supplied with carbohydrate reserves, and if the blood sugar falls too low its functions are quickly impaired. The brain of the infant responds to lack of glucose by a discharge through the splanchnic nerves, leading to liberation of catecholamines and a release of glucose from the liver.

EFFECTS OF HYPOXIA OR ASPHYXIA ON THE CNS

Placenta previa, premature separation, prolapse of the umbilical cord, maternal toxemia and utero-placental insufficiency are frequently associated with fetal hypoxia or asphyxia. Children of mothers with these complications show a two- to three-fold increase in low one-minute and five-minute Apgar scores compared to the overall population. Among the survivors, there are significant differences in neurologic abnormalities which can be observed during the neonatal period and at 4 months of age. These differences typically disappear by 1 year of age but may lead to cerebral palsy or mental retardation. The degree of neurologic abnormality in children with low Apgar scores varies with the body weight. At 1 year of age, children born with low Apgar scores and weighing between 1000 and 2000 gm. have an incidence of neurologic impairment of 10.4 per cent, as compared to 1.5 per cent in babies weighing 2500 gm. or more.

Fortunately, the nervous system of the fetus or newborn appears to be more resistant to hypoxia than that of the older child or adult. This is particularly so if the circulation can be maintained so that substrates for metabolism can be provided and the products of anaerobic metabolism can be removed.

FETAL METABOLISM

Intrauterine nutrition and growth of the fetus have been discussed in Chapter 9. Aspects of fetal metabolism deserve further elaboration in view of certain critical changes which occur at birth. At this time, the gastrointestinal tract of the newborn must switch from minimal activity *in utero* to full fledged absorption of virtually all the nutrients needed by the rapidly growing extrauterine organism. The liver must be capable of synthesizing protein and glycogen and titrating the delicate balance of blood glucose. Bilirubin must be synthesized and conjugated. The kidney must assume its excretory function and maintain acid-base balance. Last, but certainly not least, the nervous system must coordinate and direct the variety of biological processes in the functioning independent human. All of these processes begin early in fetal life but must be developed at birth to a point where they can support the independent existence of the newborn.

Gastrointestinal Tract

Contrast media injected into the amniotic sac as early as four months' gestation are swallowed by the human fetus. In spite of the early development of the deglutitive reflex, the newborn has a marked tendency to aspirate its oral contents. The neonate, especially the premature infant, frequently cannot fully integrate its swallowing mechanism. This, together with an increased tendency to vomit, presents a potential hazard to the proper functioning of the respiratory tract.

Within the stomach, ingested food must be kneaded and digested. Throughout gestation, this extraordinary organ has been preparing for its unique role. Gastric pits are visible at seven weeks' gestation, and parietal cells at 11 weeks. Probably acid secretion begins shortly after the appearance of the parietal cells. The chief cells differentiate at 12 to 13 weeks. Proteolytic activity of the fetal stomach is low but is definitely present by the third month. The full development of proteolytic activity does not occur until stimulated after birth. Mucus is formed in the epithelial cells of the fetal stomach by the fourth month.

The small intestine is capable of peristalsis and of actively absorbing glucose by 11 weeks' gestation. The transfer of water and electrolytes also occurs. In contrast to other species, the human fetus does not defecate *in utero* except under hypoxic stimuli. The meconium of the newborn is green in color because of the presence of biliverdin. White meconium suggests biliary or small bowel obstruction.

Liver

At about three weeks, the human liver can be seen as a thickened area of cells along the endodermal canal just caudal to the heart. By four weeks, it is a diverticulum of hepatic buds and ducts. The gallbladder develops from a cluster of cells, arranged in acinar fashion deep inside the liver. At four and a half to five weeks, hepatic lobes are recognizable.

During development, the enzymic pattern of the liver changes, probably reflecting exposure to an increasing variety of new substrates. During midgestation, the liver is the major source of hemoglobin in the fetus; thus enzymes involved in its synthesis are very active at this stage. Degradation of hematopoietic cells and the formation of bilirubin begin at this period. Bilirubin is found in amniotic fluid in normal pregnancies as early as the twelfth week and reaches relatively high concentrations between 16 and 30 weeks. Any disorder with increased rates of hemoglobin degradation (such as hemolytic disease) would be reflected in an elevated concentration of bilirubin in the amniotic fluid. The bilirubin in amniotic fluid is in the unconjugated form, bound to albumin. Conversion of bilirubin to bilirubin glucosiduronate occurs in the liver and converts the non-polar, lipid-soluble bilirubin to a highly polar, water-soluble conjugate. Such conversion is necessary in the adult to prevent intestinal reabsorption of the toxic pigment. However, in the fetus, elimination occurs via the placenta, and the non-polar form is more readily transferred across biological membranes. Thus, in order for fetal bilirubin to clear the placenta, it must remain unconjuated.

Conjugation of bilirubin (and steroids) involves the binding of the glucuronic acid moiety in the molecule uridine diphosphoglucuronic acid (UDPGA) to bilirubin. The reaction is catalyzed by the enzyme glucuronyl transferase. As seen in Figure 15-3, UDPGA is formed from uridine diphosphoglucose (UDPG). This involves the abstraction of two protons, with the consequent reduction of NAD, catalyzed by UDPG dehydrogenase. Both glucuronyl transferase and UDPG dehydrogenase are relatively inactive in the fetus and neonate. This enzymatic deficiency is primarily responsible for the physiological jaundice of the newborn.

In infants born at term, the concentration of bilirubin in the blood reaches a peak at three days. If the infant is immature at birth, the peak comes later, and the delayed appearance of the peak is proportional to the degree of immaturity.

High concentrations of bilirubin in the blood may lead to its deposition in the brain and cause considerable damage to nervous tissue (kernicterus). The importance of kernicterus in association with hyperbilirubinemia as a cause of morbidity and brain damage is particularly evident in premature infants. Hemolytic disease of the newborn, excessive exposure to vitamin K, sulfonamides or novobiocin, hypoxia, and acidosis—all may cause severe hyperbilirubinemia and resulting brain damage.

Animal experiments indicate that the breakdown products of bilirubin which have been exposed to light do not enter the brain and are unable to produce kernicterus. No direct evidence is available in the human, but many newborn and premature nurseries have light cradles to provide phototherapy for hyperbilirubinemia.

Phenobarbital has been shown to increase the capacity of the liver to conjugate bilirubin *in vitro*. It has been used in the perinatal period to prevent hyperbilirubinemia when increased neonatal bilirubin formation is anticipated. The use of exchange transfusion to reduce the frequency of kernicterus in hemolytic disease has already been discussed. Levels of bilirubin should be monitored carefully because of the relative inability of the newborn to handle bilirubin loads.

Carbohydrate Metabolism

LIVER GLYCOGEN AND UDPGlucose

The UDPGlucose required for bilirubin conjugation is derived from UTP and glucose-1-phosphate (see Figure 15–3). The sugar moiety must be supplied by either the breakdown of glycogen or the conversion of intracellular glucose-6-phosphate (from glucose or gluconeogenesis) to glucose-1-phosphate. Thus, the glucosiduronization of bilirubin is dependent upon an adequate source of glucose or glycogen, or both.

Glucose is provided *in utero* by the mother (see Chapter 9), and excess carbohydrate is stored in fetal liver, muscle, and heart as glycogen. Glycogen is first detected in fetal liver at about 10 weeks' gestation and increases in amount steadily throughout pregnancy. During the last third of pregnancy, carbohydrate in the fetal liver rises to twice the adult level. Skeletal muscle attains a level three to five times that of the adult and the heart about 10 times adult level. These glycogen values fall to or below adult levels within a few hours after birth even in normal, well-fed newborns.

Although the normal infant is born with very large amounts of glycogen in the liver, in the case of infants born after intrauterine malnutrition (in cases of multiple pregnancy, maternal toxemia, and so forth) liver glycogen may be very low. If we consider the relative ratio of the brain to liver weight in normal vs. malnourished infants, we see that in the normal newborn the brain is three times heavier than the liver, whereas in the malnourished it may be seven or eight times heavier. It is of interest that malnourished infants have proportionately less reduction in brain weight than in other organs. The brain must be provided with glucose, and therefore the malnourished child must provide a considerably larger portion of its total carbohydrate reserves to sustain cerebral function.

BLOOD GLUCOSE

Immediately after birth there is a rapid drop in blood glucose, reaching a minimum concentration about two hours later. There is then a gradual increase, but it is a rather long time until the adult normal values are reached. This blood sugar fall has been called the physiological hypoglycemia of the newborn. Many infants do not develop symptoms despite blood glucose levels below 20 mg./100 ml. Hormonal and metabolic adjustments are necessary to provide the glucose essential for cerebral function until feedings are initiated as an exogenous source of carbohydrate. Growth hormone remains at relatively high values during the first few days of life (see Chapter 11). There are increases in lipolysis and in the utilization of fatty acids for energy, as shown by the fall in respiratory quotient and the rise in free fatty acids. Glycogenolysis and gluconeogenesis occur as well. Since there is a low disappearance rate of exogenous glucose in the first few days of life, it is quite possible that peripheral utilization may be inhibited.

HYPOGLYCEMIA

Hypoglycemia is defined as a concentration of glucose less than 30 mg./100 ml. in full-sized infants and less than 20 mg./100 ml. in low birth weight infants. There may or may not be symptoms such as jitteriness, cyanosis, apathy, convulsions, apnea or difficulty in feeding. With the administration of intravenous glucose the symptoms disappear. The incidence of symptomatic neonatal hypoglycemia is of the order of three per 1000 births. The affected infants are frequently of low birth weight (although those offspring of diabetic mothers are typically large for gestational age). About one-half of the low birth weight hypoglycemic infants had mothers with toxemia.

Certain newborn babies are more likely to develop hypoglycemia than others. Infants born to diabetic mothers or to toxemic mothers, those with low birth weight for gestation and those with severe erythroblastosis fetalis are considered to be at risk. The blood glucose of these individuals is usually measured at 12-hour intervals for at least three days; if values below 30 mg./100 ml. are obtained, the tests are repeated more frequently. Glucose therapy is started if two values of 20 mg./100 ml. or less are obtained in the asymptomatic infant. If symptoms are observed, blood glucose is immediately measured and therapy started if the value is low.

The cause of the severe hypoglycemia of the newborn is unknown. It does not appear to be a lack of any of the enzymes involved in the breakdown of glycogen to release glucose in the blood. One of the major enzymes involved in this transformation is glucose-6-phosphatase (see Figure 15–4). It is first demonstrable *in vitro* in livers from human fetuses of about 12 weeks' gestation. Thereafter, this enzyme increases in activity. Prior to 12 weeks, the placenta has glycogen stores and glucose-6-phosphatase, and can thus provide glucose to the fetus from its own carbohydrate reserves.

Fat Metabolism

During the first half of gestation, the body of the human fetus contains a mere 0.5 per cent of fat, but by the twenty-eighth week it may have as much as 3.5 per cent fat. During the last two months of gestation, further rapid increases in body fat are seen, reaching 16 per cent of body weight by term—a situation distinctly different from that in many animals (rat, cat, dog, pig), in which fat is laid down after birth.

By 15 weeks' gestation, the human fetal liver possesses the enzymes required to convert acetate or citrate to fatty acids. Thus, in addition to carbohydrate reserves, the newborn may call upon its fat reserves for calories during the first hours or days of life. Even in the starving condition the newborn uses little protein for calories, since its metabolism is geared primarily for anabolic processes. Respiratory quotients indicate that the unfed baby probably derives two-thirds or more of his energy from the oxidation of glycogen soon after birth, but thereafter more and more from the oxidation of fat. The concentration of free fatty acids in the blood is very low at birth, but begins to rise at six hours, and approaches adult levels by 24 hours. By the end of the first day, at least two-thirds of the newborn's energy is coming from fat. About 60 calories per 24 hours per kilogram of body weight is expended by the normal newborn in heat production.

PERINATAL MORTALITY

Despite improved medical knowledge and care, perinatal death rates have not been reduced significantly over the past years. Figures of mortality are usually expressed as perinatal deaths per 1000 births or neonatal deaths per 1000 live births. In Table 15–4, these numbers are listed for the years 1966 through 1970 at the Boston Hospital for Women. In general, if such figures are broken down for sex and color, mortality is higher for non-white offspring and for male offspring. Overall perinatal mortality rates (fetal deaths of 20 weeks or more plus neonatal deaths per 1000 live births plus fetal deaths) are similar in Boston (33) to the national average (34) but higher than comparable figures from Seattle (28) and lower than figures from New York (46) or Norfolk (57).

In examining the causes of perinatal mortality, it is important to remember that frequently more than one abnormal condition may be found at autopsy of the fetus or newborn. It may be difficult to ascertain which, if any, of these conditions was the actual cause of death. In Table 15–5, we have listed the diagnoses made by Dr. Shirley Driscoll on 155 autopsies performed at the Boston Hospital for Women in the year 1970. In each case, Dr. Driscoll has listed each diagnosis not only as the most probable cause of death but also as an associated factor that might have contributed to the demise. One of the most common causes of neonatal death is hyaline membrane disease. If we consider only premature babies, this disorder is probably the commonest cause of death, with increasing risk as the maturity of the infant decreases.

The majority of deaths labeled intrauterine asphyxia in Table 15–5

Table 15–4. BIRTHS AND DEATHS AT THE BOSTON HOSPITAL FOR WOMEN

	1966	1967	1968	1969	1970
Total number of births	5655	6116	6644	7132	7010
Born dead*	71	91	102	94	103
Born alive	5584	6025	6544	7038	6907
Perinatal deaths†	153	186	200	177	180
Perinatal deaths† per 1000 births	27	30	30	25	26
Neonatal deaths‡	82	95	98	83	77
Neonatal deaths‡ per 1000 live births	14	16	15	12	11

*Excluding nonviables.
†20 weeks' gestation through 28 postnatal days.
‡Died within 28 postnatal days.

Table 15–5.   PERINATAL DEATHS AUTOPSIED AT THE BOSTON HOSPITAL FOR WOMEN (1970)

Diagnoses	Fetal Deaths		Neonatal Deaths		Total
	Major Cause	Associated Diagnosis	Major Cause	Associated Diagnosis	
Intrauterine asphyxia	20	—	4	2	26
Hyaline membrane disease	—	—	19	1	20
Placental insufficiency	14	2	1	—	17
Erythroblastosis fetalis	13	1	4	—	18
Immaturity	—	—	9	5	14
Infections	4	1	8	5	18
Malformations	—	10	7	5	22
Intracranial hemorrhage	—	—	2	6	8
Miscellaneous specific	10	—	11	4	25
Pulmonary hemorrhage	—	—	1	—	1
Unknown	26	—	2	—	28
	87		68		

represent pathological findings typical of asphyxia secondary to abruptio placentae or to compressed cord. These babies are usually stillborn, or else they live for just a few hours after birth. This category represents the major cause of perinatal mortality. Babies born before 28 weeks and weighing less than 1000 gm. have been placed in the category of extreme immaturity—a condition which is incompatible with extrauterine life for most newborns.

In general, one finds that in a large general hospital that services a significant number of deprived or low-income patients, infections and asphyxia are more prominent as causes of death. In a hospital like the

Table 15–6.   CAUSES OF NEONATAL DEATHS (DEATHS BEFORE THE TWENTY-EIGHTH DAY AFTER BIRTH) IN THE UNITED STATES*

	1969	1972	1973
Total neonatal deaths	56,085	44,432	40,664
Pneumonia	1803	1122	971
Infections, other	1367	1428	1449
Congenital anomalies	7813	6444	6142
Complications of pregnancy and childbirth	12,768	10,834	10,179
Immaturity	8890	6208	5339
Hyaline membrane disease and respiratory distress syndrome	8868	8818	8433
Asphyxia of newborn	9559	6688	5364
Infant of diabetic mother	248	166	141
Hemolytic disease	884	496	370
Hemorrhagic disease	431	444	451
All other causes	3454	1784	1825

*Compiled from data from the National Center for Health Statistics, United States Department of Health, Education and Welfare.

Boston Hospital for Women, where the referral rate is high for women with diabetes mellitus or Rh incompatibility, prematurity, congenital defects and erythroblastosis fetalis are relatively more important. These considerations are important in assessing figures on perinatal mortality from any one hospital.

In obstetric services where the fetuses of all high risk pregnancies are carefully monitored during the last weeks of gestation, a significant decrease in perinatal mortality has been obtained. On the service of Dr. Edward Quilligan at Los Angeles County Hospital, where essentially all of the high risk pregnancies are now monitored, the perinatal mortality rate has decreased from 55 in 1964 to 21 per 1000 births in 1974.

The Appalachian white has as poor a perinatal mortality rate as the ghetto Negro, so the socioeconomic state of the mother appears to be of more significance than race. There tends to be an inverse relationship between the amount of prenatal care and the number of premature babies. There is also an inverse relationship between the amount of prenatal care and the number of perinatal deaths.

An examination of the causes of neonatal deaths in the United States in recent years (Table 15–6) reveals that significant progress has been made even during this brief period in reducing the number of deaths due to pneumonia, hemolytic disease, asphyxia of the newborn, immaturity and deaths among the infants of diabetic mothers. However, no improvement has been made in reducing mortality due to congenital anomalies, respiratory distress syndrome, hyaline membrane disease and the large group included in "complications of pregnancy and childbirth."

It seems obvious that many fetuses cannot be salvaged even with the best of perinatal care. There are the problems of unexplained prematurity, stillbirths without obvious cause and congenital malformations. However, increased concern for perinatal welfare, coupled with the use of the tools for better evaluation and care of the perinate (analysis of potential Rh antigens, amniocentesis, intrauterine transfusions, electronic fetal heart rate monitoring, intensitive care units), should produce encouraging results.

The future for the perinate looks better than ever as geneticists, embryologists, perinatologists, obstetricians, pathologists, radiologists, and pediatric surgeons join together to study and care for the developing human organism in the uterine cavity.

SUMMARY

Our considerations in this chapter have dealt largely with the development of the normal human fetus and his adjustment to extrauterine life. If the sequential differentiation and maturation of the various organ systems occur on time, then the probability of neonatal distress or abnormality is minimal. The normal newborn can quickly adapt to the extrauterine environment. Some of these adaptive mechanisms have been discussed, particularly those related to the cardiovascular and pulmonary systems.

But suppose that the developing organism sustains an intrauterine insult, either genetic or environmental. He may be exposed to abnormal concentrations of glucose (maternal diabetes mellitus), bilirubin (erythroblastosis fetalis), oxygen (premature separation of the placenta), hydrogen ion (acidosis) or drugs. His hormonal balance may be abnormal because of any one of a number of endocrinopathies (maternal or fetal). He may have been exposed to microorganisms *in utero*—most importantly at the critical period of organogenesis. He may possess defective genetic material which is responsible for biochemical aberrations of metabolism, or structural abnormalities. Can he survive such insults?

Much depends on how much of an insult, how long, and at what period in development these problems arise. A glance at the causes of fetal death shows that all too often we don't know why these fetuses die. Do some of these offspring lack the "biochemical know-how" to survive beyond the 20-week period? There are many unanswered questions in this area of investigation. Intrauterine asphyxia is obviously a major cause of fetal demise, as is placental insufficiency. But even where we can put a name on the problem, our ignorance is evident as regards the causation of the precipitating events.

Once the fetus has been delivered, the question of maturity becomes of paramount importance for survival—particularly the maturity of the cardiovascular and pulmonary systems. Premature infants run a high risk of hyaline membrane disease, probably because of immaturity of the alveolar cells. This immaturity of the pulmonary system may be assessed prenatally by measuring the concentration of phosphatidylcholine in amniotic fluid. Gluck suggested that babies with low levels of this lipid may escape the threat of hyaline membrane disease if kept *in utero* for a longer period of time, to permit the enzymes for phosphatidylcholine synthesis to develop. This is but one example of the possibilities for prenatal diagnosis of fetal maturity.

The newborn infant may be full-term or even postmature and still show signs of intrauterine malnutrition—low birth weight, loss of subcutaneous fat and so forth. Such babies are often termed dysmature. Just why the baby lacks the calories necessary for normal growth and deposition of fat is not always evident. A more extensive knowledge of placental physiology and pathology and of the precipitating causes of placental insufficiency should provide some understanding in this area.

Recently, the incidence of drug addiction in newborns has increased as more young mothers expose their offspring to drugs such as heroin *in utero*. Though this subject could not be dealt with in detail in this chapter, the student of perinatal medicine should be constantly aware of the possibility of exogenous agents *in utero* altering the physiology of the newborn.

Early diagnosis (preferably prenatal) of abnormalities, coupled with intensive care units for those babies in jeopardy, offers some hope for reducing perinatal mortality. Even so, until research has provided us with a better understanding of the fetus and his environment, we cannot hope to eliminate perinatal wastage and morbidity.

SUGGESTED SUPPLEMENTARY READING

Barnes, A. C. Intrauterine Development. Lea & Febiger, Philadelphia, 1968.
This multiauthored book covers conception, placentation and the development of a variety of organ systems. The last chapters discuss abnormalities of development, including chromosomal aberrations, drugs, infections and hemolytic disease.

Dawes, G. S. Fetal and Neonatal Physiology. Year Book Medical Publishers, Inc., Chicago, 1968.
An excellent book on perinatal physiology, with special emphasis on the respiratory and circulatory systems. Gas exchange, asphyxia, and resuscitation are covered in some detail.

Jonxis, J. H. P., Visser, H. K. A., and Troelstra, J. A. Aspects of Praematurity and Dysmaturity. H. E. Stenfert Kroese N. V., Leiden, 1968.
This book contains the proceedings of a nutrition symposium held in Groningen in May, 1967. Assessment of fetal development and various aspects influencing fetal growth are covered. The low birth weight infant is discussed in detail.

Kelly, J. V. Diagnostic techniques in prepartal fetal evaluation. Clin. Obstet. Gynecol., 17:53, 1974.

Moore, K. L. The Developing Human: Clinically-oriented Embryology. W. B. Saunders Co., Philadelphia, 1973.
A beautiful description of early human development, with excellent illustrations. Each system is discussed in terms of normal development and aberrations from the normal.

Philipp, E. E., Barnes, J., and Newton, M. Scientific Foundations of Obstetrics and Gynecology. F. A. Davis Co., Philadelphia, 1970.
The sections on the fetus and newborn in this book are well worth reading. Chapters on brown adipose tissue in the newborn, respiratory distress syndrome, and fetal circulation as well as chapters on early development of the fetus are particularly helpful.

Pitkin, R. M. Calcium metabolism in pregnancy: a review. Am. J. Obstet. Gynecol., 121:724, 1975.
An excellent review of maternal and fetal calcium homeostasis. Discussions of parathormone and calcitonin are included. The relationships of calcium to phosphorus and to albumin during pregnancy are presented graphically. Clinical implications are covered at the end of the article.

Timiras, P. S. Developmental Physiology and Aging. The Macmillan Co., New York, 1972.
The approach in Part I of this book is developmental and considers fertilization, embryogenesis, and neonatal and childhood changes in various organ systems. Endocrine aspects are covered in those sections where appropriate. Part II is devoted to the physiology of aging. An excellent reference book.

Villee, C. A., Villee, D. B., and Zuckerman, J. Respiratory Distress Syndrome. Academic Press, New York, 1973.
The proceedings of a conference held in May, 1973, in Boston. Chemists and clinicians joined together to discuss fetal lung development and function in terms of the respiratory problems of the newborn.

Waisman, H. A., and Kerr, G. R. Fetal Growth and Development. McGraw-Hill Book Co., New York, 1970.
This book is based on the proceedings of a conference held at San Diego in November, 1968. The various papers cover structural and biochemical maturation of fetus and placenta, followed by sessions on methodologic approaches to evaluation of the fetus and abnormalities of growth with postnatal sequelae. The discussions of the papers are included.

Chapter 16

SOME MAJOR COMPLICATIONS OF PREGNANCY

Not all pregnancies are "normal," of course; most are marred by minor complications. The question of just what event is "abnormal" or "major" is an interesting one which is rarely discussed. From the patient's standpoint, the word "normal" usually conjures up the Platonic concept of "ideal," and herein lie many difficulties of communication between patient and physician. The latter, through his scientific training, is more likely to consider a normal pregnancy as one which is average, plus or minus two standard deviations. But the statistical definition is devoid of value judgment and suffers from the fact that what is average or common (like the common cold) may not be desirable. Labor free from pain without drugs would be distinctly abnormal from the standpoint of a biostatistician, yet—like Plato's perfect circle—should be considered ideal. Perhaps a suitable definition of normal gestation is a pregnancy which results in the desired outcome without the intervention of any untoward event which is actually or potentially deleterious to the mother or to her offspring.

Utilizing this definition, a major complication of pregnancy is one which is both undesirable and uncommon, and in practice the degree of uncommonness turns out to be something that occurs in less than 7 per cent of pregnancies. With this in mind, we have selected certain abnormalities which should be understood by all physicians, even those who may not be responsible for the delivery of obstetrical patients.

MATERNAL MORTALITY

The number of deaths associated with pregnancy is a sensitive index of the efficiency of medical care in any developing country. Once this rate has fallen to a very low level, as in most highly developed civilizations, then the perinatal mortality rate becomes a more delicate index of the quality of medical services for any given area.

Definition

The Committee on Maternal Mortality of the International Federation of Gynecologists and Obstetricians (F.I.G.O.), with the approval of the World Health Organization, defines maternal death as "the death of any woman dying of any cause while pregnant or within 42 days of termination of pregnancy, irrespective of the duration and the site of pregnancy." The rate is calculated per 100,000 pregnancy terminations, and a live birth is defined as the birth of a live infant weighing 500 gm. or more, or (if the weight is not known) born after an estimated duration of 20 weeks or more from the onset of the last menstrual period.

Until such a definition is adopted by all states and all countries, it is difficult to compare maternal mortality rates, because some countries exclude deaths not directly attributable to the pregnancy itself. Since 1933, some 30 states and the District of Columbia have adopted a similar definition (although the postpartum period was extended to 90 days) so that comparisons are possible.

Rates

In 1933, the maternal mortality rate was 620 per 100,000 births, but by 1973 this had fallen to 15 per 100,000 births. That further reduction is possible is indicated by the fact that in 1973 the maternal death rate for whites was only 10.7, whereas for non-whites the rate was 34.6 per 100,000.

Causes

In the United States during 1967, the leading causes of maternal mortality were hemorrhage (about 15 per cent), infection (about 23 per cent, of which about one-half are associated with illegal abortion), toxemia (19 per cent), and anesthesia (8 per cent), followed by ectopic pregnancy, heart disease and thromboembolism. In all but four states, maternal mortality committees investigate each death to determine the cause, to decide whether the mortality was preventable or not, and if preventable whether the responsibility can be assigned to the physician, the patient or her family. In general, about three out of four maternal deaths are judged to be preventable, and so long as that situation exists there is little point in talking about an "irreducible minimum."

Prevention

The reasons for the 95 per cent decline in maternal mortality during a brief three decades are multiple, and include the establishment of blood banks, the discovery of antibiotic drugs, the improved training and certification of obstetrical specialists, the increased numbers of women seeking antenatal care, the increased percentage of women utilizing hospitals for delivery, and the establishment of maternal mortality committees in virtually every state.

In a developing country, the importance of antenatal care cannot be overemphasized. In Malaysia from 1953 to 1962, for example, the death rate was 13 for registered patients and 150 for unregistered patients. In the United States, the current problems include an increasing need for qualified obstetricians, who now deliver less than 60 per cent of patients, consolidation of obstetrical units into larger and more efficient operations, the more efficient use of ancillary personnel, the appropriate distribution of medical services, proper nutrition for all pregnant women, the reduction of illegal abortions and the abolition of poverty.

PREECLAMPSIA AND ECLAMPSIA

Diagnosis

Preeclampsia is a syndrome occurring after the twenty-fourth week of pregnancy and arbitrarily defined as a rise in blood pressure (of at least 30 mm. Hg systolic and 15 mm. Hg diastolic) and proteinuria, with or without generalized edema. Eclampsia is preeclampsia with the addition of one or more generalized convulsions. In the presence of hydatid moles, preeclampsia may have its onset prior to the twenty-fourth week.

The old term, toxemias of pregnancy, includes all cases of hypertension in pregnancy, with or without proteinuria or edema. Thus, chronic hypertensive disease, whether secondary to a nephropathy or idiopathic, is a subdivision of "toxemia," a term which should be abandoned. Inasmuch as chronic hypertension itself predisposes to the development of preeclampsia, there may be a mixture of essential hypertension and superimposed preeclampsia, although such a diagnosis might require a renal biopsy for full confirmation. Since the common denominator of all toxemias is hypertension, we prefer the term "hypertensive disorders of pregnancy." Such disorders affect about 6 per cent of all pregnancies.

Preeclampsia, a disease peculiar to women and higher primates, occurs predominantly in first pregnancies. Among the prominent predisposing factors are diabetes mellitus, twins, obesity, some ill-defined types of malnutrition, antecedent hypertension and hydatid moles. In 100 consecutive cases of hypertension in late pregnancy, preeclampsia will be diagnosed in about 60, essential hypertension or renal disease in about 30, a chronic hypertension with superimposed preeclampsia in 5, and eclampsia in 2; in the remaining 3 the diagnosis will be quite obscure. This by no means suggests that a clinical diagnosis is correct, because when a final diagnosis has been based upon the interpretation of a renal biopsy taken during the course of the disease, there has been disagreement with the clinical impression in about a third of the cases. This is particularly true when the diagnosis of preeclampsia is made in a multipara, or is made in an obstetrical patient with acute hypertension in the absence of proteinuria.

Clinical Correlates

With preeclampsia, the usual sequence of events is a rapid weight gain resulting from the retention of sodium and water, then a rise of both systolic and diastolic pressures and then the appearance of proteinuria. In the more severe forms, there may be cerebral edema, resulting in a blurring of vision, or a severe generalized headache, which is sometimes the immediate forerunner of a convulsion. The deep tendon reflexes are often hyperactive. A boring epigastric pain, the result of a swollen liver with stretch on the capsule, is an ominous sign.

There is invariably a decrease in glomerular filtration rate and renal blood flow. Indeed, if the creatinine clearance is normal for the corresponding period of pregnancy (i.e., above the non-gravid rate), the diagnosis of preeclampsia is doubtful. Almost parallel with the decrease in filtration rate is an increase in the serum uric acid concentration, but rarely is there any significant increase of blood urea nitrogen or serum creatinine. The ability to concentrate the urine is maintained. In severe degrees of preeclampsia, there may be oliguria which does not respond to diuretic agents.

In normal pregnancy, the vascular system is relatively resistant to the pressor effects of such agents as angiotensin, norepinephrine and vasopressin. The reasons for this are not yet understood. This decreased responsiveness is lost in preeclampsia. Anterospective studies of pregnant teenagers, who are more subject to preeclampsia, revealed that this increased response to angiotensin precedes the development of hypertension. Similar studies in Dallas also showed that placental function relative to steroid metabolism, as indicated by the maternal clearance of labeled dehydroepiandrosterone sulfate, is *higher* than average during the weeks prior to the development of preeclampsia.

Gant and his co-workers observed that women who have an exaggerated response to angiotensin prior to the development of clinical preeclampsia also demonstrate a rise in diastolic pressure when they assume the supine position. The "rollover" test is conducted between 28 and 32 weeks of gestation. While the patient lies on her side the blood pressure is taken repeatedly over a 15 to 20 minute period until the diastolic pressure stabilizes. After turning to the supine position, if the diastolic pressure rises more than 20 mm. Hg the test is considered positive. This simple procedure may be predictive of the group at risk for the development of preeclampsia.

Renin substrate, plasma renin activity and plasma angiotensin concentrations are all elevated in normal pregnancy, and similar high levels are noted in preeclampsia. The generalized arteriolar constriction which causes the hypertension might well be the result of the increased sensitivity of blood vessels to circulating pressor substances, which are present in no higher concentrations than in normotensive pregnancies. The hypertension of preeclampsia and eclampsia is very labile, and intra-arterial tracings reveal rapid fluctuations from second to second. The vasoconstriction may be observed directly in the ocular fundi or in the vessels of the nail bed. As the hypertension increases, the plasma volume diminishes, suggesting that the vasoconstriction affects the postcapillary vessels as well, because most of the blood volume is contained on the venous side of the circulation. In severe preeclampsia and eclampsia, the loss of plasma volume may be reflected by a sharp rise in the hematocrit.

Blood studies suggest that with severe preeclampsia and in eclampsia, there is often a disseminated intravascular coagulation. Fibrin split products are elevated and platelet counts are decreased.

The concentration of serum albumin is reduced. Serum electrolytes ordinarily remain within normal limits. Plasma aldosterone concentrations fall below the levels noted in normal pregnancy.

With preeclampsia, uterine blood flow is decreased, and this may lead to chronic fetal distress with intrauterine growth retardation. The myometrium becomes more sensitive to oxytocin.

The occurrence of convulsions is not necessarily related to the severity of the preeclampsia. Some women, because of hereditary cerebral dysrhythmia, may be more subject to convulsive episodes than others, although this is not known for certain. Chesley has shown that there is a familial predisposition to the disease. The daughters and granddaughters of eclamptic patients have a significantly higher incidence of both preeclampsia and eclampsia.

Pathology and Pathologic Physiology

The most characteristic and constant lesion of preeclampsia is found in the glomerulus (Figure 16–1). There is a marked swelling of the endothelial cells to an extent that the capillary lumina are almost obstructed. By electron microscopy, the cytoplasm shows droplet formation and vacuolization and the deposition against the basement membrane and within the endothelial cytoplasm of amorphous material, which has been identified by immunofluorescent antibody studies as a derivative of fibrinogen. For the sake of simplicity, we shall call this amorphous material fibrin, although it is more likely akin to cryoprofibrin, and may be the result of a slow, prolonged intravascular coagulation process. The renal lesion is certainly responsible for the proteinuria, the reduced renal blood flow and the reduced glomerular filtration rate, but its relationship to the hypertension is not known. All of the renal lesions of preeclampsia are believed to be reversible.

Both eclampsia and severe preeclampsia may lead to maternal death. The most common terminal events are cerebral hemorrhage, acute cardiac

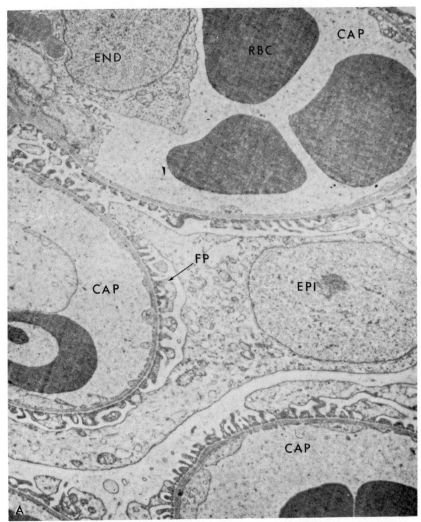

FIGURE 16–1.   *A*, Electron microscopic appearance of a normal glomerulus. *END*, endothelial cells. *RBC*, red blood cells in capillary lumen (*CAP*). *FP*, foot processes of the epithelial cell (*EPI*).

failure with pulmonary edema and hepatic failure associated with extensive periportal necrosis. According to McKay, at autopsy the brain invariably reveals intravascular platelet thrombi with perivascular hemorrhages; these may well be the primary cause of the convulsions.

The placentas of patients with preeclampsia and eclampsia show an increased number of lesions, no one of which is characteristic of the disease. True infarcts are twice as frequent as in other women, and retroplacental hematomas eight times more common, according to Fox. Cytotrophoblastic proliferation, observed in 1 per cent of normal placentas, is

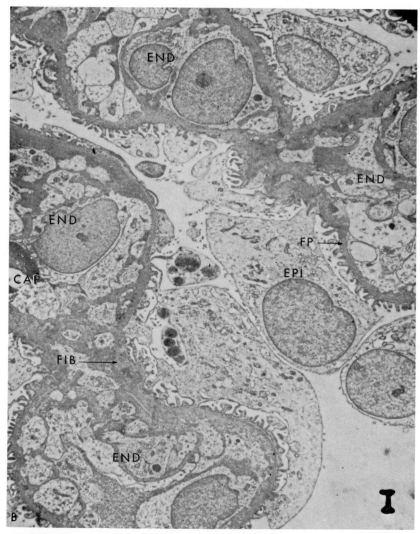

FIGURE 16-1. *Continued. B,* Electron microscopic appearance of the glomerulus in severe preeclampsia. *FIB,* deposition of a fibrin-like material against the basement membrane of the endothelial cell. The capillary lumina are almost obliterated by the swollen endothelial cells. Note that the epithelial foot processes are intact. (Courtesy of Dr. Henry Moon.)

noted in 23 per cent of preeclamptic cases. A thickened trophoblastic basement membrane, present in only 4 per cent of normal placentas, is seen in 50 per cent of preeclampsia cases, and a fibrinous necrosis of the villi, almost never observed in the placentas of normotensive subjects, is found in 12.5 per cent of preeclampsia cases. There is also an excess of syncytial knots on the villous surface and an increased rate of deportation of trophoblast to the lungs. Many of these alterations are known to result from ischemic hypoxia, supporting the concept that there is an impaired utero-placental circulation in preeclampsia. The impaired maternal flow to the placenta also contributes to the increased perinatal mortality.

Pathogenesis

Once preeclampsia is established, there appears to be some type of vicious circle which is interrupted completely only after delivery of the placenta. One concept of what this vicious circle may be is shown in Figure 16–2. The pathologic physiology may begin at any point, and the many known predisposing factors which favor the development of pre-eclampsia and eclampsia may operate at different segments of this circle. For example, with pre-existing hypertensive disease, there is already a generalized vasoconstriction. A reduced uterine blood flow may occur more readily in primigravidous than in multiparous patients because of the more extensive vascular apparatus in the multiparous uterus. Uterine ischemia would be favored by the vascular lesions of diabetes mellitus, by uterine distention with twins, or by increased uterine contractions (as with approaching term). Hydatid moles are known to be associated with an increased transport of trophoblastic elements to the lungs.

Certain abnormal diets, such as one high in oxidized fats and low in antioxidants (tocopherols, bioflavinoids, and so forth) will, in pregnant rats, lead to increased placental pathology, followed by disseminated lesions similar in many respects to those of human eclampsia. An excessive sodium intake favors sodium retention. Prior chronic renal disease (such as chronic pyelonephritis or glomerulonephritis) produces renal lesions which may render the glomeruli more susceptible to the effects of preeclampsia. The idea that uterine ischemia *per se* may lead to generalized vasoconstriction

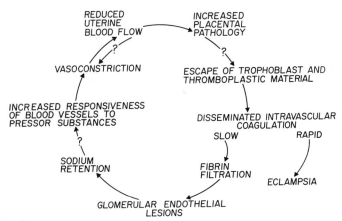

FIGURE 16–2. A concept of the sequence of events which may constitute a "vicious circle" in the pathogenesis of preeclampsia and eclampsia. The D. I. C. may actually be secondary to the disease.

and hypertension was proposed by Page in 1939 and supported by subsequent experiments with pregnant dogs. If this is so, then there would be an "inner vicious circle" which could only be overcome by improving the uteroplacental circulation (as with bed rest) or by terminating the pregnancy.

Principles of Therapy

The early detection of preeclampsia remains one of the compelling reasons for prenatal observations with increasing frequency as term is approached. A common mistake is the failure to attach significance to a rise in blood pressure from a prior level of 110/60 to 135/85 simply because the pressure did not attain the magical number of 140/90. Once the diagnosis of preeclampsia is made, prompt hospitalization becomes mandatory. Bed rest in the lateral recumbent position is in itself one of the most important therapeutic measures. Mild to moderate sedation is universally employed. Antihypertensive drugs, such as intravenous hydralazine, are ordinarily reserved for severe degrees of hypertension. In the presence of generalized edema, sodium intake is restricted, but diuretic drugs should not be used, because they further deplete the blood volume. Blurring of vision, headache or markedly exaggerated tendon reflexes, all suggestive of an impending convulsion, indicate the use of an anticonvulsive agent such as intravenous or intramuscular magnesium sulfate, which is also an effective means of controlling convulsions should they occur.

The most successful results in the treatment of eclampsia were reported by Pritchard in 1975. His treatment consists of intravenous and intramuscular magnesium sulfate to control convulsions, intermittent intravenous hydralazine to lower the diastolic pressure when it exceeds 110 mm. Hg and steps to effect delivery as soon as the woman has regained consciousness. With this simple regimen, 158 consecutive cases of eclampsia were treated without a maternal death. All fetuses that were alive when the diagnosis was made, and that weighed 4 pounds or more, survived.

Termination of the pregnancy, either by the induction of labor or by cesarean section, is at present the only curative measure. Fortunately, preeclampsia and eclampsia occur most commonly during the last month of gestation, so that fetal loss from prematurity is not common. When the typical syndrome of preeclampsia occurs prior to the twenty-fourth week, a hydatid mole may be strongly suspected.

Prevention

In our present state of ignorance about the etiology, there is a serious question as to whether anything that we might do during the antenatal period would prevent preeclampsia in *all* pregnant women. Clinical experiences in many parts of the world suggest that the avoidance of excessive sodium ingestion and the provision of a diet high in protein and vitamins reduce the frequency of preeclampsia. Of one thing we may be certain: the early detection of preeclampsia and prompt hospitalization will prevent progression to those more serious stages which lead to maternal death.

DISORDERS OF THE PLACENTA

Developmental Variations

At times there may be an accessory or *succenturiate lobe* of the placenta, connected by fetal vessels which course under the chorion. Such an anatomic arrangement is normal in the rhesus monkey, but when this occurs in women, the extra lobe may not be delivered. Careful inspection of the delivered placenta may reveal the torn fetal vessels at the margin, and a prompt manual removal of the succenturiate lobe may prevent a serious postpartum hemorrhage.

A *circumvallate* placenta (Figure 16–3) contains a whitish ring consisting of a double layer of amnion and chorion with degenerated decidua

FIGURE 16–3. Circumvallate placenta.

between, and this ring covers a variable portion of the normal chorionic plate. The mode of formation of this variant is in dispute but its presence increases the likelihood of abruption.

If the placenta develops over areas deficient in decidua, the anchoring villi may penetrate deeply into the myometrium, resulting in a *placenta accreta*. A total accreta is very rare, but small areas of deep attachment are not uncommon, and lead to the retention of placental fragments which have to be removed manually or by curettage.

Placental Lesions

The "normal" term placenta is a veritable museum of microscopic pathology. A diligent search will usually reveal areas of calcification, fibrin deposition, intervillous coagulation, ischemic necrosis, obliteration of fetal vessels within the villi, small areas of true infarction, syncytial "knots" and degeneration of the syncytium which normally covers each villus.

Inasmuch as the trophoblast depends upon the maternal rather than the fetal circulation for its survival, many of these lesions are the result of interferences with the flow of maternal blood through the intervillous space. The placenta has a considerable reserve, however, and it is only when such lesions involve a major portion of the placenta, leading to "placental insufficiency," that the fetus is in jeopardy.

An obliteration of a major maternal artery leading to a cotyledon may result in a true infarct. This will appear as a white or yellowish area of degenerated villi. The so-called "red infarcts" are in reality large areas of intervillous coagulation in various stages of organization. Such lesions are more common in the presence of preeclampsia.

Except for the trophoblastic diseases discussed in Chapter 10, neoplasms of the placenta are rare. Benign chorionic cysts are not uncommon, and in rare instances a large chorioangioma of the chorionic plate may occur.

Abruptio Placentae

Premature separation of the normally implanted placenta during the second half of pregnancy (also called "accidental hemorrhage" in the British literature) occurs in about 1 per cent of obstetrical patients. Such an event is common in the earlier months of pregnancy but is simply classified as abortion.

ETIOLOGY AND PATHOLOGY

In the majority of instances, degenerative changes in the small arteries which supply the intervillous space are believed to result in thrombosis, degeneration of the decidua and rupture of the vessel, resulting in a retroplacental hematoma. The continued arterial pumping of blood will further separate the placenta from its decidual attachment, so that in the more serious cases a complete abruption may occur. When half or more of the maternal surface is involved, fetal death is inevitable. In most instances, some of the blood is forced downward behind the membranes and escapes externally. When the hemorrhage is concealed, red cells and serum exuding from the clot may be forced through the myometrium and spread under the serosa of the uterus, producing a purplish discoloration, referred to as a Couvelaire uterus. Hypertensive disorders which predispose to arterial lesions, such as preeclampsia, chronic hypertension or chronic renal disease, are present in about 10 per cent of abruptio placentae cases. Trauma, such as a direct blow to the abdomen, is rarely a factor but may occur in about 5 per cent. The circumvallate placenta is a causative factor in about 15 per cent.

The Hibberds in England have claimed that folic acid deficiency is found with great frequency in patients with abruptio placentae, and imply that this is of etiologic importance. This has not been confirmed by Pritchard and other American workers. Furthermore, the frequency of abruption is not increased among women with established megaloblastic anemia secondary to folic acid deficiency. Nevertheless, abruption is more common in women of high parity from lower socioeconomic levels, and some form of dietary deficiency has not been eliminated as a predisposing cause.

When the separation of the placenta occurs at the most dependent edge, or when the margin of the intervillous space is torn (rupture of the marginal sinus), the bleeding is external rather than concealed, and the systemic effects associated with retroplacental hematomas do not occur.

GRADES OF SEVERITY

The following classification of severity has proven useful from the standpoint of both prognosis and therapy:

GRADE 0. A retrospective diagnosis made by examination of the placenta, with the finding of an organized retroplacental clot. No clinical symptoms prior to delivery.

GRADE 1. External bleeding without concealed hemorrhage. There is neither pain nor fetal distress. The total blood loss, which may be heavy at times, can be accounted for by the visible loss. The attachment of the placenta is usually low but does not encroach upon the lower uterine segment, as with true placenta previa. Rupture of the marginal "sinus" of the intervillous space may be included in this category.

GRADE 2. Concealed hemorrhage, with or without external bleeding. There is uterine tenderness, but no maternal shock. There are usually signs of fetal distress, or there may be fetal death. If delivery is not promptly accomplished, the abruption frequently proceeds to Grade 3.

GRADE 3. Extensive concealed hemorrhage with or without external bleeding. There is fetal death *in utero*. The uterus is tetanic, often of boardlike consistency. There is commonly evidence of maternal shock, and a coagulation defect frequently occurs.

MATERNAL COMPLICATIONS

The serious maternal complications occur in Grade 3 or severe abruption, and consist of death from hemorrhage and shock, hypofibrinogenemia with deficiency of the coagulation system, and renal failure due to bilateral cortical necrosis.

A rapid fall of fibrinogen concentration is accompanied by a drop in the platelet count and a reduction of prothrombin and proaccelerin. This is what one would expect if the coagulation process were set into motion by thromboplastin. A possible explanation for these hematologic changes is the release of thromboplastic material from the retroplacental hematoma and disrupted decidual tissue directly into the lesser circulation. Nilsen, on the other hand, believes that all the hematologic changes can be accounted for by retroplacental clotting, fibrin deposition and a re-entry of serum into the maternal circulation. In some cases there is an increased activity of plasmin (fibrinolytic enzyme), but we believe that this is a secondary effect which is not responsible for the lysis of fibrinogen. The critical level of fibrinogen, below which clotting does not occur or is grossly deficient, is about 100 mg. per cent. If blood drawn by venipuncture from a patient with Grade 3 abruption fails to clot, it may be necessary to administer a minimum of 4 gm. of human fibrinogen before delivery. The product should not be given prophylactically in anticipation of a coagulation defect because of the real danger of subsequent hepatitis.

Severe abruptio placentae is the most common cause of bilateral renal cortical necrosis. Incomplete degrees of cortical necrosis are probably common, leading to temporary oliguria with eventual recovery. Fortunately, severe degrees are rare, but they must be strongly suspected when abruptio placentae Grade 3 is followed by prolonged anuria. In modern times, a few lives may be saved by repeated hemodialysis, ultimately followed by a renal transplant.

PRINCIPLES OF THERAPY

Grade 1 abruption is ordinarily treated by the induction of labor and transfusions as needed. The fetus must be carefully monitored; if acute distress occurs, immediate cesarean section is performed.

In Grade 2 abruption, should the fetal heart tones be present, an immediate cesarean section is frequently performed in the interests of fetal salvage.

Abdominal delivery is usually recommended for Grade 3 abruption even though the fetus has expired, because the frequency of serious maternal complications increases with the passage of time. Adequate blood replacement, treatment of maternal shock and correction of a coagulation defect, if present, are essential prerequisites before a surgical procedure is per-formed. When the diagnosis of concealed retroplacental hemorrhage is suspected, prompt rupture of the membranes is indicated, with the hope that this may lessen the possibility of amniotic fluid embolism or the extrusion of thromboplastic substances into the maternal veins which drain the uterus. At the time of cesarean section, even though a Couvelaire uterus is present and the myometrium appears to be grossly infiltrated with blood, hysterec-tomy is not required. Such a uterus will contract firmly in the postpartum period, if maternal shock has been corrected.

Placenta Previa

Placenta previa may be defined as an implantation of the placenta in the lower segment such that at least a portion of a fully dilated (10 cm.) cervix would be encroached upon. Inasmuch as the determination of the degree of placenta previa can only be made definitely by palpation through the cer-vix, and the latter is commonly only 2 or 3 cm. dilated at the onset of bleed-ing, the degree of previa is estimated by the distance of the placental edge from the center of the internal os. If the os is completely covered, the degree is more than 50 per cent and it is considered clinically as a central placenta previa. If the edge should be found 3 cm. from the center, then it may be calculated that with full dilatation, 20 per cent of the cervical opening would be covered. Should a placental edge not be palpable within 5 cm. of the central point, a diagnosis of placenta previa cannot be made. Such a digital examination is ordinarily not carried out unless the patient is within three or four weeks of term and is in an operating room where a cesarean section may be done immediately if necessary (the so-called "double set-up"). Other means of diagnosis are therefore necessary when painless antepartum bleed-ing occurs three or more weeks before term.

ETIOLOGY

In the eighteenth century, it was believed that placenta previa resulted when an ovum implanted directly in the lower segment of the uterus. A more likely explanation is that at the stage of the chorion frondosum (see Chapter 8) the vascular development of the decidua in the uterine fundus is deficient for one reason or another, and attachment in the lower segment is therefore "preferred." This is supported by the fact that the total maternal area of the placenta is considerably larger with placenta previa, the insertion of the umbilical cord is more likely to be marginal or even velamentous and a few studies have actually documented the fact that decidual development in the fundus itself is deficient. We have observed several cases of central placenta previa which developed when conception occurred within two weeks after a surgical curettage, lending support to the deficient endometrium theory.

DIAGNOSIS

When significant painless bleeding occurs in the third trimester, placenta previa is assumed to be present until the possibility is ruled out. Although the condition of low attachment exists all during the earlier months, bleeding rarely occurs until the process of effacement (i.e., a retraction of the internal os) takes place. This parallels the increasing uterine motility as term is approached; therefore, the onset of symptoms is more frequent with each advancing week in the last trimester, reaching a peak with the onset of labor itself.

When the bleeding begins before 37 weeks, the initial hemorrhage is ordinarily self-limiting and usually subsides following hospitalization and absolute bed rest. Under these circumstances, a vaginal or rectal examination is contraindicated. A radiographic study ("soft tissue placentography") may be helpful in locating the placenta. Filling the bladder with radiopaque dye and measuring the distance between the bladder and the fetal head is useful in the diagnosis of central placenta previa, but is less useful in the partial degrees. Ultrasound scanning is a reasonably reliable and harmless method of placental detection and should be given first choice. One of the most reliable methods is the use of a gamma camera following the intravenous injection of technetium-99, or a scan over the uterus following the intravenous injection of radioactive iodine–labeled serum albumin (RISA). The concentration of radioactivity within the intervillous space permits localization of the placenta with a high degree of accuracy.

Placenta previa occurs more often in multiparas and its frequency is about one in 200 obstetrical patients.

TREATMENT

Any significant bleeding in the second half of pregnancy is an indication for prompt hospitalization. About half of the patients with placenta previa have their first episode of bleeding prior to the thirty-sixth week, and the majority of these are candidates for the expectant treatment. That is, they are kept at absolute bed rest without vaginal or rectal examination until they are about three weeks from term or until the fetal weight is judged to be over 2500 gm. At that time, vaginal examination may be carried out with the double set-up, and if the diagnosis is confirmed, the usual treatment is cesarean section, provided that a study of the amniotic fluid indicates maturity of the lungs.

On occasion, placenta previa will first become manifest during labor. If the edge of the placenta is marginally situated (i.e., 20 per cent or less of encroachment), the membranes may be ruptured and the fetal head may descend and control the bleeding. Methods such as scalp traction, the insertion of a hydrostatic bag and internal version have become obsolete.

With hospital management, maternal death from placenta previa should not occur, but the uncorrected perinatal mortality will be about 10 per cent, due largely to prematurity.

HYDRAMNIOS

The accumulation of amniotic fluid in excess of 2000 ml. is not necessarily a disorder of the fetal membranes, although in many instances the cause is totally unknown. When the event occurs rapidly over the course of days (acute hydramnios), there is a high association of gross fetal abnormalities, especially anencephalus and hydrocephalus. Studies have shown that fetuses with major anomalies of the central nervous system do not imbibe the usual amounts of amniotic fluid *in utero,* and inasmuch as fetal urine formation continues at the usual rate, the hydramnios is attributed to this imbalance between swallowing and urination. The occurrence of an atresia of the esophagus, for example, is almost invariably accompanied by hydramnios; conversely, the congenital absence of both kidneys is invariably associated with a virtual absence of amniotic fluid (oligohydramnios). Such major fetal anomalies occur in about a third of all cases of true hydramnios, and under these circumstances it is reasonable to attribute the phenomenon to a "drinking disorder."

By the same token, one might surmise that fetal polyuria would produce hydramnios, and it is possible that this is the reason why the complication is more common in mothers with diabetes mellitus. That is, recurring episodes of fetal hyperglycemia (reflecting the maternal hyperglycemia) might lead to an osmotic diuresis on the part of the fetus, but this remains hypothetical. Twinning and erythroblastosis fetalis are also predisposing causes.

Chronic hydramnios, occurring over a period of several weeks and associated with a normal infant, is not uncommon. Its cause is quite obscure.

Marked hydramnios may cause severe maternal distress because of pressure of the grossly enlarged uterus upon the diaphragm, the venous return from the lower extremities or the gastrointestinal tract. Its occurrence is, of course, an indication for a radiographic study of the fetus to rule out gross fetal abnormalities or multiple pregnancy. The presence of a fetal monster is ordinarily an indication for termination of the pregnancy. Should twins be present, or if the fetus appears to be normal, efforts should be made to carry the pregnancy to term, although premature labor frequently intervenes. Repeated amniocentesis for removal of fluid has been used successfully in some cases.

PROLAPSE OF THE CORD

There are two types of cord prolapse: (1) an occult or a forelying cord with intact membranes, and (2) a frank prolapse of the cord into or out of the vagina after rupture of the membranes. In either case, the etiology involves a poor adaptation of the presenting part to the lower uterine segment. An unengaged head, a shoulder presentation, a single or double footling breech or the presence of twins are among the important predisposing causes. The diagnosis of an occult or forelying cord during labor is suspected when the fetal heart rate shows a pattern of variable deceleration. The diagnosis of a frank cord prolapse, of course, is made by seeing or feeling a loop of cord in the vagina. Should the patient be in the second stage of labor, immediate delivery is indicated either by forceps or breech extraction. If the cervix is only partially dilated, however, and the fetus is still alive, an assistant should displace the presenting part to relieve cord compression and hold it there until an immediate cesarean section can be accomplished.

RUPTURE OF THE UTERUS

With good obstetrical care, uterine rupture is a rare event, occurring in less than 1 in 1000 deliveries. In some underdeveloped countries, however, this tragic event may be common because of neglected cephalopelvic disproportion. The three types of rupture are: (1) spontaneous, (2) traumatic or (3) rupture of a previous cesarean section or myomectomy scar. Spontaneous rupture may occur with any type of obstructed labor, is more common in grand multiparas and may result from overstimulation with oxytocic drugs. Traumatic rupture may result from ill-advised attempts at delivery before full dilatation or from perforation of the uterine wall with a forceps blade or by manual manipulations. As a general rule, it is wise to explore the uterine cavity with care after any difficult delivery. In developed countries, from a quarter to a half of all uterine ruptures are caused by the separation of a previous cesarean section (or myomectomy) scar. On occasion, the dehiscence is silent and is only discovered at the time of a repeat cesarean section.

The symptoms of a complete uterine rupture include both internal and external hemorrhage, acute abdominal pain, the cessation of labor and varying degrees of shock. The treatment, of course, is immediate laparotomy.

POSTPARTUM HEMORRHAGE

Although the loss of blood has been the leading cause of maternal death in the United States, essentially every loss of life was needless and should have been prevented by prompt recognition and appropriate treatment.

Postpartum hemorrhage may be defined as the loss of blood in the amount of 1 per cent of the body weight or more during the first 24 hours after delivery. After this period of time, delayed hemorrhage may occur, but its etiology is ordinarily different from that of the more immediate forms. The average blood loss in the delivery room, especially with the performance of episiotomy and repair, is between 300 and 400 ml. Losses in excess of 500 ml. occur in about 5 per cent, and losses exceeding 1000 ml. occur in 1 or 2 per cent of parturients. The recognition of excessive loss is not always simple unless all towels, drapes and sponges are weighed before and after delivery, and amniotic fluid collected prior to delivery is removed prior to the third stage of labor. The recognition of excessive loss which occurs slowly over the first 24 hours is even more difficult because of its insidious nature and the changing shifts of attendants. Even rapid hemorrhage in the delivery room is frequently underestimated, hence the rule, "Look around and guess the loss, then double it!"

Causes

In nine out of 10 cases, the immediate cause of postpartum hemorrhage is uterine atony. The remaining cases are chiefly due to lacerations of the vagina and cervix, including blood loss from the episiotomy incision. The retention of placental fragments and puerperal infection are the primary causes of delayed hemorrhage.

The uterus controls loss of blood from the site of placental attachment by firmly constricting itself around the vessels which course through the myometrium. This is facilitated by the proper management of the third stage and the use of oxytocic drugs (see Chapter 14). Atony may be the result of general anesthesia, prolonged labor, overdistended uterus (as with twins or hydramnios), intrapartum uterine infection, or hypovolemic shock itself.

A vicious circle is sometimes initiated when a laceration of the lower uterine segment or a laceration of the vagina goes unrecognized. During the first postpartal hour, the patient may lose in excess of a liter of blood and develop hypovolemic shock. During shock, the uterus will not contract firmly, the attendant palpates the fundus, diagnoses uterine atony, and treats the patient with vigorous uterine massage, more oxytocic drugs and blood transfusions. The blood pressure is restored to normal, the artery within the laceration resumes pumping, hemorrhage and shock recur, the uterus becomes atonic again and the physician begins to think about hysterectomy as a life-saving measure. If every woman who bleeds excessively has a careful examination by direct visualization of the cervix and vagina, and a manual exploration of the uterus if indicated, such a chain of events will not occur. Whenever bleeding continues in the presence of a firmly contracted fundus, a laceration should be suspected.

Treatment

The immediate treatment of uterine atony following delivery of the placenta is to bring the uterus out of the true pelvis, compress the lower uterine segment with one hand and massage the fundus with the other. Should this not suffice, bimanual compression of the uterus should be done. Packing the uterine cavity is irrational inasmuch as this distends the uterus (which is the reverse of what should occur) and conceals further hemorrhage.

At the first indication of hemorrhage, bank blood is cross-matched as rapidly as possible. This is facilitated by the prior typing of all obstetrical patients. In the meantime, an intravenous infusion should be running rapidly and a plasma expander should be on hand for emergency use should the supply of blood be delayed. Vasopressor drugs should not be used to treat hypovolemic shock. If the patient is in shock, a central venous catheter should be promptly introduced, because this will be the best guide as to the extent of blood replacement. Wrapping the legs and thighs with elastic bandages and tilting the patient in head-down position are useful adjuvant measures. Maintenance of the blood pressure within reasonably normal limits is most important because of some complications of shock which may only be manifested later. These include renal tubular necrosis and necrosis of the anterior hypophysis (Sheehan's syndrome).

The degree of anemia which may be anticipated on the third postpartum day from any given blood loss on the day of delivery may be estimated by reference to a nomogram (Figure 16–4). Even though shock has not been

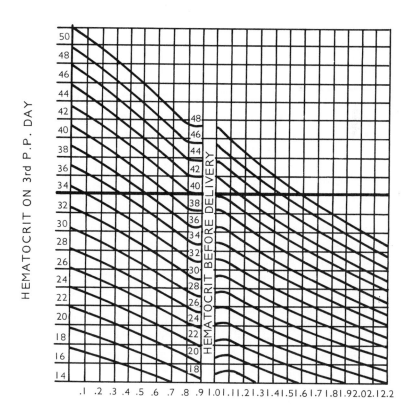

BLOOD LOSS IN PER CENT OF BODY WEIGHT

FIGURE 16–4. Anticipated hematocrit on the third postpartum day, as related to blood loss expressed in percentage of body weight. Follow vertical line representing measured blood loss to intersect with sloping line, representing predelivery hematocrit, then move horizontally to the left to the anticipated hematocrit. (Redrawn from Pastore, Amer. J. Surg., *35*:419, 1937.)

manifested, if the anticipated hematocrit on the third postpartum day should be below 34, transfusion therapy is advisable because of the predisposition of the anemic patient to puerperal infection.

PUERPERAL INFECTION

Prior to the twentieth century, puerperal sepsis, commonly called child-bed fever, was the scourge of all maternity hospitals and the leading cause of maternal death. Its occurrence in epidemic form and its relationship to infections of the examiner's hands or to the physician's participation in autopsies just prior to his attendance at deliveries led a number of observers in the eighteenth century to suggest that puerperal infection was contagious and might be carried by the physician or midwife. The idea was rejected by most obstetricians even after Oliver Wendell Holmes read his celebrated paper, entitled "The Contagiousness of Puerperal Fever," in 1843, and after Semmelweis of Vienna demonstrated a reduction of maternal mortality from 10 per cent in 1847 to less than 2 per cent in 1850 by the simple expedient of having all attendants disinfect their hands in chlorine water.

Today, by the use of rigid aseptic techniques in the delivery room and the exclusion of individuals with respiratory or other infections from the care of parturients, exogenous infections have been reduced to a minimum. Endogenous infection from the bacterial flora of the vagina is the most common cause of puerperal infection. Over half the cases of uterine infection are due to anaerobic streptococci, which are common inhabitants of the vagina. Other common causative organisms are *E. coli,* aerobic non-hemolytic streptococci and anaerobic staphylococci.

Puerperal morbidity is defined as a temperature of 38°C. or more on any two of the first 10 days postpartum, excluding the day of delivery, and with temperatures being recorded four times daily. The frequency of such morbidity is about 3 per cent, and unless it can be clearly shown to be due to urinary tract infection, mastitis, pneumonitis or other infection, it must be assumed to represent an infection of the reproductive tract.

Premature rupture of the membranes is the primary cause of intra-partum infection, whereas hemorrhage and trauma are the chief predis-posing causes of puerperal infection. Retained membranes or placental tis-sue may form a nidus of necrotic tissue for bacterial growth. Episiotomies and lacerations may become infected, of course, but fortunately the peri-neum is highly resistant to coliform organisms, so that fecal contamination, which occurs so frequently, is an uncommon cause of episiotomy infection. The site of placental attachment, now a ragged area some 5 cm. in diam-eter, is the natural focus of intrauterine infection, and from here the or-ganisms may spread into the veins, causing pelvic thrombophlebitis, or through the lymphatic vessels, resulting in parametritis, or by direct exten-sion through the myometrium or tubes, with concomitant peritonitis or multiple pelvic abscesses. Infection with *Clostridium perfringens,* although rare, may be catastrophic, and septicemia with gram-negative organisms may lead to septic endotoxic shock.

The management of mild puerperal infections in which the temperature does not exceed 38.5°C. and the pulse rate remains below 100 may consist of watchful waiting. The occurrence of a chill and a sharp rise of temperature is an indication for immediate uterine and blood cultures. Should anemia exist, transfusion therapy becomes urgent. Pending culture and sensitivity tests, vigorous antibiotic therapy should be instituted. With obvious puerperal infection, as with cases of septic abortion, it is our custom to begin intravenous penicillin infusions at a rate of 40 to 60 million units per 24 hours. Undetected gonorrheal infections of the cervix commonly remain latent for a week or longer, but may then ascend and cause salpingitis and pelvic peritonitis after the patient has returned to her home.

MULTIPLE PREGNANCY

In the United States, the frequency of twins is one in 90 births, but twinning is more common among blacks than whites and the rate increases with maternal age. As an approximation, first stated by Hellin in 1895, the frequency of triplets is one in 90^2 and of quadruplets one in 90^3. Human litters exceeding three in number have recently resulted from overstimulation of ovulation when human gonadotropins were administered to women with ovulatory failure.

Single ovum or monozygotic twins make up about one-third and double ovum or dizygotic twins two-thirds of the total number of twins. The means of distinguishing between identical and fraternal twins were discussed in Chapter 6.

When the combined weight of the fetuses reaches 7 or 8 pounds, which is about 32 weeks with twins and earlier with triplets, the rate of growth of the individual fetus slows because of inadequacies of the main nutritional supply line. The mean birth weight of a twin is about 2500 gm., of a triplet 2000 gm., and of a quadruplet 1500 gm. There is frequently a marked disparity between the birth weight of twins. With monozygotic twins, vascular anastomoses between the placental circulations may result in a gross deficiency of the fetal circulation in one and excessive circulation in the other, and this may lead to a variety of developmental anomalies.

Diagnosis

The diagnosis of twins is occasionally missed until the first infant has been delivered. It is for this reason that palpation of the fundus must be done after delivery before an oxytocic drug is administered. During the second half of pregnancy, multiple pregnancy should be suspected if the uterus is larger than anticipated, or if its rate of growth seems faster than normal. Palpation of two heads or hearing two fetal heart tones which differ by more than 10 beats per minute are positive means of diagnosis, but palpation is notoriously unreliable because of the usual tenseness of the uterine wall. Whenever twins are suspected, a single x-ray examination of the abdomen may be performed for confirmation. An ultrasound B-scan of the uterine contents is a highly reliable method and avoids radiation of the fetus. Furthermore, the method of sonography can detect two gestational sacs as early as the tenth week (Figure 16–5).

Complications

The complications associated with multiple pregnancy are the six P's: prematurity, preeclampsia, pressure symptoms, primary (and secondary) anemia, placenta previa and postpartum hemorrhage.

About 30 per cent of twins are delivered prior to the thirty-seventh week. The mean duration of pregnancy for twins is 260 days, for triplets 246 days and for quadruplets 236 days, according to McKeown and Record. The Dionne quintuplets, born in 1934, were the first set of five ever to survive longer than 15 days. The hazards of prematurity can be reduced if the diagnosis of multiple gestation is made early and a program of markedly restricted activity instituted.

The frequency of preeclampsia in twin pregnancy is 20 per cent, or some four times the expected incidence with singletons. Because of the larger area of placental attachment, the frequency of placenta previa is

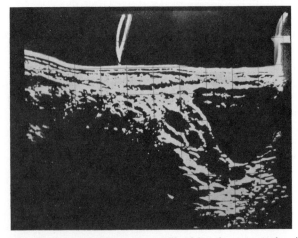

FIGURE 16–5. Two gestational sacs are outlined by ultrasonography eight weeks after conception. Markers above the abdominal wall point to umbilicus and symphysis. The filled bladder may be seen on the right. (Courtesy of Dr. L. M. Hellman.)

doubled. Hydramnios occurs 10 times more often with twins, usually involving only one sac. Its occurrence increases the risk of premature labor and adds to the pressure symptoms which already exist because of the increased uterine size.

The increased fetal demands for folic acid and for iron lead to the development of megaloblastic anemia if the dietary intake of folic acid is deficient and to iron deficiency anemia in almost every case of multiple pregnancy when iron supplements are not given. As soon as the diagnosis of twins is made, therefore, the patient should be placed upon an increased protein intake, iron supplements and multivitamin preparations which contain at least 1 mg. of folic acid daily, and she should avoid excessive ingestion of sodium chloride.

At the time of delivery, the possibility of a postpartum hemorrhage should be anticipated because of the uterine overdistention, which predisposes to atony. The patient's blood, of course, should be typed and cross-matched prior to delivery, and an intravenous solution of dextrose in water should be running during the second and third stages of labor to facilitate the use of oxytocic drugs or the use of blood transfusion if needed. General anesthesia should be avoided, especially if the labor is premature. The first twin will ordinarily present by vertex or breech, but malpresentations are common with the second twin. The interval between births should not exceed 15 or 20 minutes. Vaginal examination should be carried out immediately after delivery of the first twin. The second sac will be noted (except in the very rare instance of monoamniotic twins with a single sac) and the presenting part palpated. If the vertex or breech presents, the management is watchful waiting, but oxytocin stimulation is frequently needed to reinstitute effective contractions and the fetal heart rate of the second twin must be monitored continuously. Should the second twin present abnormally, if there is a prolapse of the cord or an arm, or if bradycardia is detected, preparations are made for a rapid delivery, usually by internal podalic version and extraction.

CHRONIC DISEASES COMPLICATED BY PREGNANCY

Pregnancy may coexist with almost every known chronic disease (other than senility), and essentially every known acute disease has been reported in pregnant women. In a book such as this, which concentrates on the core content of obstetrics and gynecology, only a few relatively common disease processes of special concern in pregnancy can be discussed. The physician or student who encounters neurologic or psychiatric disorders, collagen diseases, neoplasms, tuberculosis, syphilis, or any other chronic disease process complicated by pregnancy should consult one of the monographs noted in the references.

We have selected six entities to discuss in some detail because pregnancy may have unusual effects upon the course of the disease or because the disease process may vitally affect the outcome of pregnancy. These are diabetes mellitus, heart disease, pyelonephritis, glomerulonephritis, essential hypertension and hematologic disorders.

In general, there are very few absolute contraindications to pregnancy, and even these are gradually disappearing with modern medical advances. The difficulty, indeed the sheer impossibility, of predicting when a pregnancy might threaten the life of the mother caused such an impasse with our old abortion laws that several state supreme courts declared them unconstitutional. As pointed out by Guttmacher, any absolute contraindication to pregnancy is also an absolute indication for sterilization of the wife or husband. A few diseases, such as endometriosis or rheumatoid arthritis, are actually improved during gestation.

Diabetes Mellitus

Prior to 1923, when insulin was first used in a pregnant diabetic, conception occurred only rarely because of a high infertility rate. The maternal mortality during pregnancy was 25 per cent and the fetal mortality 50 per cent in the pre-insulin era. Since then, the maternal mortality has dropped below 1 per cent, but the perinatal mortality is barely below 20 per cent for mothers with pre-existing diabetes. The discovery of insulin, incidentally, permits the unfettered reproduction of a hereditary disorder, so that the incidence of diabetes mellitus in the general population is steadily increasing.

The nomenclature used in describing diabetes mellitus is sometimes confusing to the non-endocrinologist. A "prediabetic" is an individual who is genetically predisposed to the disease but who does not demonstrate any abnormality of carbohydrate metabolism. The term is only useful in a retrospective sense, but is frequently used in the obstetrical literature because it is known that mothers who subsequently develop diabetes may have an obstetrical history of overweight infants and an increased perinatal mortality. The term "gestational diabetes" refers to the existence of diabetes only with the stress of pregnancy, with an apparent return to normal thereafter.

The classification of diabetes in pregnancy proposed by White is widely employed and is useful in terms of the prognosis for both mother and fetus:

GROUP A.   Diagnosis based upon abnormal glucose tolerance test only ("chemical diabetes").

GROUP B.   Onset of clinical diabetes after age 20, duration less than 10 years, no demonstrable vascular disease.

GROUP C.   Onset between ages 10 and 20, duration 10 to 19 years, no x-ray evidence of vascular disease.

GROUP D.   Onset before age of 10 or duration over 20 years, x-ray diagnosis of vascular disease in legs, retinal changes on funduscopy.

GROUP E.   Same as D with addition of calcifications of pelvic arteries by x-ray examination.

GROUP F.   Diabetic nephropathy (Kimmelstiel-Wilson syndrome).

GROUP R.   Active retinitis proliferans.

In most modern obstetrical units, Group A diabetics will constitute over half of the total number. Their discovery during pregnancy is of prognostic importance, because within five years some 25 to 30 per cent and within 10 years some 50 to 60 per cent will have developed overt clinical diabetes.

There has been controversy for years about the relative usefulness of oral versus intravenous glucose tolerance tests, but after comparing the results of both tests administered to the same patients, we have concluded that the intravenous test, although less sensitive, is more reliable and gives few false positives. The test is indicated in pregnant women who (1) have a family history of diabetes, (2) have given birth to an infant weighing over 9 lbs. in a previous pregnancy, (3) have had a stillborn or anomalous infant, (4) are excessively obese, or (5) have had one positive test for glucosuria on a prenatal visit.

The intravenous test consists of giving a load of 25 gm. of glucose after drawing a control blood sample, then sampling frequently and plotting the results on semi-log paper to determine the half-time ($t_{1/2}$) for disappearance of half the added load. The slope, K, which is literally the rate of disappearance in per cent per minute, is determined by applying the formula $K = 69.3/t_{1/2}$. In normal pregnancy, K is greater than 1.2. Class A diabetics have a prenatal mortality only slightly above control levels, and there does not appear to be any advantage in delivery prior to term, as there is in Groups B through F.

Pregnant patients with diabetes mellitus have increased insulin requirements, especially in the second half, increased episodes of hypoglycemia in the first trimester, largely because of nausea and vomiting, and an increased tendency to acidosis. Despite adequate care, some 5 per cent of pregnant diabetics will have a period of acidosis during late pregnancy, and about 30 per cent of fetuses will die *in utero* during the episode.

Diabetes mellitus (Groups B through R) affects pregnancy in a number of ways. Among the chief *complications* are the following: (1) the frequency of preeclampsia is 24 per cent, almost five times the control incidence; (2) hydramnios occurs in 20 per cent, and is a serious complication for the fetus (q.v.); (3) oversized fetuses are many times more common, and this may lead to mechanical dystocia; (4) intrauterine death, especially during the last two or three weeks of pregnancy, is not uncommon. Spellacy, in a review of the literature between 1953 and 1967, found an overall fetal death rate of 11.4 per cent and a neonatal death rate of 8.5 per cent, for a total perinatal loss of 19.8 per cent; (5) the frequency of congenital fetal anomalies is 5.8 per cent, or almost triple the control figure; and (6) due to recurring episodes of fetal hyperglycemia, there is hypertrophy of the fetal pancreatic beta cells, and many newborns of diabetic mothers have severe hypoglycemia shortly after birth. The newborns of diabetic mothers in Groups A, B, and C are more subject to hyaline membrane disease (respiratory distress syndrome).

The management of the pregnant diabetic requires a team approach. Ideally, the internist responsible for managing the disease should see the patient jointly with the obstetrician at each prenatal visit. Pregnant diabetics should be seen every week during the second half of pregnancy, and there must be rigid control of the blood glucose levels, which requires occasional blood samples. The quantity of insulin needed will probably increase by 50 to 75 per cent in late pregnancy. The mothers should be delivered before term but after fetal lung maturity is attained. Most obstetricians will select 37 weeks as the optimal time. Because induction of labor at this time is not suitable for some patients, cesarean section may rarely be employed.

Group F diabetes, which includes all mothers with nephropathy, has essentially 100 per cent perinatal loss and an appreciable maternal mortality. Group R patients may develop blindness during pregnancy. For these reasons, a woman in Group F or R should not become pregnant. With this exception, diabetes mellitus and pregnancy are quite compatible, albeit hazardous for both mother and fetus.

Heart Disease

Organic cardiac disease occurs in only 1 or 2 per cent of pregnant women, yet it ranks fourth or fifth as a cause of maternal death in the United States. About 85 per cent of cases are the result of rheumatic fever, 10 per cent are due to congenital defects and most of the remainder are secondary to chronic hypertension.

Many of the signs and symptoms of normal pregnancy, such as dyspnea, mild systolic murmurs, tachycardia and ankle edema, simulate those observed in heart disease. Any one of the following four signs, however, would be diagnostic of cardiac disease: a severe arrhythmia, a diastolic murmur, a Grade III systolic murmur, or unequivocal cardiac enlargement.

Of greatest importance is the functional classification, and that formulated by the New York Heart Association is widely utilized. In abbreviated form the groupings are as follows:

CLASS I.   The cardiac disease does not impose any limitation of physical activity.

CLASS II.   There is slight limitation of physical activity.

CLASS III.   There is marked limitation of physical activity.

CLASS IV.   There is inability to carry on any physical activity without discomfort. Symptoms of cardiac insufficiency occur even at rest.

Fortunately, half of the pregnant women with heart disease remain in Class I and have an excellent prognosis. About one in 10 will fall into Classes III and IV, and most of these patients must be hospitalized for the duration of their pregnancies in order to prevent cardiac failure. Indeed, the primary purpose of prenatal management of women with heart disease is to prevent maternal death.

The greatest periods of danger are (1) from the fourth through the eighth month of pregnancy, when the physiologic changes (increased cardiac output, increased blood volume and increased heart rate) are maximal and (2) during labor and the first several days of the puerperium. If the patient survives the pregnancy and postpartum period, it is not believed that pregnancy affects the course of her heart disease or reduces her longevity.

The management of the pregnant woman with heart disease should be a joint venture between the obstetrician and a cardiologist. She must be observed at least every two weeks in order to recognize the early signs of heart failure, such as rales at the lung bases. Every effort must be made to avoid infection, and the occurrence of any febrile illness is an indication for absolute bed rest. Most patients with rheumatic valvular disease receive prophylactic penicillin therapy for the prevention of subacute bacterial endocarditis, but not all obstetricians are in agreement with this custom. It is especially important to avoid anemia and abnormal fluid retention. Adequate rest and the avoidance of undue exertion are mandatory, and each patient should have written instructions, varied according to her functional classification, as to the extent of her activities. Cardiac surgery, especially for those with a tight mitral stenosis, may be safely undertaken during the first half of pregnancy. Digitalization should not be used prophylactically, but should be undertaken at the earliest sign of failure.

Of particular importance is the management during labor and delivery. Cardiac disease is not an indication for cesarean section. Labor should be conducted with a minimum of pain and anxiety for the patient, preferably with the employment of regional anesthesia. The patient should not bear down, so elective low or outlet forceps deliveries are recommended. Ergot-like drugs should be avoided because they may produce prolonged rises of venous pressure.

Therapeutic abortion in the first trimester is frequently recommended for Class III and IV patients because of the high maternal risks, and sterilization is liberally employed under these circumstances.

Urinary Tract Infections

Both cystitis and infections of the upper urinary tract are common during pregnancy and the puerperium, largely because of the ureteral dilatation and urinary stasis which accompany normal pregnancy, as well as the trauma which accompanies delivery and the frequent use of the catheter (which should be discouraged). On the other hand, pregnancies may sometimes represent episodes in the long course of chronic pyelonephritis, which is a serious disease and a common cause of death.

Asymptomatic bacteriuria occurs in from 3 to 12 per cent of pregnant women, most series reporting about 5 per cent. The definition, established by Kass in 1960, is the presence of 100,000 colonies per milliliter when a clean midstream specimen of urine is cultured. A colony count of 10,000 per ml. is probably significant when special precautions are used, such as washing the vulva with soap and water, inserting cotton in the vaginal introitus, and immediate culture or refrigeration pending culture of the specimen. *E. coli* is the offending organism in three out of four cases. There are several inexpensive chemical screening agents which will detect the presence of significant bacteriuria in a high percentage of cases, and if positive will indicate those patients who should have culture and sensitivity tests if such are not routinely performed.

Women with asymptomatic bacteriuria have an increased incidence of anatomic abnormalities of the upper urinary tract, as revealed by pyelography, a higher incidence of impaired renal function, a higher incidence of previous overt urinary tract infections by history, and about one-third will develop acute pyelonephritis during pregnancy or the puerperium. At least three out of four cases of pyelonephritis will occur in the group with prior bacteriuria, and herein lies the chief reason for routine screening, because the eradication of asymptomatic bacteriuria will prevent symptomatic infections. Indeed, Kincaid-Smith believes that persistent bacteriuria of significant degree actually represents chronic pyelonephritis in the majority of instances.

The relationship between asymptomatic bacteriuria and premature labor is *sub judice,* about half the reports indicating a significant relationship and half showing none. It is possible that this represents differences in the frequency of underlying renal disease with impairment of function, which is known to be associated with premature labor and an increased incidence of preeclampsia. Be that as it may, there is sufficient reason to attempt an eradication of bacteriuria (with sulfonamides or with nitrofurantoin or ampicillin) in order to prevent acute pyelonephritis.

Acute pyelonephritis is a serious and occasionally fatal disease, from gram-negative bacteremic shock. The onset is marked by chills, fever, pain in the flanks, tenderness over one or both kidneys and pyuria. The absence of pyuria raises the question of ureteral obstruction and is cause for immediate urologic investigation. The disease occurs in 1 or 2 per cent of pregnant or puerperal women and therefore constitutes one of the most common serious complications of pregnancy. Fortunately, the response to antibiotic therapy (e.g., 2 gm. of ampicillin per 24 hours) is usually dramatic. After seven to 10 days of therapy, it is important to obtain follow-up cultures of the urine to be sure that it remains sterile. A recurrence of symptomatic infection suggests the existence of some abnormality of the urinary tract (e.g., stones, vesicoureteral reflux, or congenital anomaly) which calls for a thorough investigation. For those women with chronic pyelonephritis who become pregnant, small daily suppressive doses of sulfonamides throughout pregnancy and the puerperium may be useful.

Glomerulonephritis

The term "chronic nephritis," used so frequently in the older obstetrical literature, simply referred to chronic proteinuria with a reduction in renal function, with or without hypertension. Obviously, such series of cases included examples of glomerulonephritis, systemic lupus erythematosus, idiopathic nephrosis, chronic pyelonephritis, and other less common renal lesions. Even today, a renal biopsy has become essential for a precise diagnosis. Regardless of the imprecise nature of many reports of chronic nephritis in pregnancy, one rule of thumb has emerged: chronic renal disease, even with impaired renal function, is quite compatible with pregnancy and good fetal outcome *provided that there is no associated hypertension.* Primary renal disease with hypertension carries an exceedingly high perinatal mortality, with intrauterine growth retardation being the rule rather than the exception, and there is an excessively high incidence of superimposed preeclampsia. As an example, Tillman in 1951 reported 14 patients with the nephrotic stage of glomerulonephritis who were normotensive and who had 21 pregnancies. There was one stillbirth, due to a true knot in the cord, but all other pregnancies produced healthy living children. By contrast, in 16 women with chronic nephritis and hypertension in 21 pregnancies, only eight proceeded to term, six had premature infants and there were seven stillbirths. Eleven of the 16 developed preeclampsia, one other had eclampsia, and all had deteriorating renal function. One must conclude that primary renal disease with hypertension is a contraindication to pregnancy and an indication for therapeutic abortion.

Glomerulonephritis, systemic lupus and most cases of the nephrotic syndrome are believed to be immunologic disorders in which appropriate quantities of antigen, antibody and complement form a complex which attaches to the glomerular basement membrane. This complex attracts leukocytes, which release proteolytic enzymes and basic proteins, which in turn injure and fragment the basement membrane. In the case of lupus, the antigen may be DNA; in glomerulonephritis, it is frequently a protein of the streptococcus which cross-reacts with a basement membrane protein, producing an antibody which is the immediate cause of the glomerular lesions. Unlike the renal lesion of preeclampsia, these antigen-antibody complexes may be demonstrated by fluorescent microscopy in renal biopsies. The nephrotic syndrome is rare in pregnancy and may be mistaken for preeclampsia if it occurs in the third trimester. It may be distinguished, however, by the usual absence of hypertension, the frequent lipiduria, the very marked hypoproteinemia and marked edema and the notable elevation of serum lipids.

Essential Hypertension

This term refers to a sustained hypertension about which we are virtually ignorant. The level of 140/90 mm. Hg blood pressure is arbitrarily used to define the abnormal, but it must be confirmed on several occasions under reasonably basal conditions. In most instances, the hypertension antedates the pregnancy, and with this knowledge at hand we may be reasonably secure in our diagnosis. When hypertension is noted for the first time during late pregnancy, we are frequently in a quandary about differentiating essential hypertension from preeclampsia. The absence of proteinuria and edema favors the former diagnosis, provided that we eliminate primary renal disease, pheochromocytoma and other causes of elevated blood pressure.

Essential arteriolar hypertension is a disorder with a complex multifactorial etiology. There is a strong hereditary influence, and the majority of women with sustained hypertension have at least one hypertensive parent or sibling. Pregnancy is quite definitely a provocative episode which will reveal a latent hypertension even in young primigravidous patients. Such women may be normotensive after the puerperium, have recurring hypertension with each pregnancy, eventually to become chronically hypertensive only in their later years. Analogy with diabetes suggests the term "gestational hypertension" for such cases.

When experimental hypertension is induced in dogs, cats, rabbits, sheep and rats, by renal ischemia, pregnancy results in a lowering of the blood pressure, and the reduction of arterial pressure occurs near the end of gestation. The same effect is noted in women, except that the hypertension lowers in the middle months, then returns and is frequently exaggerated in the third trimester. In the absence of any prior observations, therefore, the observance of normal pressures during the middle trimester does not eliminate the diagnosis of essential hypertension. The disorder is one of the predisposing causes of preeclampsia, which occurs in from 15 to 32 per cent of patients with essential hypertension. The higher the blood pressure, the greater the likelihood of the development of preeclampsia and the greater the likelihood of an unfavorable fetal outcome. Secondary renal involvement with nephrosclerosis is a serious complication in pregnancy, leading to intrauterine growth retardation and a high perinatal loss.

Moderate degrees of chronic hypertension do not require any specific therapy during pregnancy, and the outcome is usually successful. Severe hypertension should be managed with a combination of antihypertensive drugs in the same manner as in non-gravid patients. This may protect the mother from cerebrovascular accidents and other sequelae, but in all likelihood does not improve the welfare of the fetus.

Hematologic Disorders

HEMOGLOBINOPATHIES

Hemoglobin S occurs in about 8 per cent of American blacks. The sickle cell trait (S/A) is not deleterious in pregnancy except for some increase in the frequency of urinary tract infections. About 1 in 250 blacks will have either Hb S/S (sickle cell disease) or Hb S/C (sickle cell-hemoglobin C disease), and both conditions are extremely serious in pregnancy. The maternal mortality is high (10 to 20 per cent), as is the perinatal mortality (about 50 per cent). No other medical disorder complicating pregnancy carries such a high risk. About 3 per cent of American blacks carry the Hb C trait, but C/C disease is rare. Hb C/C disease is manifested by a mild hemolytic anemia, but except for the use of folic acid no therapy is needed.

The β-thalassemia trait mimics iron deficiency anemia in pregnancy, because the mean hemoglobin level is about 9 gm./100 ml. with a microcytic hypochromic anemia. If the serum iron level is normal or elevated, iron should not be administered because of the risk of hemosiderosis. Pregnant women with thalassemia minor should receive folic acid but should rarely be given transfusions. Homozygous thalassemia (Cooley's anemia) is so often fatal in childhood that its association with pregnancy is extremely rare.

ANEMIAS

Because of the variable increases in plasma and red cell volumes, it is difficult to arrive at a precise definition of anemia in pregnancy, but a hemoglobin level of 11 gm./100 ml. is generally accepted as the lower limit of normal.

The diagnosis of *iron deficiency anemia* is ordinarily not difficult. The mean corpuscular hemoglobin concentration (MCHC) is reduced, the serum iron is usually below 60 μg./100 ml. and the iron binding capacity is usually more than 300 μg./100 ml. With mild degrees of anemia, the easiest way to confirm iron deficiency is to give the patient ferrous sulfate (*e.g.,* 300 mg. three times daily), and in two or three weeks there should be an increase in the reticulocyte count and the hemoglobin. Other than iron deficiency, no anemia responds to the administration of iron. When for some reason the patient cannot tolerate or refuses to take iron orally, its parenteral administration may be justified. Folic acid deficiency sometimes coexists with iron deficiency, requiring the administration of both substances before the anemia can be corrected.

Blood transfusions should be reserved primarily for the treatment of acute hypovolemia secondary to blood loss, not for the treatment of uncomplicated anemia.

Even though *leukemia* is the second most common cause of death from malignant disease in females aged 15 to 34, its association with pregnancy is not common. Although examinations of the peripheral blood are helpful, bone marrow studies are required for precision. Pregnancy itself does not appear to have any deleterious effect upon the course of leukemia, but the cytotoxic drugs used to produce maternal remissions may be teratogenic for the fetus in early pregnancy.

Despite the changes in the coagulation system in pregnancy, antepartal *thrombophlebitis* is not common, occurring in less than 1 per cent of gravidas; but it is twice as common during the postpartal period. Because 1 in 6 patients with thrombophlebitis may develop pulmonary emboli, immediate anticoagulant therapy is indicated, with heparin as the initial drug of choice.

THE USE OF CESAREAN SECTION

The first account of the use of cesarean section in living women was published in Paris by Rousset in 1580, and he stated that the first successful operation was performed at Siegenhausen by a cattle-gelder named Alespachen on his own wife about the year 1500. Ramsbotham, in his obstetrical textbook of 1845, described cesarean section as a "dreadful operation." It was later in the century that the introduction of aseptic methods, suture materials and general anesthesia made the operation reasonably feasible.

A vertical incision in the uterine fundus, referred to as the classical cesarean, was utilized until 1926, when Kerr introduced the low transverse incision in the lower segment. This latter method is generally employed because the danger of rupture in subsequent pregnancies is reduced and because the incision is covered by the bladder flap, which obviates adhesions to the intestines. With modern techniques, speed is only essential when the indication is acute fetal distress, and the operation carries little more maternal risk than normal delivery.

In the United States, cesarean section is utilized in from 5 to 11 per cent of all deliveries. The variability is due in part to the nature of the obstetrical population served and in part to the philosophies of the staff. In the past decade, there has been an increased use of rapid abdominal delivery for fetal distress, and the introduction of fetal monitoring equipment and use of fetal scalp blood sampling has improved the accuracy of the diagnosis of fetal hypoxia during labor. There is also a significant increase in the use of cesarean section for breech presentations.

About 45 per cent of all cesarean operations are "repeat sections." When the original indication for cesarean section no longer exists, it is possible to permit labor and vaginal delivery, providing that the low transverse incision was utilized and there was no postoperative morbidity to suggest wound infection. There is still a 1 per cent risk of uterine rupture and an additional 1 per cent occurrence of dehiscence of the uterine scar, i.e., separation with bulging but not ruptured membranes. In practice, therefore, the dictum, "Once a cesarean, always a cesarean," is generally followed.

About one-third of all *primary* cesarean sections are performed because of some type of cephalopelvic disproportion or failure to progress. Antepartum hemorrhage (abruptio and placenta previa) and fetal distress (including prolapse of the cord) each accounts for about 15 per cent, whereas uterine dysfunction, malpresentations or some type of maternal disease each accounts for about 10 per cent. Miscellaneous indications account for the remaining 7 per cent.

When the infant is delivered from the uterus, it is customary to hold it below the level of the placenta while aspirating any fluid from the nose and mouth, to permit some transfer of fetal blood from the placenta before the cord is clamped. This is routinely done with vaginal delivery, and may prevent hypovolemia, which some pediatricians believe is a predisposing cause of the respiratory distress syndrome.

A classical cesarean section followed by total hysterectomy is indicated when the uterus is the site of disease, such as myomata, gross infection, carcinoma-in-situ of the cervix, uterine rupture or placenta accreta. Some obstetricians prefer cesarean-hysterectomy to tubal ligation as a means of sterilization, especially after several prior cesarean sections.

All of the major complications discussed in this chapter, except of course those occurring after delivery, increase the danger to the fetus. Some, but by no means all, can be anticipated, thus creating a group of high-risk mothers who should receive the special attention of a qualified obstetrician. There is a growing trend to assign routine prenatal care to nurse obstetrical assistants, which is entirely feasible and even desirable in some areas, but only the low-risk mothers should be so delegated. The past obstetrical history is one of the major clues in the assessment of risk because many types of reproductive failure tend to be repetitive.

SUGGESTED SUPPLEMENTARY READING

Barber, H. R. K., and Graber, E. A. Surgical Disease in Pregnancy. W. B. Saunders Co., Philadelphia, 1974.
Multiple authorities contribute 38 chapters for a broad coverage of the subject.
Greenhill, J. P., and Friedman, E. A. Biological Principles and Modern Practice of Obstetrics. W. B. Saunders Company, Philadelphia, 1974.
This book, based upon the classical De Lee textbook, has nevertheless been completely rewritten.
Pritchard, J. A., and MacDonald, P. C. Williams' Obstetrics, 15th Edition. Appleton-Century-Crofts, New York, 1976.
This is a definitive textbook for obstetrical specialists.
Reid, D. E., Ryan, K. J. and Benirschke, K. The Principles and Management of Human Reproduction. W. B. Saunders Company, Philadelphia, 1972.
This 900-page textbook has particularly helpful discussions of obstetrical pathology.
Rovinsky, J. J., and Guttmacher, A. F. (Eds.). Medical, Surgical and Gynecologic Complications of Pregnancy, 2nd Ed. Williams and Wilkins Co., Baltimore, 1965.
Stander, R. W. (Ed.). Blood dyscrasias in pregnancy. Clin. Obstet. Gynec., *17*, 126, 1974.
This is a collection of six authoritative reviews of the various hematologic disorders which may occur during pregnancy.

Chapter 17

COMMON GYNECOLOGIC DISORDERS OTHER THAN NEOPLASMS

The ten most common gynecologic complaints are (1) irregular, profuse or prolonged bleeding, (2) the absence of menses, (3) vaginal discharge or vulvar irritation, (4) pelvic pain, (5) leakage of urine, (6) a protrusion from the vagina, (7) some type of sexual complaint, (8) hot flashes and night sweats, (9) a mass in the abdomen, and (10) a fear of cancer. Collectively, these comprise so large a number of the medical complaints of women that each physician, irrespective of his specialty, should have a basic understanding of the common gynecologic disorders which underlie these symptoms.

Some of the common gynecologic disorders discussed in previous chapters are dysmenorrhea (Chapter 2), the menopause (Chapter 3), sexual problems and psychogenic pelvic pain (Chapter 4), infertility (Chapter 7), abortion, ectopic pregnancy and trophoblastic diseases (Chapter 10). Important topics which remain include menstrual disorders; endometriosis; vaginal, pelvic and venereal infections; pelvic relaxations and urinary incontinence. A discussion of malignant gynecologic neoplasms will be reserved for Chapter 18.

THE GYNECOLOGIC HISTORY

When a new patient consults her physician she should be asked at the outset why she came, perhaps with the question, "Do you have a problem?" Her reply constitutes the *presenting complaint,* which may or may not turn out to be the *chief complaint.* She must then be permitted to describe the present illness in her own way, with the physician asking some leading questions but never suggesting the answers. It is helpful to listen attentively with a minimum of note-taking at this time, so that he may later write a detailed description of her complaints in the proper sequence.

When the complaint involves some irregularity of uterine bleeding, it is worth the time to prepare a graphic pattern month by month (see Figure 17–1), because this may reveal at a glance the most likely causes. When the complaint involves pain, urinary incontinence or some type of genital prolapse it is important to record quantitative data indicating the degree of disability. Does the symptom interfere with her usual daily work in any way? Does it interfere with sexual activities? (Her response may indicate the need for a further sexual history.) If there is leakage of urine, how often and how much? Does she wear protection, or has the leakage resulted in social embarrassment? Occasionally a patient may remark that her presenting symptom really doesn't bother her much at all, but she was "worried about cancer." This, of course, calls for a thorough physical examination and a lot of reassurance.

The student, in particular, must not be led astray by the patient's organ language. When she says "my breast hurts," it may be that the intercostal muscles of the chest wall are tender; when she states "my left ovary is painful," she could have a radicular pain of vertebral origin or a diverticulitis of the sigmoid. Nevertheless, with a puzzling symptom-complex it is often revealing to ask her what she thinks the cause of her problem may be. She might give you the correct diagnosis!

Following the history of her current complaints, the usual past history is obtained but with particular emphasis on complete details of the menstrual history, of all pregnancies, all surgical procedures and all vaginal or pelvic infections. It is also of importance to record the nature of current and past contraceptive practices, and a brief note as to the health of her husband or "significant other."

THE PELVIC EXAMINATION

No complete physical examination omits the examination of a woman's pelvis (the euphemistic term is "pelvic examination deferred"). Furthermore, it is almost becoming a matter of negligence not to make a cytologic examination of cervical or vaginal cells for cancer screening unless this has been done within the previous year.

The bladder should be emptied prior to the examination. With the patient in lithotomy position, suitably draped, the external genitalia are carefully inspected. The patient may be asked to strain downward to determine whether there is any obvious cystocele or rectocele. Bartholin's glands are then palpated between the thumb and index finger, and the urethra is gently stripped downward to determine whether there may be any exudate from the meatus or from Skene's ducts.

A non-lubricated, warm vaginal speculum is then inserted to expose the cervix. If there is an abnormal amount of vaginal discharge, a drop mixed with warm saline solution should be examined microscopically for *Trichomonas* organisms or for the hyphae of *Candida albicans*. In addition, a culture from the endocervical canal may be obtained and placed in Thayer-

Martin carrying medium for transmission to the laboratory as a screening test for gonorrhea. Material from the posterior fornix is spread on a slide and immediately fixed in alcohol and ether. The cervix may now be painted gently or sprayed with an iodine solution. Should there be an area which stains poorly, indicating deficient glycogen, a direct scraping with a spatula or tongue blade is made from that area and fixed immediately. These slides are submitted for cytological screening for cancer cells. If required, a smear from the lateral vaginal wall may also be obtained for an estimation of the woman's estrogen status. Any suspicious lesions of the cervix should be biopsied for tissue diagnosis.

After removing the speculum, two gloved fingers are lubricated and gently inserted into the vagina. The corpus uteri is brought between these fingers and those of the opposite hand pressing deeply in the lower abdomen. With the uterus in an anterior position, the size, shape and consistency of the fundus may ordinarily be described with ease, unless the abdominal wall is unusually obese. With retrodisplacements, the uterine fundus may be found resting in the cul-de-sac, posterior to the cervix. Both adnexal areas are now palpated for masses or ovarian enlargement. Normal fallopian tubes cannot be felt, and normal ovaries can be outlined about half the time, but only in non-obese, fully cooperative patients. The parametrial areas and lateral pelvic side walls can best be palpated by a recto-vaginal examination, utilizing the examiner's middle finger in the rectum and his index finger in the vagina.

If the patient is pregnant, additional maneuvers are performed to estimate the shape and capacity of the bony pelvis, as described in Chapter 13.

CYTOLOGY OF THE VAGINA AND CERVIX

Since the introduction of vaginal cytology for the detection of cervical cancer in 1943 by Papanicolaou and Traut, the "Pap smear" has occupied an important role in gynecologic practice. It is not only useful as a screening device for the detection of cervical dysplasia or carcinoma, but when properly employed vaginal cytology is useful for detecting endometrial carcinoma and even malignancies of the tube or ovary (on rare occasions), for determining the estrogen status of the woman and for assisting in the diagnosis of vaginal infections.

In the evaluation of cytologic specimens for cervical malignancy, Papanicolaou originally proposed a classification which read as follows:

CLASS 1: Absence of atypical or abnormal cells.
CLASS 2: Atypical cytology but no evidence of malignancy.
CLASS 3: Cytology suggestive of, but not conclusive for, malignancy.
CLASS 4: Cytology strongly suggestive of malignancy.
CLASS 5: Cytology conclusive for malignancy.

Some laboratories still utilize this system in their reports, but today most cytopathologists prefer a verbal description, utilizing terms similar to those employed for histologic preparations. The problem with the original classification is not with Classes 1 and 2, which are clearly benign, nor with Classes 4 and 5, which are clearly malignant. Problems arise with the interpretation of Class 3, which some cytologists employ to mean "inconclusive" or "I don't know," while to others it means "probable dysplasia" and to still others, "possible carcinoma in situ." A descriptive system is therefore preferable.

Maximal information from the vagino-cervical smear depends upon sampling the vaginal pool in the posterior fornix and from both the ectocervix and endocervix. For economy, the various samples may be spread on a single slide, as long as immediate fixation is accomplished. The vaginal pool portion is useful for cytohormonal evaluation, for assistance in identifying various vaginal infections, for detecting adenocarcinoma of the endometrium (in about 75 per cent of cases) or even cancer of the ovary (in about 25 per cent of cases). The ectocervical and endocervical scrapings detect cervical dysplasia, carcinoma-in-situ or invasive cancer with an accuracy of 98 per cent.

It is believed that every woman over the age of 18 should have cytologic screening for cervical cancer, and ideally this should be repeated annually for the rest of her life. Women so screened should not die of cancer of the cervix. Obviously, vaginal pool cytology cannot be depended upon to rule out endometrial, tubal or ovarian cancer, although in some cases this may give the first indication that such lesions exist.

Cytohormonal evaluation is best accomplished by obtaining a light scraping of the upper mid-vaginal mucosa, with immediate fixation. Probably the best estimate of the woman's hormonal status is an expression of the degree of maturation of the vaginal cells, such as the *maturation index* (M.I.). The M. I. is the percentage of mucosal cells which are parabasal, intermediate or superficial, in that order. For example, vaginal atrophy due to an absence of estrogen would yield an index of 100/0/0, whereas a maximal estrogen effect (rarely achieved) would be represented by a M. I. of 0/0/100. When progesterone dominates, as in pregnancy, almost all of the cells are intermediate, for example, an M.I. of 0/95/5. With experience, the physician may recognize some of these patterns even in unstained hanging-drop preparations from the vaginal pool.

Cytology for demonstrating the presence or absence of the female sex chromocenter (Barr body) is better accomplished from buccal smears than from vaginal specimens.

UTERINE BLEEDING DISORDERS

During the reproductive period, women may experience variations in the quantity of menstrual flow (hypermenorrhea or hypomenorrhea), in the duration of flow (short or prolonged), and in the frequency of cycles (polymenorrhea or oligomenorrhea); in addition, there may be acyclic intermenstrual or postmenopausal bleeding. In obtaining the history, it is very important to obtain the actual dates of bleeding, if possible, together with the patient's estimate of the quantity of flow at each time. Unless she has marked the bleeding on a calendar, this process of anamnesis may be rather tedious. Having recorded the information in prose form, it is then a valuable step to draw a diagram of the bleeding pattern, such as shown in Figure 17–1. Some of these are sufficiently characteristic to suggest the differential diagnosis, as illustrated in the charts, and this will be helpful in planning further diagnostic steps.

As a rule of thumb, an *endocrine disturbance*—whether it involves the hypothalamic-pituitary system, the thyroid, the adrenal cortex, the ovaries or the placenta—*will alter the ovarian cycle and therefore the frequency of the menstrual flow. Organic lesions* of the vagina, cervix, myometrium or endometrium may cause alterations in the amount and duration of flow but *do not ordinarily disturb the ovarian cycle.*

The term *dysfunctional uterine bleeding* simply refers to prolonged and/or excessive flow, usually with irregularities of the cycle as well, occurring in women who have no obvious organic lesions. Most of the time, endometrium obtained at the time of diagnostic dilatation and curettage will show either a proliferative stage or a true hyperplasia, suggesting a failure of ovulation with persistent, unopposed estrogen effect. This type of dysfunctional bleeding, secondary to an anovulatory cycle, occurs most frequently in adolescents and in premenopausal women. Occasionally, if endometrial samples are obtained on the third, fourth or fifth day of bleeding, a mixture of proliferative and secretory endometrium is revealed, suggesting

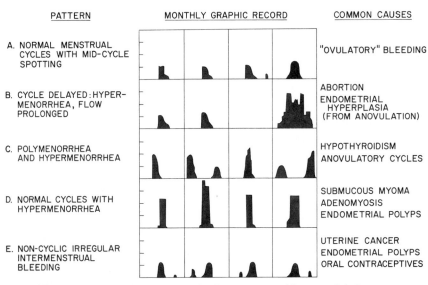

FIGURE 17–1. Varieties of uterine bleeding patterns, with some of their most common causes.

a diagnosis of *irregular shedding of the endometrium.* This is secondary either to abnormal corpus luteum function or to a failure of the end organ to reflect the hormonal cycle accurately.

It must be remembered that failure of ovulation may produce amenorrhea, polymenorrhea or acyclic bleeding, or may typically result in a delayed cycle followed by prolonged and profuse bleeding, such as the pattern shown in Figure 17–1B. Should such a pattern be manifested shortly after the menarche, and if pregnancy can be excluded, there is essentially no other diagnosis to consider; in such cases diagnostic curettage is both unnecessary and quite possibly psychologically damaging to an adolescent. The administration of progesterone or a synthetic progestin is specific therapy. Should such a pattern occur in the premenopausal woman, however, an organic lesion must be ruled out by endometrial biopsy or curettage before resorting to endocrine therapy.

Mild degrees of hypothyroidism most commonly result in polymenorrhea with increased flow, due to a shortening of the preovulatory phase, and the administration of thyroid substance or triiodothyronine ordinarily restores the cycle to a normal pattern. Conversely, hyperthyroidism most commonly leads to oligo- and hypomenorrhea.

In the bleeding patterns shown in Figure 17–1D and E there is a basic monthly cycle, so in accordance with our rule of thumb, one would expect to find some organic lesion rather than an endocrine disturbance. An exception to this would be the commonly encountered break-through bleeding which occurs with oral contraceptives containing low amounts of estrogen. True postmenopausal bleeding should always be regarded as abnormal, even when the patient is known to be taking estrogens.

It must be kept in mind that excessive or prolonged uterine bleeding may be a manifestation of a systemic disease, such as leukemia, a coagulation disorder or a deficiency of ascorbic acid.

AMENORRHEA

Failure to menstruate by the age of 18 is called *primary amenorrhea,* and in some 40 per cent of cases it is found to be due to a congenital anomaly, such as the disorders of sex chromosomes, the XY androgen-insensitivity syndrome ("testicular feminization"), or a congenital absence of the uterus and vagina. If such anomalies are ruled out, and there has been some delay in the development of secondary sex characteristics, a diagnosis of delayed puberty is appropriate, although this is only substantiated by the subsequent occurrence of the menarche.

Secondary amenorrhea is defined as the cessation of menses for three or more months. Assuming that a functioning endometrium is present, this symptom is always due to some alteration in the production of gonadal hormones. This may result from extra-ovarian sources (placenta, adrenal cortex), from primary ovarian disease or, more commonly, from insufficient or disordered gonadotropic stimuli. Because of the multiple etiology, some classification of causes such as the following may prove to be convenient.

During the reproductive years, the most common cause of secondary amenorrhea is pregnancy. The next most common cause of an abrupt cessation is psychogenic. This may result from a variety of stressful situations, such as the fear of pregnancy or even an insatiable desire for pregnancy (pseudocyesis), or a change of locale (e.g., "boarding house amenorrhea") or a death in the family. Amenorrhea may accompany a period of acute undernutrition, such as a crash diet, or may result from prolonged hypothalamic suppression, as in 1 or 2 per cent of oral contraceptive users.

Many of these episodes of secondary amenorrhea are self-limited and respond to reassurance and psychotherapy. In others, clomiphene citrate may be used to induce ovulation. This drug is an estrogen analogue which is believed to antagonize the inhibitory effect of estrogen upon the LH-releasing center in the hypothalamus. When there is true pituitary insufficiency, ovulation may be induced by the parenteral injection of human gonadotropins prepared from the urine of postmenopausal women, but

Table 17-1.   THE CAUSES OF SECONDARY AMENORRHEA

I. Physiologic
   A. Pregnancy
   B. Puerperal
   C. Menopause
II. Failure of uterine response
   A. Hysterectomy
   B. Cryptomenorrhea (due to acquired stenosis of cervix or vagina)
   C. Destruction of endometrium
      1. Tuberculosis
      2. Traumatic (Asherman's disease)
III. Ovarian insufficiency
   A. Destruction (surgery, irradiation, infections)
   B. Neoplasms, including the rare functioning tumors
   C. Autoimmune disease (premature menopause)
IV. Central nervous system disorders
   A. Pituitary
      1. Insufficiency (unknown cause)
      2. Destruction (e.g., Sheehan's syndrome)
      3. Neoplasms (various adenomas)
   B. Primary CNS disorders
      1. Idiopathic (?) hypothalamic dysfunction (e.g., polycystic ovary syndrome)
      2. Psychogenic and environmental (common)
      3. Tumors
      4. Inhibition of prolactin inhibiting factor (Chiari-Frommel syndrome)
   C. Secondary hypothalamic dysfunction
      1. Iatrogenic
         a. Steroid administration (e.g. continuous progestin, "post-pill amenorrhea")
         b. Psychotropic drugs (e.g., phenothiazines)
      2. Thyroid disease
      3. Adrenocortical disease
      4. Nutritional (sudden weight loss from dieting, anorexia nervosa, malnutrition)
      5. Severe chronic systemic disease

multiple pregnancies frequently result from overstimulation. Under some circumstances, continuous adrenocortical suppression with a corticosteroid will result in a restoration of ovulatory cycles.

Among the laboratory tests which are useful in an investigation of the cause of secondary amenorrhea are: the 24-hour urinary output of 17-ketosteroids and 17-hydroxycorticosteroids; serum LH and FSH concentrations; a serum T_4 or other measure of thyroid function; an examination of the cervical mucus and vaginal cytology for estrogen status; and an endometrial biopsy. In some instances, it is helpful to test the end-organ response by the administration of progesterone (e.g., 50 mg. in oil i.m. or 10 mg. orally of norethindrone acetate daily for five days) to determine if bleeding occurs. The absence of bleeding indicates pregnancy, very low or absent estrogen, or absence of a responsive endometrium.

Determination of the cause of secondary amenorrhea may at times be difficult, but it is obvious that treatment depends upon an accurate diagnosis. The induction of ovulation in a woman who does not want to become pregnant is ordinarily not indicated, and the production of periodic withdrawal bleeding (as with the oral contraceptives) under the guise of "restoring the menstrual periods" is rarely useful except for psychiatric reasons.

BENIGN DISEASES OF THE VULVA

Condylomata Acuminata

Condylomata acuminata are warts of varying size, usually multiple, appearing on the perineum, vulva, mucosa of the introitus and occasionally within the vagina and on the cervix. They are believed to be due to the same virus that causes ordinary skin warts, and are definitely more common in the presence of chronic leukorrheal discharges or poor hygiene, or in association with pregnancy. The first step in therapy is to eradicate any associated infections, such as trichomoniasis, candidiasis, gonorrhea or *H. vaginalis* infections (q.v.). When the lesions are small, they may be carefully painted with 20 per cent podophyllin suspended in tincture of benzoin. The medication should be washed off with soap and water 6 or 8 hours after the application. Several hours later, a reaction which may be quite painful occurs, so it is best to treat only four or five lesions at a time. Such applications are done weekly until the lesions have cleared. Large condylomata exceeding 1 cm. in diameter should not be treated in this manner, but should be either surgically removed or desiccated by electrocoagulation followed by removal with a sharp curet or knife blade.

Cysts

Cysts of the vulva are quite common, and most of them are Bartholin duct cysts, traumatic inclusion cysts, or epidermal or "sebaceous" cysts.

Cysts occurring behind the walls of the vagina usually arise from remnants of the mesonephric (Gartner) or paramesonephric (Müllerian) ducts (see Chapter 2).

Cysts of the Bartholin duct due to some form of occlusion occur in 1 or 2 per cent of all gynecologic patients. Small asymptomatic cysts should be left undisturbed; large or symptomatic ones are best treated by marsupialization. Traumatic inclusion cysts are usually found near the scars of prior episiotomy incisions or obstetrical lacerations in which small bits of squamous epithelium have become buried. Epidermal cysts also contain bits of epidermis, often within the occluded ducts of sebaceous glands. Most inclusion or epidermal cysts cause no distress and require no treatment.

An acute abscess of Bartholin's gland is another matter, because this is a most painful lesion. Most of these are caused by mixed bacterial flora which somehow gain access to a duct which is blocked. When fluctuant, the abscess may be incised under local anesthesia and packed with iodoform gauze for several days to maintain drainage. If a large, non-infected cyst re-forms, it may then be marsupialized.

Dermatoses

Almost any of the dermatoses may occur on the vulva, including psoriasis, lichen planus, intertrigo (especially in very obese women) and contact dermatitis. The last is sometimes due to nylon or to soaps or to medications used in the vagina or on the vulva, such as the vaginal deodorant sprays. Acute vulvitis is most often secondary to vaginal candidiasis or trichomonas infestation.

Dystrophies

Chronic vulvar dystrophies are common, especially in the post-menopausal period, and may result in a very disturbing persistent pruritus. Jeffcoate believes that the term dystrophy should be used to include all such entities as leukoplakia, leukoplakic vulvitis, lichen sclerosis et atrophicus, kraurosis, and atrophic and hypertrophic vulvitis. This, of course, simplifies the terminology, because it is extremely difficult to differentiate between some of these varieties by simple inspection. The important issue is whether or not the lesion constitutes an atypical dysplasia which may be considered premalignant, or may in itself be some type of intraepithelial carcinoma. Quite frequently, hyperplastic and atrophic lesions coexist, and since the hyperplastic areas are more likely to be dangerous, they should be biopsied.

Treating chronic vulvar lesions with a variety of topical agents over a long period of time *without biopsies* is a common failing among physicians. A biopsy can be readily done in the clinic or office. After raising a wheal with a local anesthetic solution, a sample of the lesion may conveniently be removed with a small skin biopsy punch, which resembles a tiny cork borer. The various dystrophies must be distinguished from localized neuro-dermatitis (lichen simplex chronicus), which may often be relieved by application of corticosteroid-containing lotions.

Herpes Genitalis

Herpes genitalis appears to be occurring with increased frequency, even allowing for the fact that our recognition of the disease has improved in recent years. The lesions are caused by the herpesvirus hominis, type II, although we suspect that some type I viruses may cause primary vulvar lesions as the result of oro-genital contact. The type II virus is ordinarily introduced at the time of coitus, resulting in infection of the cervix followed by secondary lesions of the vulva. Cytologic specimens obtained by scraping the lesion may be diagnostic, because there are multinucleated giant cells, bizarre nuclear changes and vacuolated cytoplasm in the epithelial cells.

Primary, asymptomatic herpes of the cervix must be quite common, because 15 to 20 per cent of adult women have antibodies to the herpesvirus type II in their sera. In contrast, almost 90 per cent of women with carcinoma of the cervix have such antibodies, and it is strongly suspected that herpesvirus type II is a major etiologic factor for cervical dysplasia, carcinoma *in situ* and invasive cervical cancer.

The initial attack of herpes of the vulva results in very painful lesions. During the first 24 hours, there are tender vesicles which rupture a day or two later, leaving shallow ulcers which persist for another week or so. During the initial phase there may be a viremia with high fever and even signs of a viral encephalitis. During pregnancy, the virus has been known to infect the fetus, with fatal outcome.

The course of herpes infection may be shortened and the pain alleviated by photoinactivation of the virus. This is accomplished by painting the raw lesions with a solution of 0.1 per cent neutral red or proflavin and then exposing them to a cold light source for 10 or 15 minutes. The wisdom of using such dyes has been questioned because their addition to cell cultures followed by exposure to ultraviolet light has led to mutagenic cell changes. An alternative method of affording immediate relief is freezing the lesions with an ethyl chloride spray.

Untreated herpes genitalis tends to recur at intervals even without exposure to reinfection because the virus may remain viable within the cells of the original lesion. The recurrences are less painful and of shorter duration because of the development of humoral antibodies. Herpes may be extremely serious in those persons whose immune mechanisms are being suppressed because of organ transplants or other reason.

INFECTIONS OF THE VAGINA

Vaginitis, with or without an associated vulvitis, is one of the most common gynecologic complaints, particularly among young women. The usual triad of infectious agents are *Candida albicans* or *tropicalis, Trichomonas vaginalis* and *Haemophilus vaginalis.*

Candidiasis

The most frequently observed vulvovaginitis is candidiasis (often referred to by the older term moniliasis). The reasons that this fungal infection has surpassed the venereal infection trichomoniasis in frequency include the increasing use of antibiotic drugs, the widespread use of oral contraceptive hormones and the use since 1960 of an effective oral trichomonacide which is steadily reducing the number of carriers.

All of the vaginal fungi which form hyphae or filaments belong to the genus Candida, which is indigenous to man and many animals. The organism may be found in the vaginas of about 5 per cent of nongravid women and 15 to 20 per cent of pregnant women, although in only about half will there be a symptomatic vaginitis. The fungus may be cultured from feces about half the time and is generally ubiquitous. The problem, therefore, is not where it comes from, but what the predisposing factors are which permit the fungus to take over and cause vaginitis.

Candida species thrive on glycogen. It would be expected, therefore, that pregnant women are highly susceptible to candidiasis because of the unusually high glycogen content of the vaginal cells. To a lesser extent, the oral contraceptives have the same effect, and the frequency of infection is several times higher among the Pill-takers than in those who do not use the Pill, although this view has been recently challenged. Antibiotic drugs produce candidiasis either by reducing the normal bacterial flora of the vagina, permitting the "weeds" to take over, or by increasing by several hundredfold the fungal content of the intestinal tract. Women with diabetes mellitus are particularly prone to candidiasis, and one of us has observed acute infections following a "candy spree" sufficiently often to suggest that there is a cause and effect relationship. Coitus is probably unimportant as a source of infection.

Women with acute "monilial" vaginitis complain primarily of vulvar burning and itching, rarely of discharge. Upon introducing a speculum, one may see thick white patches of discharge adherent to the vaginal walls. These are called "thrush patches," because this is the same disease that causes oral thrush in newborn infants, but they are observed in only about half of the non-pregnant patients with symptomatic candidiasis. The diagnosis is made by scraping the lateral vaginal wall with a tongue blade and tapping the discharge into three or four drops of 20 per cent potassium hydroxide solution on a slide, then adding a cover slip. The strong alkali lyses all cells, but leaves the hyphae and conidia, which may be readily seen under the microscope (Figure 17–2).

When the vulvovaginitis is particularly acute, rapid relief may be obtained by painting the vagina and vulva with a 1 per cent solution of aqueous gentian violet. This is messy, and the dye stains everything, but women will put up with this inconvenience for the relief of intense vulvitis. Successful eradication of candidiasis, however, depends upon the prolonged use of fungicidal preparations in the vagina. Included among these are nystatin or candicidin suppositories or creams, miconazole nitrate cream (Monistat), and propionic acid jellies. A cream combining nystatin with some corticoid for relief of the itching may be used as an adjunct to the intravaginal therapy. Most failures are due to premature termination of the treatment or failure to use the medication during the menstrual period. The preparation selected should be employed twice daily for the first week and then at bedtime for two or three weeks to prevent recurrences, which are all too common despite such intensive treatment.

Trichomoniasis

Trichomoniasis is due to a vaginal infestation with *Trichomonas vaginalis,* a unicellular protozoon flagellate discovered in 1836 (Figure 17–3). If the organism is routinely looked for in every woman during the reproductive

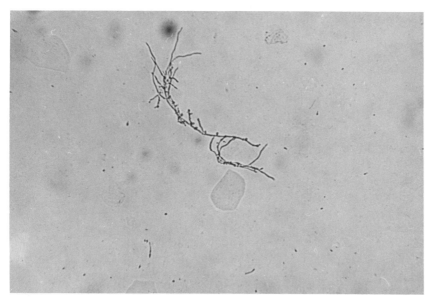

FIGURE 17–2. Hyphae of candidiasis as seen in a preparation of 20 per cent potassium hydroxide. (From Gardner and Kaufman, Benign Diseases of the Vulva and Vagina, C. V. Mosby Company, St. Louis, 1969.)

years, it will be found in about 15 per cent of patients. Most of them will admit having a malodorous discharge. In the classic case of acute trichomoniasis, there is an associated vulvitis, the discharge is greenish to gray and frothy and the upper vagina is red and may show small punctate petechiae. The diagnosis is made by mixing a drop of discharge with two or three drops of Ringer's solution and observing the motile trichomonads under the microscope.

Trichomonas vaginalis vaginitis is a venereal infection inasmuch as 99 per cent of infections originate from coitus with a male or female carrier of the protozoon. Rarely does the infection cause any symptom in men. Metronidazole (Flagyl) was introduced in 1959, and its oral use has been a highly successful means of eradicating the organism in both men and women. The usually recommended dosage is 250 mg. three times daily for 10 days, but Gardner reports equally successful results by giving 500 mg.

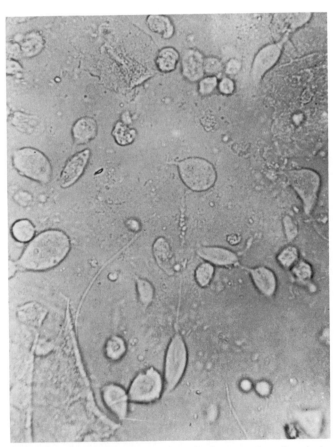

FIGURE 17–3. Trichomonads as seen in a fresh suspension of vaginal secretions in Ringer's solution. (From Gardner and Kaufman, Benign Diseases of the Vulva and Vagina, C. V. Mosby Company, St. Louis, 1969.)

(two tablets) every 12 hours for five days. When the sexual partner is not treated at the same time, he reports a recurrence rate of 27 per cent. When the male is treated, the recurrence rate is 8 per cent, but this may represent a failure of monogamous relationships.

Haemophilus Vaginalis

Vaginitis due to *Haemophilus vaginalis* was first described by Gardner and Dukes in 1954 and has since been recognized as constituting the major share of what were formerly called cases of non-specific bacterial vaginitis. Any patient who has a gray, homogenous, malodorous discharge with a pH of 5.0 to 5.5 that does not contain trichomonads probably has *H. vaginalis* infection. The organism is a minute, non-motile, gram-negative rod, again transmitted by coitus. The infection is relatively mild, but may be the source of a chronic discharge with a bad odor. In a wet mount there are few pus cells, the typical *Lactobacilli* are absent, and there are myriads of tiny rods which cling to the epithelial cells, giving them a stippled appearance (the so-called "clue cells"). A Gram stain reveals almost no bacteria other than the tiny Gram-negative rods. Cultures may be obtained but require special partial-anaerobic methods. Temporary relief may be provided by the use of sulfonamide-containing vaginal creams, but recurrences are the rule. When warranted, cures may be obtained by the use of therapeutic dosages of tetracycline or ampicillin for 10 days or so. In many instances, both sexual partners must be treated simultaneously in order to effect a cure. The essential facts about the common forms of vaginitis are shown in Table 17–2.

DISORDERS OF THE MYOMETRIUM AND ENDOMETRIUM

Adenomyosis

Adenomyosis is the result of a benign invasion of the myometrium by the endometrium, resulting in numerous islands of endometrial glands and stromal elements within the uterine wall. The nodular forms are caused by myomas containing endometrial elements (adenomyomas). The usual manifestation, however, is a diffuse infiltration resulting in a symmetrical enlargement of the corpus, most commonly in parous women over the age of 35. The invading endometrium rarely participates in the menstrual cycle, since it originates from the basal layers; however, at the time of menstruation there may be swelling and edema, so that the uterus is larger at this time than it is between menses, which is one of the diagnostic signs.

Adenomyosis uteri is not uncommon, and there is often a coincidental association with such other lesions as external endometriosis, endometrial cancer and myomas. The lesions are usually asymptomatic and are noted

Table 17-2.   COMMON VAGINAL AND/OR VULVAR INFECTIONS

Agent	Venereal	Predominant Symptoms	Appearance of Discharge	Odor	pH	Diagnosis	Treatment
Candida (Monilia)	No	Marked vulvar itching	Thick, white Often in "thrush patches"	No	<5.0	Hyphae and spores seen in 20% KOH (or culture)	Nystatin suppositories Propionic acid jelly Monistat cream Candicidin suppositories
Trichomonas	Yes	Irritating discharge	Greenish to yellow, thin and frothy	Yes (Putrid)	>5.5	Hanging drop in saline	Metronidazole P.O.
H. vaginalis	Yes	Chronic discharge	Milky, gray	Yes (Acrid)	>5.5	Hanging drop and Gram stain	Oral antibiotics (e.g., tetracycline)
Herpes virus Type II	Yes	Painful ulcers on vulva	Watery (if cervicitis present)	No	<5.0	Appearance Pap smear	Photoinactivation with 0.1% neutral red or proflavine(?)
Condylomata acuminata (Virus)	Maybe	Pointed warty growths	Due to concurrent infections	Rare	<5.0	Appearance Biopsy	25% podophyllin in Tr. Benzoin Surgery

for the first time by the pathologist who examines a uterus removed for other reasons. On the other hand, adenomyosis may give rise to progressive hypermenorrhea and dysmenorrhea, necessitating hysterectomy. Inasmuch as most of the invading endometrial glands connect with the uterine cavity, the diagnosis may be established preoperatively by hysterography, utilizing a contrast medium of low viscosity so that the glands fill, giving a tree-like appearance within the myometrium.

Sometimes the invading endometrium consists solely of stromal elements, variously called stromal adenomyosis or stromatosis; when the lesion is invasive, it is called endolymphatic stromal myosis (histologically benign) or stromal sarcoma (histologically malignant).

Polyps

The term endometrial polyp simply refers to a finger-like pedunculated overgrowth of the endometrium. Polyps may occur singly or in large numbers, filling the uterine cavity, and sometimes a polyp will protrude through the cervix. They may be totally asymptomatic or may result in intermenstrual bleeding due to erosion or necrosis of the polyp. They may also result in hypermenorrhea if they consist of functioning endometrium, which is the exception.

Endometrial polyps may develop at any age. In a series of diagnostic curettages on non-gravid women, Overstreet found the incidence of polyps to be 17 per cent. Their discovery, however, depends upon the ritual of always using a polyp forceps after dilatation of the cervix and *before* using the curet, which may fragment them beyond recognition. Their removal by polyp forceps and curet constitutes definitive treatment.

Asherman's Syndrome

Destruction of the endometrium by trauma may lead to amenorrhea or marked hypomenorrhea, as described by Asherman in 1948. The trauma may be chemical, resulting from the injection of caustic materials to produce abortion, but it most commonly results from vigorous postabortal or postpartum curettage. Intrauterine synechiae form and may be demonstrable by hysterography. The treatment is curettage, breaking up the adhesions and inserting a fluid-filled intrauterine device to prevent further synechia formation. With time, the endometrium will regenerate in the majority of cases.

Endometriosis

Endometriosis refers to benign endometrial implants or metastases outside of the uterus. The most common sites are the cul-de-sac, ovaries, uterine ligaments and the serosal surfaces of the uterus and bowel. Very rarely, endometrial implants are found in the umbilicus, perineum, vagina, cervix, bladder or other bizarre locations.

Widespread recognition of the disease awaited Sampson's classic description in 1921, at which time he ascribed the process to a transtubal regurgitation of menstrual blood containing viable bits of endometrium. This is undoubtedly the usual pathogenesis. The occurrence of implants in strange locations at which this pathogenesis is not possible probably results from lymphatic or hematogenous spread in the same manner that endometrial adenocarcinoma may be disseminated. There are still some advocates of the celomic metaplasia theory, which holds that menstrual blood in the pelvic cavity is simply the inciting cause of a transformation of peritoneal surfaces into endometrial tissue.

Serosal implants typically appear as puckered purplish papules. In the ovaries, dark cysts filled with old inspissated blood resembling chocolate present the classic appearance shown in Figure 17–4. Microscopically, the diagnosis is confirmed by the finding of endometrial elements, but these are frequently missed in random sections.

Extensive endometriosis may at times be quite asymptomatic, whereas in some patients minimal lesions may give rise to severe pain at the time of

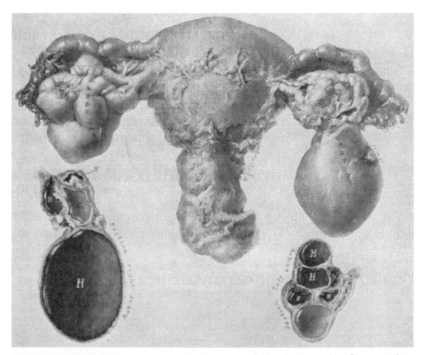

FIGURE 17–4. The gross appearance of endometriosis of the ovaries. Cross-sections of ovaries show the hematomas ("chocolate cysts"). (From Sampson, Am. J. Obstet. Gynecol., *4*:489, 1922.)

menses as well as to dyspareunia. In all likelihood, the pain is produced by peritoneal stretch, and therefore depends upon the degree of fibrosis, which limits the expansibility of the lesion when it bleeds into itself. Pelvic adhesions resulting from endometriosis are unusually dense, requiring sharp rather than blunt dissection, and it is a once-in-a-lifetime experience to be able to remove a large endometrioma of the ovary without spilling its chocolate-like contents into the peritoneal cavity.

The disease occurs most often in white nulliparous females and is frequently associated with long-standing infertility; the reasons for this are rather obscure. Pregnancy, or the pseudopregnant state induced by the continuous administration of progestins, will usually cause regression—but not disappearance—of the lesions. This forms the basis for the hormonal treatment of endometriosis with ascending doses of a progestin (preferably with a low but constant level of exogenous estrogen) in young women who desire children, and in whom the disease is not too extensive. Some pregnancies will occur after cessation of therapy, but so will they occur (and perhaps more often) after surgical eradication of the lesions with preservation of the pelvic organs.

DIAGNOSIS

The diagnosis is made on the basis of the history combined with positive pelvic findings. Pelvic pain, often referred to the rectum and sacral area and maximal during the menstrual flow, is the most common presenting complaint. Dyspareunia due to tender fixed nodules in the cul-de-sac or to a fixed uterine retroversion is also common. The palpation of such nodules and the finding of fixed adnexal masses resembling chronic pelvic inflammatory disease (with which endometriosis is often confused) completes the diagnostic picture. Laparoscopy is sometimes employed to confirm the diagnosis when pelvic palpation is equivocal. Diagnosis and treatment should not be instituted on the basis of the history alone.

TREATMENT

Surgical treatment, when indicated, consists of three types. (1) Conservative treatment means removal of endometriotic lesions by excision or fulguration, lysis of adhesions, and sometimes suspension of the uterine fundus if it is found to be fixed in the cul-de-sac. This treatment is reserved for the younger patient who desires children. Repeated, more extensive surgery is subsequently required in about 25 per cent of women so treated. (2) Semi-conservative treatment is removing the entire uterus and excising all endometriotic lesions but conserving ovarian function. (3) Radical treatment involves removing the uterus, both tubes and both ovaries. This constitutes a cure. Contrary to older teachings, continuous (not cyclic) estrogen replacement therapy may be given safely even though some lesions of endometriosis remain. The more radical treatment is preferred for most women over 35 years of age.

BENIGN DISEASES OF THE CERVIX

One-third to one-half of all women examined will reveal such benign changes of the cervix as ectropion, chronic cervicitis, Nabothian cysts, lacerations with eversion and polyps. Many of the changes are of little clinical significance, but must be differentiated from a malignant lesion of the cervix, for the cervix is the second most common site of cancer in women (the breast is the most common site). Some of the common benign lesions are illustrated in Figure 17–5.

Ectropion

An ectropion, often referred to by the misnomer *congenital erosion,* is an outgrowth of the endocervical columnar epithelium, and appears as a

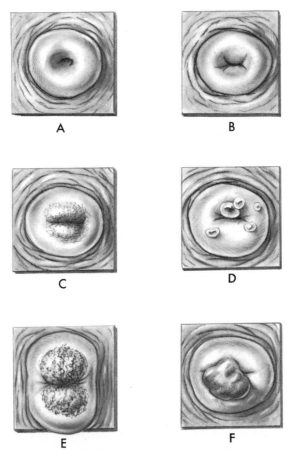

FIGURE 17–5.   Appearance of the cervix. *A,* In a nullipara. *B,* A multipara. Common benign lesions are: *C,* An ectropion. *D,* Nabothian cysts. *E,* Bilateral lacerations with eversion. *F,* An endocervical polyp.

velvety red ring surrounding the os or a red patch on the portio of the cervix. The squamocolumnar junction of the cervix (it is here that cancer so often begins) frequently shifts its location, depending to a large extent upon the pH of the cervical environment (Figure 17–6). Squamous epithelium thrives, but the columnar epithelium does not, in an acid medium. With prolonged uterine bleeding (even in slight amounts), such as may occur in the puerperium, the slightly alkaline blood may induce the columnar epithelium to cover the entire posterior half of the cervix. The use of an acid jelly buffered to a pH of 4.0 (e.g., Acijel) in the vagina each night will frequently cause the squamocolumnar junction to retreat into the endocervical canal where it belongs. That area of the portio or cervical canal formerly occupied by columnar epithelium but now covered by squamous epithelium and punctuated with crypt and tunnel openings is called the *transformation zone*. It is here that most squamous cell carcinomas of the cervix have their origin.

An *eversion* is quite similar, but is simply due to an exposure of the endocervical canal because of old lacerations which may cause the cervix to gape open when a speculum is spread. Unless chronically infected, such a cervix requires no treatment.

Chlamydial Cervicitis

Microorganisms known as Chlamydia (TRIC agents; formerly *Bedsonia*) cause such varied diseases as trachoma, psittacosis and lympho-

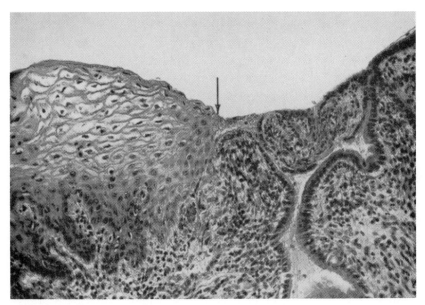

FIGURE 17–6. The squamocolumnar junction (arrow) of the cervix. (From Davies and Woolf, Clin. Obstet. Gynecol., 6:265, 1963. P. B. Hoeber, Harper & Row, New York.)

pathia venereum. Chlamydial organisms, identified by a tissue culture method, may be recovered in almost 50 per cent of men with a non-gonococcal urethritis, in over 35 per cent of women with cervicitis, but in less than 5 per cent of women without genital tract complaints. Thus chlamydial cervicitis may be the most common venereal disease. A current and effective treatment is with tetracycline, 250 mg. four times daily for 21 days.

Chronic Cervicitis

The cervical glands, which are in reality labyrinth clefts, may have their openings closed by an ingrowth of squamous epithelium, a process referred to as epidermidization or squamous metaplasia. Beneath these areas, cervical mucus may accumulate, forming shiny little hemispheres known as *Nabothian cysts.* When they become infected and fill with pus, they are yellow in appearance, giving rise to chronic cervicitis.

A chronic infection of the exocervix may cause an annoying leukorrhea but does not cause infertility. It is best treated by cauterization, either with a hot cautery tip creating radial spokes or by electrocoagulation or cryosurgery. A chronic *endocervicitis,* on the other hand, can interfere with sperm migration and result in infertility, as well as adjacent lymphangitis in the parametrial areas, causing pain on motion of the uterus. The diagnosis is made by finding thick mucus within the cervical canal packed with leukocytes during the preovulatory phase when the mucus should normally be thin and clear, like fresh egg white. Endocervicitis is more common in nulliparous women whose cervix has a pin-point external os which interferes with normal drainage. Mechanical dilatation of the external os, treatment of the cervical canal with solid silver nitrate, or even systemic antibiotic therapy in selected cases may be necessary to eradicate the condition. Endocervicitis is a frequently unrecognized cause of dyspareunia.

Endocervical Polyps

Polyps which dangle from the endocervical canal appear as soft, usually bright red, pedunculated growths of variable length. They often bleed when lightly scraped and may cause postcoital or intermenstrual spotting. Cervical polyps are easily removed in the office by twisting them off with a sponge stick and cauterizing the base. Even though they are very rarely malignant, the removed tissue should be submitted for microscopic examination. A more luxurious polypoid growth resembling an exophytic cancer may occasionally be observed in women who have been taking oral contraceptives for some time ("Pill polyps"), and only biopsies or colposcopy will reveal their true nature. These will regress after discontinuance of the hormonal stimulus.

Dysplasia

Dysplasia (disordered growth) of the cervical epithelium may be found in from 2 to 6 per cent of women, mostly between the ages of 20 and 40. The initial diagnosis is usually made by the cytologist, who will recognize the dysplasia as mild, moderate or severe. The cervix often appears reddened but may appear normal. An iodine stain will reveal non-staining areas (Schiller positive) which serve as a guide for biopsy. Dysplastic areas may also be recognized by the use of a colposcope, and a colposcopically directed biopsy facilitates a precise histologic diagnosis. If the lesion extends into the endocervical canal, a cone biopsy will be needed to rule out carcinoma.

From 10 to 35 per cent of cervical dysplasias, depending upon the severity, will proceed to carcinoma-in-situ within one to five years. The remainder either remain static (about 50 per cent) or regress. In young women who wish to retain their fertility, one or two cryosurgical treatments

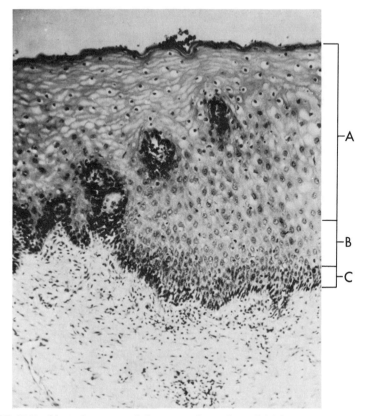

FIGURE 17-7.   Normal mucosa of the exocervix. *A*, Superficial layer. *B*, Intermediate layer. *C*, Basal cells. (From the collection of H. F. Traut.)

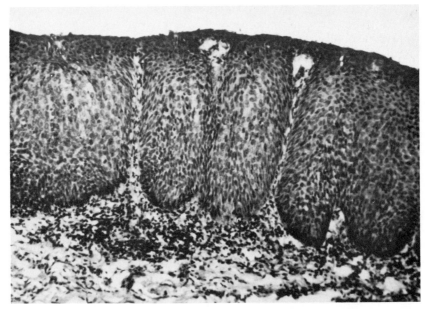

FIGURE 17–8. Severe dysplasia of the cervical epithelium. (From Hill, Am. J. Obstet. Gynecol., *95*:308, 1966. C. V. Mosby Co., St. Louis.)

will cure 3 out of 4 cases of dysplasia, but obviously all patients, irrespective of the treatment employed, should have frequent "Pap smears" for an indefinite period of time.

FUNCTIONAL CYSTS OF THE OVARY

At this point it is pertinent to describe enlargements of one or both ovaries, which are the result of non-neoplastic or functional changes. Simple follicle cysts and corpus luteum cysts are the most common. Bilateral polycystic ovaries associated with anovulation (the Stein-Leventhal syndrome) is an unusual disorder which may also be described under the heading of functional enlargements. We should also consider the sequence of events when a cyst (or tumor) undergoes torsion.

Follicular and Corpus Luteum Cysts

A *follicular cyst* is simply an exaggeration of an event which begins before every ovulation. Occasionally these simple serous cysts may become easily palpable and cause concern on the part of the physician. In women under 40 we recommend the "tennis ball rule"; that is, so long as any cyst does not exceed the size of a tennis ball, or 7 cm. in diameter, it is unlikely to be neoplastic and should simply be observed. In women

gether with an estimate of the risk of regional lymph node involvement and also an approximate five-year survival rate. The staging of carcinoma of the cervix is shown in Figure 18–2.

DIAGNOSIS

The diagnosis of cervical carcinoma is established by single, multiple or cone tissue biopsy. Painting the cervix gently with Lugol's solution (Schiller test) is of value in indicating the best areas for biopsy or for inclusion in the cold-knife conization procedure. Abnormal epithelium fails to take the iodine stain as deeply as normal epithelium. Periodic cytology as well as colposcopy is invaluable as a guide to the indication for a biopsy even when the cervix appears grossly healthy.

TREATMENT

Carcinoma *in situ* may be treated by conization alone, provided that there are adequate margins of normal epithelium completely surrounding the lesion, as revealed by the cone specimen, and provided that subsequent

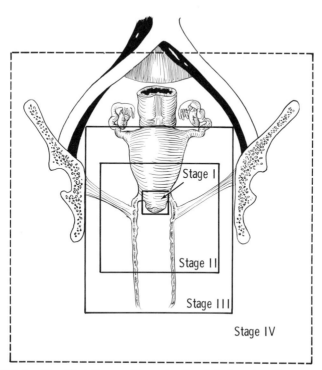

FIGURE 18–2. Simplified diagram illustrating the clinical staging of cancer of the cervix. (From Benson, Handbook of Obstetrics and Gynecology, 4th Ed., Lange Medical Publications, Los Altos, Calif., 1971.)

smear tests are negative. Women who do not want to have more children are usually treated by total hysterectomy.

Most gynecologists treat Stage IA lesions (early stromal invasion) by extra-fascial total hysterectomy without lymph node dissection. Stage IB and some Stage IIA lesions may be treated by radical hysterectomy and bilateral pelvic lymph node dissection or by a combination of radium and external radiation therapy. The five-year survival rate is about the same for both methods, so that other considerations determine the choice between radical surgery and radiation therapy.

Stages IIB, III and IV carcinoma of the cervix are almost universally treated by radiation therapy. Centrally recurring cancer following radiation therapy and occasional Stage IV lesions which do not involve the pelvic side walls are sometimes treated by a total pelvic exenteration, which involves removal of the bladder, uterus, vagina and rectum.

Deaths from carcinoma of the cervix are preventable if all women over the age of 20 receive cytologic screening at least annually and if appropriate and adequate treatment is carried out for the detected early stages of the disease. Such treatment is best conducted in a medical institution where large numbers of cancer patients are managed.

Adenocarcinoma of the Cervix

For every 15 or 20 squamous cell carcinomas of the cervix one may encounter a primary adenocarcinoma arising from the cervical canal. The histology reveals an atypical glandular pattern, occasionally mixed with patches of benign squamous metaplasia. To be considered primary there must be an absence of endometrial cancer, as determined by differential curettage.

The clinical staging, treatment and prognosis are essentially the same as for squamous cell carcinoma.

TUMORS OF THE MYOMETRIUM

Myomas

Myomas are by far the most common tumor of the uterus and the most frequent indication for hysterectomy. After the age of 35, about 20 per cent of uteri contain myomas of varying size, and this figure is much higher in black races. They are commonly referred to as fibroids, but this term is a misnomer, because they arise from immature smooth muscle cells, and therefore consist primarily of myometrial cells arranged in bundles or whorls. They may vary in size from tiny seedlings to mammoth tumors and are usually multiple. The reasons for their growth are obscure, but they are dependent on estrogen and therefore cease to grow and usually decrease in size after the menopause.

All uterine myomas are intramural in location to begin with, but as they enlarge they must either encroach upon the uterine cavity and become submucous or bulge from the peritoneal surface and become subserous in location (Figure 18–3). The shifting position of these hard balls of smooth muscle is due in part to contractions of the surrounding myometrium. Sometimes they are extruded into the peritoneal or intrauterine cavities on pedicles, becoming pedunculated myomas. On rare occasions, their blood supply through such pedicles becomes compromised, and they may attach themselves to surrounding intraperitoneal structures and become parasitic.

Depending upon the adequacy of the blood supply to these tumors, a variety of pathologic changes may occur within the myomas. Hyaline changes are found in most, and there may be necrosis with hemorrhage, cystic degeneration, or calcification, and very rarely (less than 1 per cent of myomas) a *sarcoma* may develop. The fact that a sarcoma can develop does not influence our management of myomas during the reproductive period of life.

The majority of myomas are asymptomatic and require no treatment other than periodic examinations. Degeneration or hemorrhage within a

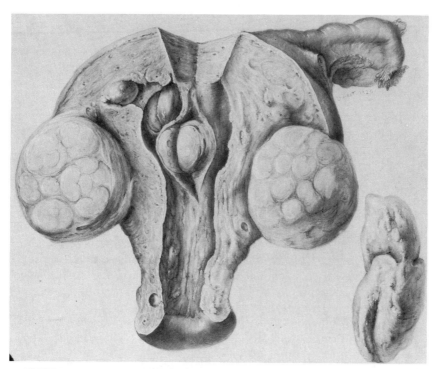

FIGURE 18–3. Drawing of a uterus containing small submucous myomas and a medium-sized intramural myoma. The isolated myoma (lower right) was removed from the broad ligament. (From Morton, in Davis' Gynecology and Obstetrics, Vol. II., Harper & Row Publishers, Inc., Medical Dept., (Loose-leaf Section), Hagerstown, Md., 1970.)

myoma, especially common during pregnancy, may cause the tumor to become quite painful. Subserous myomas may become so large that they exert pressure on the rectosigmoid, the bladder or even the ureters. Excessive menstrual flow is characteristic of submucous myomas, sometimes leading to profound secondary anemia.

The mere existence of myomas that are readily palpable on bimanual examination of the uterus is not an indication for surgery. If the rate of growth is slow and the tumors remain asymptomatic, they may simply be observed two or three times a year until the menopause. Under these circumstances, patients should be warned that the use of oral contraceptives may cause more rapid growth of the tumors.

In younger women who desire children, who are unable to conceive despite the absence of any other infertility factors, or who repeatedly abort, the individual tumors may be removed by multiple myomectomy. Such patients must understand, however, that there is about one chance in three that recurrent myomas may necessitate hysterectomy later, and that if the uterine cavity is entered during myomectomy, future pregnancies must ordinarily be delivered by cesarean section. When myomas become truly symptomatic because of their size or because of hypermenorrhea, the appropriate treatment is total hysterectomy.

Subtotal or supracervical hysterectomy is no longer performed for benign diseases of the uterus, because there is no point in leaving a useless structure (the cervix) which is subject to a 2 per cent risk of subsequent cancer. By the same token, after the age of 45 or thereabouts, there is little reason for leaving the ovaries when a hysterectomy is performed, because ovaries are subject to a 1 per cent risk of subsequent cancer and the mortality from ovarian cancer when it does occur is about 80 per cent—several times higher than that for cancer of the cervix, which may now be readily detected in its preinvasive stage by periodic examinations.

Sarcomas

Uterine sarcomas may arise *de novo* from any type of connective tissue cell in the uterus, or they may occur as the result of a malignant change in a pre-existing leiomyoma. As noted above, this latter event is rare and may occur in only one in 500 or so cases of uterine myomas; nevertheless, because myomas are so common, a leiomyosarcoma is the most common histologic type encountered. Altogether, the various types of sarcomas constitute less than 5 per cent of uterine malignancies.

A major difficulty has been the differentiation between highly cellular myomas and "low grade sarcomas," and this accounts for the wide range of reported salvage rates following surgery, from nearly zero to as high as 40 per cent. The differential diagnosis depends primarily upon the number of mitotic figures observed per high power field. The prognosis varies inversely with the number observed. Sarcomas metastasize through the blood stream, and often the first symptoms result from pulmonary lesions. The prognosis is essentially hopeless in the presence of metastases because sarcomas are notoriously resistant to radiotherapy and chemotherapy.

A sarcoma may be suspected when a uterine myoma, previously static, suddenly increases in size, especially after the menopause. Often the diagnosis is initially made by the pathologist upon examining a uterus removed because of a clinical diagnosis of large myomas.

CANCER OF THE ENDOMETRIUM

Adenocarcinoma of the endometrium is the second most common gynecologic malignancy, and appears to be increasing in frequency. The great majority of cases occur between the ages of 50 and 70, and the increase is generally attributed to the progressively longer life expectancy of women, which is now 75 years.

As with all other cancers, the expectancy of a cure depends primarily upon the extent of the disease. The International Federation of Gynecology and Obstetrics (FIGO) recommends the following classification:

Stage 0. Carcinoma *in situ* of the endometrium.

Stage I. Carcinoma confined to the corpus.

Stage II. The carcinoma has extended downward to involve the cervix.

Stage III. The tumor has extended outside the uterus but is still confined to the pelvis.

Stage IV. The carcinoma has obviously involved the mucosa of the bladder or rectum or has extended beyond the confines of the true pelvis.

To a lesser extent, the prognosis is affected by the degree of differentiation. A well-differentiated adenocarcinoma (Figure 18–4) has a better prognosis, stage for stage, than a poorly differentiated tumor. When benign-appearing squamous cells are intermingled with the glandular elements, the lesion is called an *adenoacanthoma* and is usually classified with the well-differentiated group. When the squamous elements are malignant, however, the lesion is called an adenosquamous cancer, which has a poorer prognosis.

PATHOGENESIS

A voluminous and spirited gynecologic literature has accumulated relative to the pathogenesis of endometrial adenocarcinoma, with particular reference to the role of estrogen. There is little doubt that a protracted estrogen effect unopposed by progesterone favors the development of endometrial carcinoma. Thus, young women with the polycystic ovary (Stein-Leventhal) syndrome, women with feminizing ovarian tumors and patients with recurring episodes of anovulatory endometrial hyperplasia have an increased incidence. So do women who are obese, nulliparous, hypertensive or diabetic, for reasons which are somewhat obscure. Atypical adenomatous hyperplasias of the endometrium observed in the perimenopausal

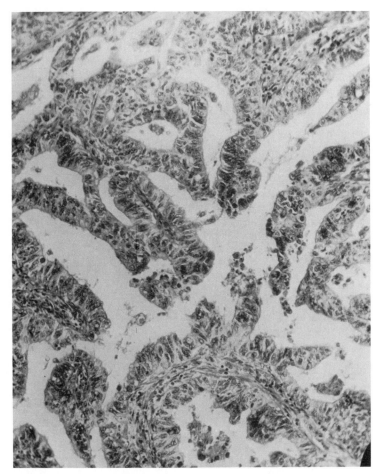

FIGURE 18–4. Photomicrograph of a well-differentiated adenocarcinoma of the endometrium.

period of life are definitely predisposing lesions. It is quite possible that the primary carcinogenic agent, whatever it may be, must remain in the endometrium for years before the cancer develops, and that the periodic shedding of the endometrium which occurs in normal ovulatory cycles discourages its development. If this is true, then there would be a theoretical reason for the administration of progestins, perhaps once or twice a year (the so called "hormonal curettage") to postmenopausal women who are on prolonged estrogen therapy, as well as to younger women with anovulatory syndromes.

The primary symptom of endometrial cancer is abnormal uterine bleeding. Fortunately, this ordinarily occurs before the lesion extends beyond Stage I (providing that there is no cervical stenosis to conceal the event). Any bleeding that occurs six months or more after the menopausal woman has stopped menstruating must be regarded as possible endometrial cancer even when she is on cyclic estrogen therapy. Women who have scanty bleeding "according to plan" when they interrupt estrogen therapy are in a low suspicion group, and under these circumstances an endometrial biopsy or endometrial washings submitted for cytology, if clearly negative, may be sufficient evidence for the absence of cancer. A negative Pap smear taken from the cervix or vagina is *totally unreliable* for the exclusion of endometrial adenocarcinoma. Postmenopausal women who are not receiving exogenous estrogens in any form and who bleed are in a high suspicion group and should have a differential curettage under anesthesia. The term "differential" means that the cervical canal is curetted first and that any tissue obtained is submitted separately to differentiate between clinical Stages I and II.

TREATMENT

The most important part of the treatment of endometrial carcinoma is complete removal of the uterus, tubes and ovaries. Obviously, this is the only treatment needed for Stage O. In early Stage I, when the uterus is normal in size and the carcinoma is well differentiated, primary surgery is also adequate and should yield a five-year survival rate of 85 to 90 per cent. It has been well established, however, that any further extent of the disease (Stage I with an enlarged uterus, Stage II and Stage III) is best treated by a combination therapy (either intracavitary radium or external radiation therapy to the entire pelvis), followed later by surgery. Stage IV lesions and recurrent metastases are currently treated with massive doses of intramuscularly administered long-acting progesterone preparations. About a third of such patients respond with a satisfactory remission of the disease for a year or longer.

In the case of endometrial cancer, hope for the future depends upon continued early recognition of the warning symptoms of irregular or postmenopausal bleeding by all women and immediate diagnostic steps by the physician who is consulted. Periodic shedding of the endometrium by the use of oral progestins may be of prophylactic value in anovulatory or postmenopausal women.

CANCER OF THE FALLOPIAN TUBE

Papillary adenocarcinoma arising in the fallopian tube is a rare lesion, and there are less than 1000 cases reported. Its importance lies in the fact that it is rarely considered in the differential diagnosis of an adnexal mass, yet there may be a characteristic yellow watery discharge, intermittent, often bloody, and containing cancer cells. Because the diagnosis is rarely made early, survival after radical surgery is less than 40 per cent.

NEOPLASMS OF THE OVARY

Tumors of the ovary may occur at any age, even *in utero,* but the peak incidence for benign tumors is between 20 and 44 years of age, and that for ovarian cancer is between 45 and 64 years of age. The chance that a woman will develop a malignant ovarian tumor during her lifetime is only 1 in 100, compared with 1 in 50 for cervical cancer or 1 in 65 for carcinoma of the endometrium. Nevertheless, more women die of ovarian cancer (about 10,000 annually in the United States) than of cervical or endometrial cancer because the five year survival rate in all forms of ovarian cancer is as low as 20 per cent. Over the last 40 years there has been a steady decline in the death rates for uterine cancer, but a steady increase in the death rates for cancer of the ovary. The problems of prophylaxis, early detection and adequate therapy of ovarian cancer are perhaps the foremost challenges in the field of gynecology.

DIAGNOSIS

The diagnosis of functional ovarian cysts during the reproductive period of life was discussed in Chapter 17. The detection of an ovarian neoplasm in its earlier stages is ordinarily incidental to a routine, pelvic examination because the lesions are usually silent, unless there is torsion or rupture. Prior to the menopause, any adnexal mass greater than 5 cm. in diameter which progressively enlarges under observation, or any adnexal mass greater than 9 cm. in diameter, should be considered neoplastic. After the menopause, any ovary that is palpable on bimanual examination should be suspected as neoplastic. Once a mass is palpated, other diagnostic methods, such as sonography, x-ray and laparoscopy, may be utilized for differential diagnosis.

PROPHYLAXIS

Removal of both tubes and ovaries at the time of hysterectomy performed for any reason provides, of course, complete protection against ovarian neoplasms for that individual. This is becoming increasingly acceptable after the age of 40, especially since we have learned that essentially all estrogen after the menopause originates from the peripheral conversion of adrenocortical precursors, not from the ovaries. Lest this seem radical, it must be remembered that removal of the ovaries will save more lives from cancer than will the routine removal of the cervix at the time of hysterectomy, a procedure regarded as mandatory. About 10 per cent of women with ovarian cancer have had a prior hysterectomy.

CLINICAL STAGING OF OVARIAN MALIGNANCY

Unlike the staging of many malignancies, that for ovarian cancer is based upon surgical exploration. Unexplored cases which are clinically diagnosed as ovarian carcinoma are separated into a special category. The international system of staging is shown below. In the variants of Stage I, note should also be made as to whether the capsule of the tumor is ruptured or not ruptured.

Stage I: Growth is limited to the ovaries.
Ia: Limited to one ovary; no ascites.
Ib: Limited to both ovaries; no ascites.
Ic: Limited to one or both ovaries, but ascitic fluid containing malignant cells is present.
Stage II: Growth involving one or both ovaries with pelvic extension.
IIa: Extension or metastases to uterus or tubes only.
IIb: Extension to other pelvic tissues.
Stage III: Growth involving one or both ovaries with widespread intraperitoneal metastases (omentum, bowel, peritoneum, and so on.)
Stage IV: Distant metastases outside of the peritoneal cavity.

The Classification of Ovarian Neoplasms

The bewildering array of over 60 different primary ovarian tumors which have been described is a source of dismay to the student and even to the gynecologic pathologist. Innumerable classifications have been proposed, but one which is based upon the probable histogenesis is perhaps the easiest to remember. The following system is adapted from a monograph by Janovski and Paramanandhan (1973) and a review by Scully (1970). The percentages which appear in parentheses after various malignant tumors are Scully's estimates of the comparative incidence of ovarian cancer types found at operation. The student should consult a textbook of gynecologic pathology or a monograph on ovarian tumors for the details of gross and microscopic structure.

There are essentially five components of the human ovary which give rise to neoplasms:

I. The surface or celomic epithelium.
II. The sex-undifferentiated (gonadal stromal) mesenchyme.
III. The sex-differentiated mesenchyme, or sex cords.
IV. The germ cells.
V. Heterotopic tissues and mesonephric rests.

In addition to these, the ovary is a common site of metastatic malignancies originating in other organs.

Table 18–2.   SURFACE OR CELOMIC EPITHELIAL ORIGIN

Benign	Malignant Counterpart	
Serous cystadenoma	Serous cystadenocarcinoma	(35–50%)
Mucinous cystadenoma	Mucinous cystadenocarcinoma	(10–20%)
Endometrial cyst	Endometrioid adenocarcinoma	(15–20%)
Cystadenofibroma	Malignant cystadenofibroma	(rare)
Brenner tumor	Malignant Brenner tumor	(rare)
Clear cell tumor	Clear cell adenocarcinoma	(4–6%)
	Undifferentiated or unclassifiable	(5–10%)

TUMORS OF SURFACE OR CELOMIC EPITHELIAL ORIGIN

The ovarian celomic epithelium during neoplasia may differentiate via tubal epithelium (forming the serous tumors), via endocervical epithelium (forming the mucinous tumors) or via the endometrium (forming the endometrioid and clear cell varieties). The Brenner tumor is ordinarily solid and resembles a fibroma, except that it contains characteristic islands of epithelial-like cells (Table 18–2).

About 90 per cent of all ovarian cancers fall into this first group, the most common being the papillary serous cystadenocarcinoma, which makes up over 40 per cent of ovarian malignancies and which unfortunately has the poorest prognosis.

TUMORS OF SEX-UNDIFFERENTIATED MESENCHYMAL ORIGIN

With the exception of ovarian fibromas, which make up about 5 per cent of benign ovarian tumors, all of the tumors in this group are rare (Table 18–3). Any solid ovarian tumor, but notably the fibromas, may be associated with hydrothorax, with or without ascites. This strange phenomenon, as yet unexplained, is referred to as the Meigs syndrome.

TUMORS OF SEX-DIFFERENTIATED MESENCHYMAL ORIGIN

These tumors, although uncommon, are of particular interest because of the marked systemic symptoms which may result from their hormonal

Table 18–3.   SEX-UNDIFFERENTIATED MESENCHYMAL ORIGIN

Benign	Malignant Counterpart
Fibroma	Fibrosarcoma
Myxoma	Myxofibrosarcoma
Leiomyoma	Leiomyosarcoma
Neurofibroma	Neurofibrosarcoma
Mixed mesodermal tumor	Malignant mixed mesodermal tumor

Table 18–4. SEX-DIFFERENTIATED MESENCHYMAL (SEX CORDS) ORIGIN

  I. Stromal luteoma (e,a).
 II. Granulosa cell tumor (e).   Malignant varieties (5–10%)
III. Theca cell tumor (e).
 IV. Granulosa-theca cell tumor (e).
  V. Androblastoma (Sertoli-Leydig cell tumor; arrhenoblastoma) (a).
 VI. Gynandroblastoma (mixed cell types) (e,a).
VII. Hilar cell tumor (a).
VIII. Gonadoblastoma (also contains elements derived from germ cells) (e,a).

Note: All of these tumors are potential producers of estrogens (e) or androgens (a), or both (e,a). All except the luteoma are potentially malignant.

activity (Table 18–4). Granulosa cell tumors are the most common and constitute about 70 per cent of the group. There are many varieties of histologic patterns, and about one-third demonstrate malignant behavior, although the prognosis is relatively good (85 to 95 per cent five-year survival rates). Granulosa cell tumors may occur in children, but over half of them arise in postmenopausal women. Both granulosa cell and theca cell tumors are associated with a high incidence of hyperplasia, anaplasia and adenocarcinoma of the endometrium. Theca cell and hilar cell tumors and gynandroblastomas are almost always benign. Androblastomas, on the other hand, are frequently malignant. Even small androblastomas and hilar cell tumors may cause extensive virilization.

TUMORS OF GERM CELL ORIGIN

With the exception of the common adult cystic teratoma, which may rarely be malignant, all of the tumors in this group are predominantly malignant, and for the most part occur in children and young adults (Table 18–5). As a whole, this group accounts for about 16 per cent of all ovarian neoplasms, primarily because of the prevalence of benign cystic teratomas, commonly referred to as dermoids. Cystic teratomas constitute about 10

Table 18–5. GERM CELL ORIGIN

  I. Dysgerminoma (5%)
 II. Teratoma group
   A. Extraembryonal forms
     1. Endodermal sinus tumors
     2. Primary choriocarcinoma
   B. Embryonal teratomas, solid and cystic (1–2%)
   C. Adult teratomas
     1. Solid
     2. Cystic ("Dermoids") (1–2%)
   D. Struma ovarii (thyroid tissue)
   E. Carcinoid

per cent of all benign ovarian neoplasms and occur mainly during the first two decades of life. Indeed, cystic teratomas are by far the most common ovarian tumors occurring in childhood and adolescence. They may contain any type of adult tissue such as skin, cartilage, bone, teeth, brain tissue, thyroid and so forth, but in most tumors the predominant mass consists of sebaceous material and hair. They are bilateral in from 10 to 20 per cent of cases.

Dysgerminomas constitute 5 per cent of ovarian cancers, are almost always unilateral, occur primarily before the age of 30 but have a relatively good prognosis following removal (75 to 90 per cent five-year survival rates).

The endodermal sinus tumor is almost invariably fatal. Histologically, it resembles the yolk-sac structure found in the rodent placenta and known as an endodermal sinus.

METASTATIC TUMORS

The ovaries are hospitable hosts to a large variety of metastatic malignancies arising in other organs, notably the breast, stomach and colon. It is estimated that about three-fourths of all gastric cancers, a third of all breast cancers and a quarter of all malignancies of the rectosigmoid, colon and appendix metastasize to one or both ovaries. Altogether, metastatic tumors including lymphomas constitute about ten per cent of ovarian cancers found at operation. When the tumor contains mucin-secreting cells with a signet ring formation (usually but not always secondary to stomach cancer) it is commonly referred to as Krukenberg's tumor.

Treatment of Ovarian Tumors

The primary treatment of all ovarian neoplasms is surgical. The extent of the surgery depends upon the age of the patient, the degree of malignancy, if any, and the clinical staging of the disease. In young women, for example, such unilateral Stage I tumors as dysgerminomas or granulosa cell tumors may simply be removed with no further therapy. In most instances of Stages I and II ovarian cancers, all pelvic organs are removed and postoperative radiotherapy to the pelvis is administered. Ovarian cancer in Stages III and IV is treated by radical surgical removal, often accompanied by omen-

Table 18–6.   HETEROTOPIC OR MESONEPHRIC REST ORIGIN

  I. Malignant tumors arising from endometriosis
 II. Pheochromocytoma (benign)
III. Adrenal cell rest tumor (benign or malignant)
IV. Mesonephric adenoma (benign)
 V. Mesonephroma (malignant)
     Note: These tumors are all rare. The adrenal cell tumors may cause masculinization.

tectomy, followed by external irradiation to the pelvis for selected patients, and systemic chemotherapy for most patients. In some instances of Stage III ovarian cancers, a repeat "second look" laparotomy is performed following irradiation or chemotherapy to determine whether further radical removal of any residual tumor is feasible. Every woman with ovarian cancer can receive the best possible therapy only by the collective judgment and close cooperation of the gynecologist, pathologist, radiotherapist and chemotherapist.

GYNECOLOGIC ASPECTS OF PREVENTIVE MEDICINE

In the last two decades, the obstetrician-gynecologist has found himself placed in the role of a primary physician insofar as the care of women is concerned. This began with the annual Pap smear, but he soon found out that he could not ignore the general physical and emotional well-being of his patient, nor refer patients to other specialists for every examination or ailment that was not strictly gynecologic in nature. As a result he became interested in preventive medicine and comprehensive medical care. Gynecology has evolved into the care of women, not just the diseases of women.

The periodic examination of women, whether carried out by a generalist, an internist or a gynecologist, has become part and parcel of preventive medicine. The purposes are three-fold: (1) To detect cancer in its pre-invasive (or at least its early invasive) stage, (2) to detect the onset of other diseases, such as diabetes, hypertension, anemia, tuberculosis, urinary tract infections and emotional illnesses, and (3) to assist in the goals of family planning and population control and in the physical health, mental health and sex education of the individual woman. Most visits are made annually, but in the perimenopausal period and for certain women with problems a semi-annual examination is more appropriate.

About half of all cancers in men occur in the gastrointestinal system, lung or prostate, where the detection of readily curable lesions is rare. Women are a bit more fortunate in that about 60 per cent of malignancies begin in sites which are readily accessible to direct inspection, palpation or cytologic examination (Figure 18–5).

What should be done at the time of a routine annual visit? This will vary with each physician, but the following procedures should serve as minimal standards:

1. An interval history and review of systems, including, of course, the menstrual history and any emotional or marital problems.

2. Weight and blood pressure.

3. A brief general physical examination, which should include palpation of the thyroid, a careful examination of the breasts, ausculation of heart and lungs, inspection of all skin surfaces, abdominal palpation and digital rectal examination.

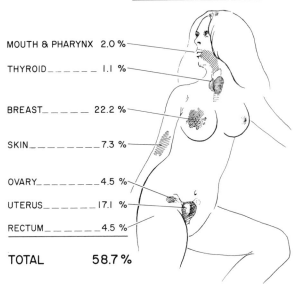

OVER HALF OF ALL CANCER INVOLVES SITES ACCESSIBLE TO DIRECT EXAMINATION

MOUTH & PHARYNX 2.0 %

THYROID _ _ _ _ _ _ 1.1 %

BREAST _ _ _ _ _ 22.2 %

SKIN _ _ _ _ _ _ _ _ 7.3 %

OVARY _ _ _ _ _ _ _ _ 4.5 %

UTERUS _ _ _ _ _ _ 17.1 %

RECTUM _ _ _ _ _ _ 4.5 %

TOTAL   58.7 %

FIGURE 18–5. Common sites of cancer which may be detected in women by physical examination.

4. Pelvic examination, to include vaginal and cervical cytology and biopsy of any suspicious lesions.

5. Testing of a voided urine sample for albumin and sugar.

Depending upon the history and findings, the woman may be referred for other diagnostic studies or to another physician. Among the common procedures utilized are mammography, x-ray examination of the chest, a complete blood count, endometrial biopsy, glucose tolerance test, barium studies of the upper or lower gastrointestinal tract, a test for blood in the stools (after a meat-free diet), sigmoidoscopy, and a mid-stream urine culture. There are many who would argue that some one of these should be routinely done, but knowing that there is no such thing as a *complete* physical or laboratory examination, some compromise must be adopted, based upon the anticipated yield of positive findings and balanced by the cost, discomfort of the procedure and degree of patient cooperation. One thing is certain, the pernicious "Pelvic examination deferred" syndrome which is endemic among members of medical and surgical house staffs, both among in-patients and out-patients, should be ended. The physical examination of a woman is grossly incomplete unless a careful and complete pelvic examination with cytology is performed.

Screening programs for breast cancer are conducted by the National Cancer Institute. These include physical examination, thermography and mammography. In women over the age of 34, cancers have been detected in 1 per cent of those screened and biopsied. Preliminary results indicate that mammography detected 92 per cent, physical examination 57 per cent, and thermography only 39 per cent of the cancers proven by biopsy. The encouraging fact is that of the cancers discovered to date, 77 per cent have had no nodal involvement.

Obstetrics and gynecology are indivisibly entwined to form a single discipline which has as its common denominator the care of women who are past the age of puberty. Almost by necessity, if not by choice, the obstetrician-gynecologist has become a primary physician for women. The preconceptional, interconceptional and even the post-menopausal phases of a woman's life are closely related to her reproductive functions and sexuality. Of necessity, therefore, a physician's training in obstetrics and gynecology not only involves such preclinical areas as genetics, embryology and gynecologic pathology, but requires knowledge and experience in sizable segments of such related clinical fields as surgery, internal medicine, psychiatry, urology, oncology, endocrinology, perinatal medicine and preventive medicine.

SUGGESTED SUPPLEMENTARY READING

Disaia, P., Morrow, C. P. and Townsend, D. E. Synopsis of Gynecological Oncology. John Wiley & Sons, Inc., New York, 1975.

Gusberg, S. B. (Ed). New Directions for research on endometrial cancer. Gynecologic Oncology 2, 113, 1974.
The manuscripts and discussion of a symposium sponsored by the American Cancer Society.

Janovski, N. A., and Paramanandhan, T. L. Ovarian Tumors. W. B. Saunders Company, Philadelphia, 1973.
A well-illustrated monograph with extensive bibliographies.

Novak, E. R., and Woodruff, J. D. Novak's Gynecologic and Obstetric Pathology, 7th Ed. W. B. Saunders Company, Philadelphia, 1974.
An authoritative, profusely illustrated text which has become the standard reference on gynecologic pathology.

Scully, R. E. Recent Progress in Ovarian Cancer. Human Pathology, 1:73, 1970.
A good review article on all of the malignant ovarian tumors.

INDEX

459